ansmitting Boundar
of Dam-Reserv

ries for Time-Domain Analysis
oir Interaction

Bene
dikt Weber

r Baustatik und Konstruktion
he Technische Hochschule Zürich

September 1994

ISBN 978-3-7643-5123-6 ISBN 978-3-0348-7751-0 (eBook)
DOI 10.1007/978-3-0348-7751-0

Rational Tr

Institut fü
Eidgenössisc

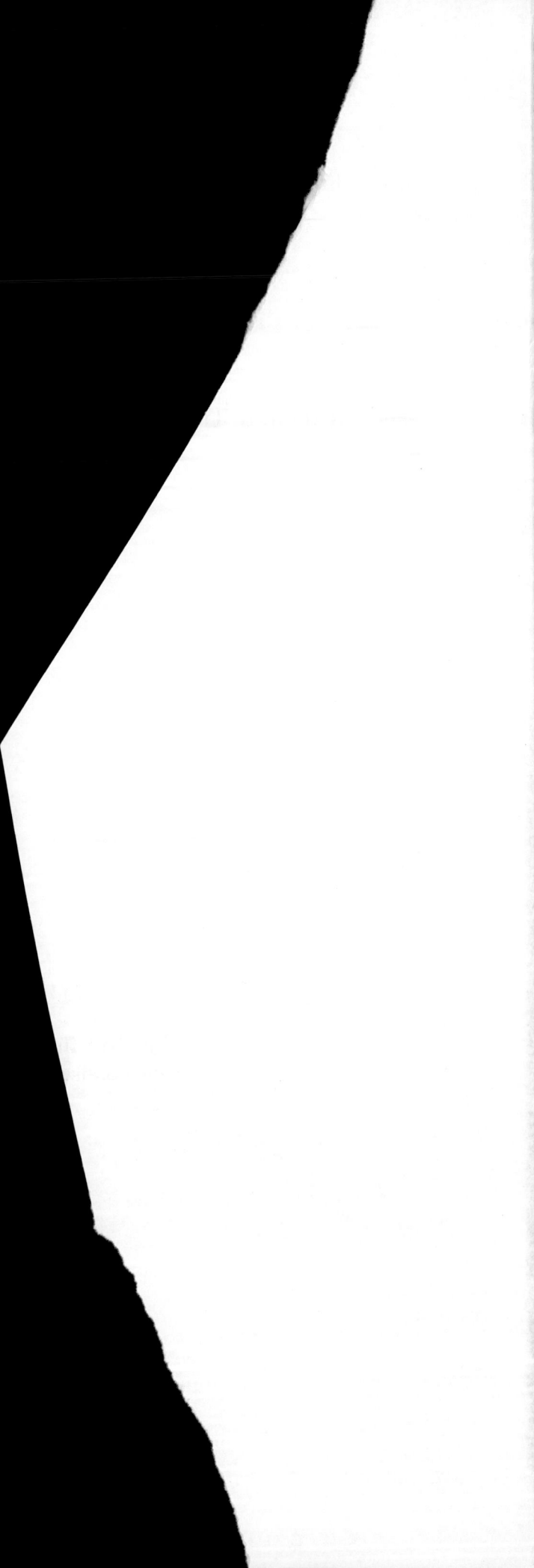

Preface

Most existing arch dams have been designed for seismic loading by static methods involving the use of seismic coefficients. Although there are no known examples of arch dams which have been seriously damaged by earthquakes, the need for more realistic seismic analyses is now well recognized, not only for new dams but especially in the context of the safety evaluation of existing dams. Fortunately, with the finite element method, engineers have a powerful tool for modeling the complex geometry and the nonlinear material behavior of a dam. However, there is still a major complication in the analysis procedure, namely the interaction of the dam with the reservoir and with the foundation during an earthquake. Interaction is a wave propagation problem involving transmitting boundaries. The State of the Art in engineering practice is to neglect wave propagation by modeling the water as incompressible and the foundation as massless.

More advanced analysis methods using compressible water and foundation with mass have been available for some time. However, these methods are restricted to linear models, because they work in the frequency domain. On the other hand, there are also advanced nonlinear models for dams, but they can only be used in the time domain, usually with simple transmitting boundaries.

In this report, which is based on an a doctoral thesis, rigorous transmitting boundaries in the time domain are developed which permit combining compressible water with nonlinear dam behavior. The new numerical model is based on a systems-theory approach. While the nearfield is modeled by conventional finite elements, the farfield is captured by a system which introduces only a few additional internal degrees of freedom. This makes the procedure very efficient. Because the farfield system is based on a rigorous frequency-domain solution, the method is also accurate.

With this very original thesis significant progress in modeling the earthquake behavior of arch dams with reservoir and foundation has been achieved. It is expected that the work will be highly regarded by specialists in the field of structural dynamics and earthquake engineering.

Zurich, September 1994 — Prof. Hugo Bachmann

Acknowledgments

This thesis is part of a larger research project on the seismic analysis of dams at the Institute of Structural Engineering of the Swiss Federal Institute of Technology under the guidance of Prof. Dr. Hugo Bachmann. On this occasion I would like to express my gratitude to Prof. Bachmann for his support in my research and for giving me all the freedom I needed to accomplish this work, and also for his comments on the thesis.

My further thanks go to the co-examiners Prof. Dr. Edoardo Anderheggen and Dr. Jürgen Halin, for reviewing the thesis and making many useful comments. I am also very grateful to Prof. Dr. Eduardo Kausel from MIT for serving as an external reviewer, and for carefully reading through the complete work and asking the relevant questions that helped me improve many formulations.

I also would like to thank all my colleagues at the Institute who helped in any way to complete this work, especially Glauco Feltrin, with whom I have had the pleasure to share an office for many years, and who has been one of the few persons with whom I could discuss my theoretical problems, Tadeusz Szczesiak for calculating the analytical solutions for the reservoir with semi-circular cross-section, and lastly to Dr. Gerald Prater for proof-reading the manuscript.

I also gratefully acknowledge the continuing generous support of the project by the Swiss Federal Office of Water Resources, and the excellent infrastructure of the Swiss Federal Institute of Technology, especially its comprehensive main library, and the almost unlimited computer resources.

Finally, I would like to express my deep appreciation to Ruth Schäffler, who accompanied me through this hard time of probation in our friendship.

Zumikon, September 1994 B. Weber

Contents

Preface i

1 Introduction 1
1.1 Transmitting boundaries . . . 1
1.2 Frequency-domain solutions . . . 2
1.3 Time-domain solutions . . . 2
1.3.1 Local, non-consistent boundaries . . . 3
1.3.2 Consistent, non-local boundaries . . . 3
1.3.3 Rational boundaries . . . 4
1.4 Rational approximation . . . 4
1.5 Dam-reservoir interaction . . . 6
1.5.1 Compressibility effects . . . 6
1.5.2 Two-dimensional solutions . . . 8
1.5.3 Three-dimensional solutions . . . 9
1.5.4 Experimental work . . . 9
1.6 Objectives . . . 10
1.7 Scope . . . 10
1.7.1 Model and restrictions . . . 10
1.7.2 Related problems . . . 11
1.7.3 Presentation . . . 12
1.8 Outline . . . 12

2 Model Problems 15
2.1 Semi-infinite rod . . . 15
2.2 Spherical cavity . . . 17
2.3 Semi-infinite rod on elastic foundation . . . 21
2.4 Semi-infinite rod on visco-elastic foundation . . . 25
2.5 Semi-infinite channel with constant cross-section . . . 26

3 Common Numerical Procedures 29
3.1 Local, non-consistent boundaries . . . 30
3.1.1 Engquist-Majda boundaries . . . 30
3.1.2 General case . . . 33
3.1.3 Paraxial boundaries . . . 34
3.1.4 Lindman boundary . . . 34
3.1.5 Extrapolation algorithm . . . 35
3.1.6 Superposition boundaries . . . 37
3.2 Boundary elements . . . 38
3.2.1 Frequency-domain boundary elements . . . 40
3.2.2 Time-domain boundary elements . . . 41

3.3 Infinite elements and mapping . . . 43
3.3.1 Spatial discretization, finite elements . . . 43
3.3.2 Infinite elements . . . 47
3.4 Mixed time-frequency domain methods . . . 49
3.4.1 Hybrid method . . . 50
3.4.2 "Recursive" method . . . 50

4 Scalar Systems, Signal Processing **51**
4.1 Rational systems . . . 51
4.1.1 Recursive realization . . . 52
4.1.2 Differential equation formulation . . . 53
4.2 Discrete-time systems . . . 53
4.2.1 z-transform . . . 54
4.2.2 Stability and causality of rational systems . . . 56
4.2.3 Hankel matrix . . . 58
4.2.4 Singular value decomposition . . . 59
4.2.5 Discrete Fourier transform (DFT) . . . 60
4.2.6 Sampling . . . 60
4.2.7 Norms . . . 62
4.3 Continuous-time systems . . . 62
4.3.1 Laplace transform . . . 62
4.3.2 Stability and causality of rational systems . . . 67
4.3.3 Markov parameters . . . 69
4.3.4 Integral Hankel operator . . . 70
4.3.5 SVD of integral Hankel operator . . . 71
4.3.6 Numerical Fourier transform . . . 71
4.3.7 Norms . . . 72
4.4 Bilinear transform . . . 72
4.4.1 Definition of mapping . . . 72
4.4.2 Singular values . . . 74
4.4.3 Discrete Fourier transform . . . 74
4.4.4 Back-transformation for rational transfer function . . . 75

5 Approximation by Scalar Systems **77**
5.1 Approximation of transfer function (frequency sampling) . . . 77
5.1.1 Discrete-time systems . . . 77
5.1.2 Continuous-time systems . . . 81
5.2 Approximation of Markov parameters . . . 82
5.2.1 Markov parameters of discrete-time system . . . 82
5.2.2 Markov parameters of mapped discrete-time system . . . 85
5.2.3 Potential problems . . . 89
5.3 Singular value decomposition (SVD) . . . 90
5.4 Methods based on SVD of Hankel matrix . . . 92
5.4.1 Carathéodory-Fejér (CF) method . . . 92
5.5 Additional topics . . . 95
5.5.1 Transfer function in the complex plane . . . 95
5.5.2 Causality and stability . . . 97
5.5.3 General remarks . . . 97

6 Multivariable Systems, State-Variable Description 100
6.1 Discrete-time systems . . . 100
6.1.1 Scalar systems . . . 100
6.1.2 Multivariable systems . . . 102
6.1.3 Similarity transformation . . . 104
6.1.4 Stability . . . 105
6.1.5 Observability, controllability . . . 105
6.1.6 Lyapunov equations . . . 109
6.1.7 Hankel matrix . . . 110
6.1.8 Balanced systems . . . 111
6.1.9 Norms . . . 113
6.2 Continuous-time systems . . . 113
6.2.1 Multivariable systems . . . 114
6.2.2 Observability and controllability . . . 114
6.2.3 Lyapunov equations . . . 115
6.2.4 Hankel integral operator . . . 116
6.2.5 Balanced systems . . . 117
6.2.6 Norms . . . 118
6.3 Bilinear transform . . . 118
6.4 Remarks . . . 120

7 Balanced Realization 121
7.1 Algorithm . . . 121
7.1.1 General derivation . . . 121
7.1.2 Full rank Hankel matrix . . . 123
7.1.3 Rank deficient Hankel matrix . . . 124
7.1.4 Reduced systems . . . 126
7.2 Examples . . . 128
7.2.1 Spherical cavity . . . 128
7.2.2 Rod on visco-elastic foundation . . . 130
7.2.3 Two-degree-of-freedom system . . . 131
7.3 Conclusions . . . 137

8 Symmetric Second-Order Systems 138
8.1 Scalar systems . . . 138
8.1.1 Symmetric systems . . . 138
8.1.2 Realizations . . . 139
8.1.3 Physical models . . . 140
8.2 Multivariable systems . . . 141
8.2.1 Vector formulation . . . 141
8.2.2 Matrix formulation . . . 143
8.3 Example . . . 146

9 Fluid Model and Analytical Solutions 148
9.1 Fluid model . . . 148
9.2 Analytical solutions for 2-D fluid problem . . . 149
9.3 Analysis of simplified models . . . 156
9.3.1 Upstream excitation . . . 156
9.3.2 Vertical excitation . . . 159
9.4 Analytical solutions for 3-D fluid problems . . . 162

9.4.1 Rectangular channel . . . 162
9.4.2 Semi-circular channel . . . 164

10 Finite Element Models **170**
10.1 General formulation for fluid . . . 170
10.1.1 Weak form . . . 170
10.1.2 Shape functions . . . 172
10.2 Fluid element matrices . . . 173
10.3 Solid element matrices . . . 175
10.4 Interface element matrices . . . 175
10.5 Infinite fluid domain . . . 177
10.5.1 Fluid element matrices for channel cross-section . . . 177
10.5.2 Semi-analytical solution for the whole channel . . . 178
10.6 Coupled system . . . 180
10.6.1 Nearfield . . . 180
10.6.2 Appending the farfield . . . 182
10.6.3 Ritz vectors . . . 183
10.6.4 Right-hand-side input . . . 184
10.6.5 Special cases . . . 185
10.7 Frequency-domain analysis, simplified modeling . . . 187
10.7.1 Two-dimensional fluid problem . . . 187
10.7.2 Talvacchia dam . . . 190
10.7.3 Morrow Point dam . . . 195

11 Time-Domain Implementation **199**
11.1 Further modifications for impedance matrix . . . 199
11.1.1 Scaling . . . 199
11.1.2 High-frequency behavior . . . 200
11.2 Coupled system with approximated farfield . . . 200
11.2.1 General case . . . 200
11.2.2 Diagonal approximation . . . 203
11.2.3 Special cases . . . 203
11.3 Time Integration . . . 204
11.4 Review of procedure . . . 205
11.5 Examples with rigid dam . . . 210
11.5.1 Two-dimensional fluid domain with constant depth . . . 210
11.5.2 Three-dimensional examples . . . 212
11.6 Examples with flexible dam . . . 216
11.6.1 Taft earthquake records . . . 216
11.6.2 Two-dimensional example: Pine Flat dam . . . 217
11.6.3 Three-dimensional example: Morrow Point dam . . . 218

12 Closure **224**
12.1 Conclusions . . . 224
12.2 Future Developments . . . 225
12.2.1 Extensions of the method . . . 225
12.2.2 Extensions of the general computer program . . . 226
12.2.3 Applicability to other problems . . . 226

A Program DANAID **228**
A.1 Input processing . . . 229
A.1.1 Format . . . 229
A.1.2 Scanner . . . 229
A.1.3 Parser . . . 230
A.1.4 Geometry definition . . . 230
A.2 Element types . . . 231
A.3 Data structures . . . 233
A.3.1 Nodal data . . . 233
A.3.2 Element data . . . 233
A.3.3 Sets . . . 234
A.4 Memory management . . . 235
A.5 Analysis procedure . . . 236
A.6 Output file structure . . . 237
A.7 Libraries . . . 239
A.8 Status of program and further developments . . . 239

Bibliography **241**

Abstract **253**

Zusammenfassung **255**

Glossary **257**

Notations **260**

Index **263**

Chapter 1

Introduction

1.1 Transmitting boundaries

For many civil engineering structures, interaction with surrounding media leads to important effects that have to be included in a numerical model. Practical examples are soil-structure interaction for buildings on soft soils and fluid-structure interaction for dams and reservoirs. To keep the effort for the analysis reasonable, only the *nearfield* consisting of the structure and a small portion of the surrounding media can be modeled directly while the rest, the *farfield*, has to be captured by an appropriate boundary condition at the *artificial boundary*. The setup for a dam-reservoir problem is schematically shown in Figure 1.1

While the nearfield is conveniently modeled by finite elements, the boundary representing the farfield needs special treatment. For dynamic problems such as earthquake analysis, waves generated in the nearfield propagate to the farfield. The effect of energy loss through the artificial boundary results in *radiation damping*, which may be substantial. If the boundary condition at the artificial boundary is incorrect, waves are reflected back to the nearfield and a wrong radiation damping results. Boundaries that do not reflect waves back to the nearfield are called in the literature by various names: non-reflecting boundaries, quiet boundaries, absorbing boundaries or *transmitting boundaries*.

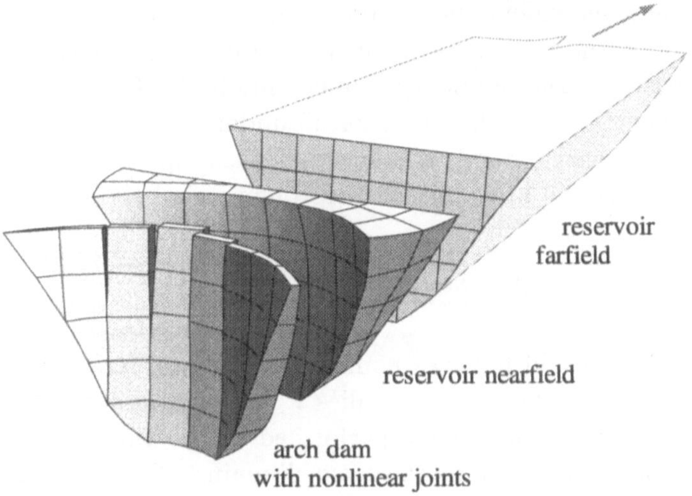

Figure 1.1: General setup for dam-reservoir problem

While various transmitting boundaries have been designed based on the concept of reflection, the rigorous procedure is to derive the boundary condition from the solution of the farfield. Transmitting boundaries that are equivalent to the farfield are called consistent boundaries. The boundary condition can be formulated as a dynamic stiffness for solid problems or generally as an impedance or a Dirichlet-to-Neumann (DtN) boundary condition [HH92]. For unbounded domains, analytical or semi-analytical solutions can be obtained in the frequency domain. These solutions have to satisfy the *radiation condition* at infinity as stated by *Sommerfeld* [Som65].

The frequency-domain solution of a channel exhibits two qualitatively different types of solutions depending on the frequency range. The two ranges are separated by the cut-off frequency which corresponds to the fundamental frequency of the channel cross-section. Below the cut-off frequency the pressure decays exponentially with distance, above the cut-off frequency the solution consists of pressure waves propagating to infinity. The radiation condition prevents waves coming into the model from infinity. Transmitting boundaries can easily be derived from the frequency-domain solution of the channel. If, however, the nearfield behaves nonlinearly, transmitting boundaries are needed that work in the time domain. In the time domain the frequency ranges are not separated and the difficulty is to find boundaries that capture both types of solutions below and above the cut-off frequency simultaneously.

1.2 Frequency-domain solutions

For linear problems, the nearfield and the farfield can be analyzed in the frequency domain. Transfer functions relating the response to the excitation can be calculated and from these, using the Fourier transform, the response in the time domain is determined. Boundaries obtained by this procedure are consistent boundaries. Besides the many analytical solutions for simple geometries [Shu87, Kot59], one of the important contributions in the literature is the semi-analytical solution for a two-dimensional layered soil on a rigid base by Lysmer an Waas [LW72]. The concepts were adapted by Hall and Chopra for two-dimensional [HC82] and for three-dimensional [HC83] reservoirs and by Lotfi et al. [LRT87] for a combination of fluid and soil layers. These methods apply to prismatic semi-infinite domains. The cross-section is discretized while the longitudinal direction is captured by an analytical solution satisfying the radiation condition, thus the name semi-analytical. Similar to semi-analytical methods are infinite elements and mapping finite elements with frequency-dependent shape functions[CC87, Nat81].

The other class of methods that has gained considerable attention are the boundary integral methods, first of all the boundary element methods. Boundary integral methods have the intrinsic advantage that the radiation boundary condition is automatically satisfied. However, if a boundary extends to infinity (for example, free surface of half-space) only part of that boundary can be modeled, introducing a substantial error in some cases.

1.3 Time-domain solutions

If the structure or the nearfield behaves nonlinearly but the farfield remains linear, the concept of transmitting boundaries still applies. The only problem is that, because of the nonlinearities, the analysis has to be performed in the time domain. This requires transmitting boundaries that work in the time domain. The problem of transmitting boundaries in the time domain is the main topic of this investigation. Several attempts to formulate transmitting boundaries in the time domain have been made in the past.

The simpler ones are the local transmitting boundaries. These are dashpots or similar frequency-independent devices which, however, are not equivalent to the farfield and are thus called here *local, non-consistent* boundaries. Consistent boundaries, on the other hand, are frequency-dependent and are usually formulated in the time domain as convolution. They are therefore called here *consistent, non-local* boundaries. We first discuss the local, non-consistent boundaries and then the consistent, non-local boundaries. Then we introduce a new class, the *rational* boundaries. Rational boundaries are consistent and, depending on the realization, local or non-local.

1.3.1 Local, non-consistent boundaries

Local, non-consistent boundaries are formulated directly in the time domain. They are local in space and time, that is, they are applied to each node and each time step individually using only information from the current node and the current time step including spatial and temporal derivatives. The influence of a few neighboring nodes and a few preceding time steps is implicitly included through the discretization of the derivatives. Local transmitting boundaries are frequency-independent and can therefore be utilized directly in the time domain analysis. They are, however, transmitting only for certain waves while for other waves they are reflecting. The simplest transmitting boundaries are the viscous boundaries [LK69], which are simple dashpots. Higher-order boundaries have been proposed for plane waves [EM77, LW84] and for cylindrical and spherical waves [BT80, BBE84].

A somewhat different approach is taken by the infinite elements based on frequency-independent shape functions [OB85b, Ana90] and by the doubly asymptotic approach [Sha87]. Although the philosophy is different, these methods are equivalent to local transmitting boundaries.

1.3.2 Consistent, non-local boundaries

Consistent boundaries are obtained by first solving the problem in the frequency domain and then transforming it to the time domain. Frequency domain solutions are calculated by analytical or semi-analytical methods or by the boundary element method. These solutions are transformed to the time domain by the Fourier transform. Therefore they have no inherent errors other than the discretization and truncation errors. The classical example are the convolution methods. This route was taken by Wepf [WWB88] where the frequency-domain solution is based on a boundary-element method for the nearfield of the reservoir and on an analytical solution for the farfield. Convolution methods use a lot of memory and computer time because each boundary node is influenced by all past values at all other boundary nodes. The effort for the convolution is comparable to that obtained by making the model large enough such that the waves do not reach the boundary within the time of the analysis. The computational effort is substantially reduced by a technique developed by Tsai and Lee [TL90] and improved by Ylang [YTL90] where the convolution is only applied to each mode rather than to each term of the impedance matrix. A computational effort similar to the convolution is necessary if the boundary element method is formulated directly in the time domain [MB82, AvE87]. Another way is to perform a Fourier transform instead of a convolution for each time step [MW89] or for a number of time steps [DW88].

1.3.3 Rational boundaries

A new class of methods for transmitting boundaries is introduced, the rational boundaries. Rational boundaries are boundaries based on rational transfer functions. A rational transfer function can easily be transformed to the time domain resulting either in a differential equation for the continuous-time case or in a recursive scheme for the discrete-time case.

The idea of rational boundaries is best demonstrated by the case of the familiar time-stepping used in structural dynamics. A system of linear differential equations can either be solved by convolution (Duhamel integral) or by a time stepping scheme (Newmark). Clearly the latter is much more efficient in terms of operations count as well as in terms of memory requirements.

Recursive methods have been used in signal processing for digital filters. A large number of publications started to appear in the electrical engineering literature in the early seventies. See, for example, [Ste70, Dec72]. Early applications of recursive digital filters in structural dynamics considered single-degree-of-freedom systems [Lee84, Mee87].

Also differential equations, visualized as systems of masses, springs and dashpots, have been used for a long time to approximate the behavior of infinite domains. Examples are the added mass for a machine foundation on a half-space [RHW70] and the more complicated models for soil-structure interaction for a rigid foundation by Wolf and Somaini [WS86]. All these models are selected on an ad-hoc basis. A more general theory, although only for scalar systems and only for the physical interpretation of the differential equations, is presented in Wolf [Wol91].

Rational boundaries are derived from the frequency-domain solution and are therefore consistent. They are, depending on the realization, more or less local in space and time. A purely recursive realization has to take into account many values from the past but can still be considered local compared to convolution. In a state variable (internal variable) realization only one or two past time steps are included and the algorithm is really local in time. If the system matrix of a state-variable formulation is diagonalized, the state variables are uncoupled and the algorithm is also local in space.

In this work a general procedure based on system theory is presented. This includes a general formulation for multivariable systems, procedures for the approximation, investigation of stability, representation by a second-order system with symmetric matrices and the coupling to the finite element system. The starting point is not the physical system of masses springs and dashpots but a more general system of differential equations. This system is then transformed to the same form as a system of masses, springs and dashpots in order to be consistent with the finite element formulation. These equations may still be interpreted as a system of masses, springs and dashpots but have, other than that, no physical significance.

1.4 Rational approximation

The rational transmitting boundary representing the farfield is modeled by a rational system. A *system* is a mathematical abstraction of a physical model. It has some input-output relationship described by a differential equation, a transfer function or an impulse response. A *rational system* has a rational transfer function. Systems are the subject of *system theory* which will be extensively used here. In this work we are concerned with *linear time-invariant systems.* Linear time-invariant systems can be described by linear differential equations with constant coefficients.

The key step in developing rational transmitting boundaries is the *approximation* of the frequency-domain solution by a rational system. An good overview of rational approximation related to systems together with an extensive bibliography is given by Bultheel and Dewilde [BD82].

Rational functions are very powerful as approximation functions because they are very flexible and can even adapt to singularities. On the other hand, there are two major problems that make the rational approximation of systems difficult. Firstly, the approximation problem is nonlinear, secondly, the resulting system has to be stable. A simple way to start is to use the linearized problem by approximating a weighted error. This method has been applied for continuous-time systems by Sanathanan [ST74]. For digital filters (discrete-time systems) the approximation has been performed either in the time domain [BP70, EF73] or in the frequency domain [Dud74]. In the discrete-time domain, the impulse response is identical to the Markov parameters and we prefer to consider the time-domain approximation to be an approximation of the Markov parameters or as a Padé approximation. Closely related to the Padé approximation are continued fraction methods [Sha76] which are, however, not considered in this work.

On the other hand, there is an extensive mathematical literature on rational approximation using the minimax criterion [Pow81]. Some modifications with regard to stability have usually been made to apply these methods to system theory and digital filters. The Remes exchange algorithm has been adapted in [Dec74], nonlinear optimization has been applied in [Ste70, Dec72].

More closely related to systems are the methods based on the singular value decomposition of the Hankel matrix. The Hankel matrix contains all necessary information of a system. Its rank is equal to the degree of the system and it is used to construct the system matrices. It is also useful to define the Hankel norm. The second ingredient, the singular value decomposition, is a powerful theoretical and numerical tool to determine the rank of a matrix and for finding the important subspaces of a matrix. Within this framework there are two major directions [KA87], the optimal Hankel-norm approximation and the balanced realization.

The *optimal Hankel-norm* approximation goes back to Carathéodory and Fejér (1911), Schur (1917) Takagi (1924) [Tak24, Tak25] and Achieser (1967) [Ach67, appendix]. The theory has been generalized for infinite-dimensional Hankel matrices by Adamjan Arov and Krein (1971) [AAK71]. The latter is a functional analysis approach which is very theoretical and may not be directly used for numerical work. However, it is useful as a theoretical background. Simplified (but still very demanding) presentations are given by Genin and Kung [GK81], Meinguet [Mei83b] and Partington [Par88].

Linear systems notation has been used by Kung [Kun80, KL81b, KL81a, LK82], Glover [Glo84] and Meinguet [Mei83a, Mei88]. The articles by Kung and Glover also include numerical algorithms but they apply to model reduction starting from a higher-order rational system. The only method that is immediately applicable to transmitting boundaries is the method by Gutknecht and Trefethen [GST83, GT80] called the Carathéodory-Fejér (CF) method. The CF method is described in Subsection 5.4.1. It can only be used for scalar systems.

The second direction is the *balanced realization*. Any system described in state-variable form can be transformed to be balanced. For a balanced system the observability and the controllability Gramians are equal and diagonal. The observability Gramian is related to the output energy obtained from the different states and the controllability Gramian describes the input energy necessary to influence the different states. Thus in a balanced system, small entries in the Gramians relate to states that are hard to influence by the input and difficult to observe in the output. These states are not important and can be neglected. The approximation is therefore obtained by taking a subsystem of the balanced system neglecting almost unobservable and uncontrollable states. The procedure does not minimize a norm but yields good results and is numerically robust. The connection between balanced systems and the Hankel matrix is that the entries of the balanced Gramians equal the singular values of the Hankel matrix. The singular vectors can be used for the balancing transformation. Model reduction by balancing is described in several references [Moo78, Moo81, SB80a, PS82, GN86]. It should also be noted that many of the optimal Hankel-norm approximations use balancing as an intermediate step [Glo84].

While balancing is useful for model reduction, the problem is to find a realization from the input-output description in the first place. A classical realization algorithm is the one by Ho and Kalman [HK66]. In this algorithm the system matrices are derived from the Hankel matrix using a decomposition different from the singular value decomposition. Zeiger and McEwen [ZM74] introduced the singular value decomposition in the Ho-Kalman algorithm. As shown by Chen [Che84, p. 270], when using the singular value decomposition, the algorithm yields a balanced realization. Taking a subsystem as approximation, the algorithm effectively performs the realization and the approximation simultaneously. A similar algorithm is due to Kung [Kun78]. These methods are directly applicable for multivariable system. The algorithm by Zeiger and McEwen has been implemented in this work. Parts of the work have been published by the author earlier [WHB89, Web90, Web92].

1.5 Dam-reservoir interaction

The main application for the transmitting boundaries proposed in this work is the dam-reservoir interaction under seismic loading. In this section we give a brief literature review of this topic.

1.5.1 Compressibility effects

Before doing so, we first address the question of compressibility effects of the water. The reason is that transmitting boundaries are only necessary for wave propagation problems and wave propagation only occurs in a compressible fluid. If compressibility effects could be neglected, an incompressible model could be used and there would be no need for transmitting boundaries. To give a first impression of compressibility effects we consider a two-dimensional model of a reservoir with a rigid dam. The exact analysis is presented in Chapter 9. If the dam is moved horizontally, a dynamic pressure arises in the reservoir in excess of the static pressure. The dynamic pressure distribution due to a sinusoidal excitation with a frequency $f = 0.8 f_w$ is shown in Figure 1.2. The frequency f_w denotes the cut-off frequency which is the fundamental frequency of the reservoir $f_w = c/4H$. The parameters are H the height of the reservoir and $c = 1440$ m/s the wave velocity of water. For a typical height of 100 m the cut-off frequency is $f_w = 1440/400 = 3.6$ Hz. Note that the excitation frequency is below the cut-off frequency. The pressure is shown as a time sequence of 4 snap shots arranged vertically. The time increment is $\Delta t = 0.25/f_w$ ($= 0.07$ s for $H = 100$ m). On the left-hand side the incompressible case is shown, on the right-hand side the compressible case. The two cases are qualitatively similar. The dynamic pressure is in phase with the dam (not evident from the figure) and is confined to a region close to the dam. If the artificial boundary is placed outside of that region, the behavior is not affected even if the boundary it is not transmitting.

A different picture arises in Figure 1.3 where the excitation frequency is chosen above the cut-off frequency, in this example $f = 1.5 f_w$. The incompressible model on the left shows a similar pattern as the previous example with the lower excitation frequency. In fact the incompressible model is frequency-independent and the difference is only due

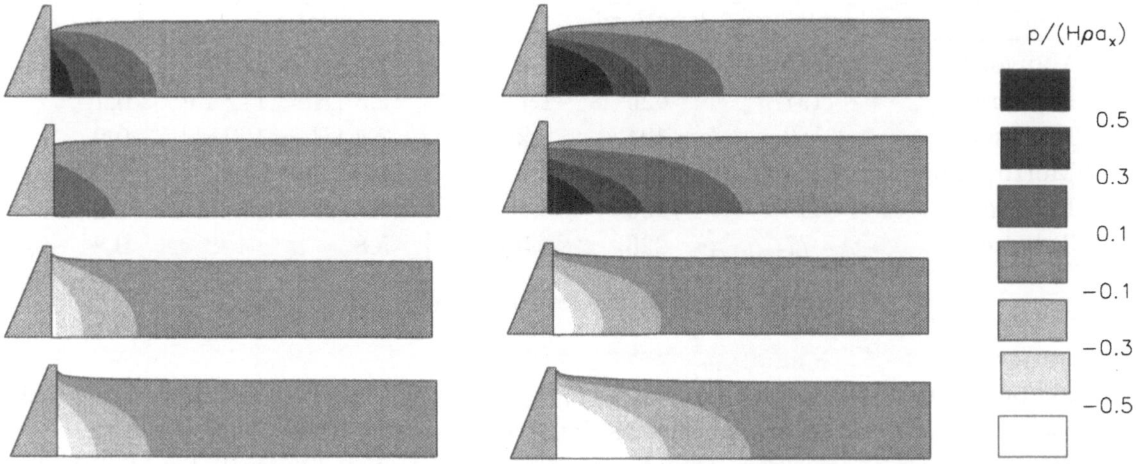

Figure 1.2: Pressure distribution in reservoir, $f = 0.8 f_w$. Left: incompressible fluid. Right: compressible fluid.

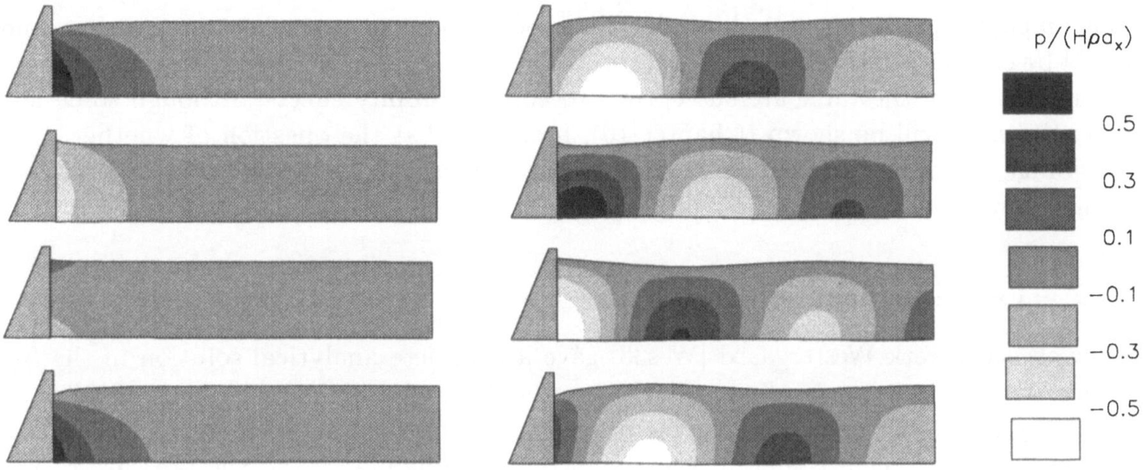

Figure 1.3: Pressure distribution in reservoir, $f = 1.5 f_w$. Left: incompressible fluid. Right: compressible fluid.

to the timing of the frames. The compressible model on the right shows a completely different behavior. The dynamic pressure is out of phase and consists of wave packages that propagate from the dam to the right. The pressure waves decay only slowly due to foundation absorption and the reservoir is influenced even far away from the dam.

For a three-dimensional problem with a flexible dam the pressure distribution is more complicated, but the basic behavior is the same as for the simple case. Compressiblity effects may be neglected if the fundamental frequency of the dam-reservoir system is well below the cut-off frequency. This criterion is not useful for practical purposes, because an analysis with compressible water is needed to determine the fundamental dam-reservoir frequency. Following Hall [Hal88], a more practical way is to compare the frequency of the first symmetric mode of the dam with empty reservoir, f_d, to the cut-off frequency of the reservoir, f_w. The symmetric dam mode has to be taken because antimetric modes are not greatly affected by compressibility. According to Hall, compressibility effects may

Dam	height [m]	length[m]	f_d [Hz]	f_w [Hz]	f_d/f_w
Mauvoisin	250	560	2.0 [num.]	2.1 [H = 190 m]	0.9
Kölnbrein	197	626	1.7 [exp.]	2.3 [H = 172 m]	0.7
Emosson	180	424	2.2 [exp.]	2.4 [H = 160 m]	0.9
Morrow Point	142	219	3.7 [num.]	3.0	1.2
Pacoima	113	180	5.1 [exp.]	4.4 [H = 90 m]	1.2
Talvacchia	77	226	4.5 [num.]	5.8	0.8

Table 1.1: Frequencies for various dams

not be expected to be important for a ratio f_d/f_w below 0.7. To give an impression of the range of this parameter, values for various dams are shown in Table 1.1.

The numbers for f_d are either taken from experiments (labeled '[exp.]') listed in [Hal88] or from numerical analyses (labeled '[num.]') presented in this report and in the literature [Kni93, ISM91]. The values for f_w are either estimated using an idealized geometry (labeled '[H = ... m]') or calculated from the numerical eigenvalue problem. Idealized geometries are the semi-circular channel with $f_w = c/(3.41H)$ and the rectangular channel with $f_w = c/(4H)$. The numbers in the table are averaged values for these two geometries using an equivalent height as indicated. The numbers depend on various assumptions and should be considered as estimates only, not as absolute values. According to the criterion, all dams listed in the table are susceptible to compressibility effects, although some less than others. It will be shown (Chapter 10), however, that the question of whether compressibility has to be included in the analysis or not, cannot be answered on the basis of a single parameter.

1.5.2 Two-dimensional solutions

In his classical work Westergaard [Wes33] gave a complete analytical solution in the frequency domain for the two-dimensional compressible reservoir model with a rigid dam. As an approximation he considered the pressure distribution for low frequencies. For low frequencies the problem is identical to the incompressible model and the pressure is proportional to the acceleration of the dam. The effects of the reservoir can thus be included by an *added mass*. The assumption of low frequency behavior was justified by the following — from today's perspective unreasonable — statement: "Since the main vibrations in the earthquake are not likely to have a period less than 1 sec., this form of resonance need not be expected in any project now contemplated". Kotsubo [Kot59] has included resonance effects and has also given results for a transient excitation using the convolution approach. He also considered three-dimensional models with different cross-sections.

The two-dimensional problem of hydrodynamic pressure on a rigid dam was taken up again by Chopra [Cho67]. Later he and his co-workers extended the models to the interaction with a flexible dam [Cho68], effects of reservoir bottom absorption [FC83], and the interaction with a flexible foundation [CC81]. All these calculations were based on analytical solutions for the reservoir combined with Ritz vectors for the dam-foundation system. Hall and Chopra [HC82] modeled the dam and the reservoir nearfield by finite elements and the reservoir farfield by a semi-analytical solution. The introduction of finite elements for the reservoir nearfield allows to investigate the influence of irregular reservoir geometry.

Boundary elements have also been used to solve the two-dimensional problem. Wepf [WWB88] used the convolution approach based on a frequency-domain solution. The reservoir nearfield was modeled by boundary elements, the farfield by an analytical method. Even though the original motivation for the boundary elements was the modeling of the infinite region, this aspect has been dropped because with boundary elements the bottom and the free surface of the reservoir cannot be extended to infinity. The effects of reservoir bottom absorption and vertical excitation were also not considered. Based on the same concepts Humar and Jablonski [HJ88] developed a frequency-domain solution including reservoir bottom absorption and vertical excitation. Antes and von Estorff [AvE87] used a time-stepping boundary element method. Vertical excitation was included and bottom absorption was accounted for by modeling the foundation.

The two-dimensional solutions referred to so far applied only to linear problems. However, the convolution approach by Wepf [WWB88] and the time-stepping boundary element method by Antes and von Estorff [AvE87] are time-domain solutions and are easily combined with a nonlinear dam. This was done by Feltrin [FWB90] where the solution of Wepf was combined with the earlier developed discrete-crack procedure for the dam by Skrikerud [SB86]. Other nonlinear analyses used some simplified treatment of the farfield [FVL88].

1.5.3 Three-dimensional solutions

Three-dimensional analytical results in the frequency domain were developed for channels with different geometries by Werner and Sundquist [WS49], Kotsubo [Kot59] and Shull'man [Shu87] assuming a rigid dam. Foundation absorption was included for a channel of semi-circular cross-section by Szczesiak and the author[SW92]. A wedge-shaped reservoir was considered elsewhere [Per73].

A finite element model of a simple arch dam was combined with analytical results for a wedge-shaped reservoir in [PC81]. Hall and Chopra generalized the two-dimensional finite element approach [HC82] to the three-dimensional case [HC83]. This method is now considered the state of the art for linear analysis.

A three-dimensional boundary element solution has been developed by Jablonski and Humar [JH90]. The reservoir nearfield and the dam were modeled by boundary elements, the reservoir farfield by a semi-analytical finite element solution.

All three-dimensional methods referred to so far work in the frequency domain and are only applicable to linear analysis. Time-domain procedures have been proposed using simple treatment of the farfield such as viscous boundaries [OB88, AO88, GO90] or neglecting the compressibility of the water [TL87, DH89, FMR92, MR93]. Tsai and Lee [TL90] presented a time-domain method which rigorously treats the farfield. Convolution was used for each mode rather than for each term of the matrix coupling the nearfield and the farfield which substantially reduces the demand on computer resources.

1.5.4 Experimental work

The validation of numerical methods by experiments is an important question but is not subject of this work. A comprehensive overview of experiments worldwide is presented by Hall [Hal88]. Only two experiments that were specially designed to investigate the compressibility effects of the water will be mentioned here.

One is the classical laboratory experiment by Hatano [Hat65]. A steel tank of 480 cm length and 10 cm width was filled with water of a depth of 150 cm and was excited harmonically by a piston of 10 cm $\times$ 10 cm. Resonance was observed at the theoretical value of the excitation frequency. This resonance was no longer observed when a 10 cm layer of sand was placed at the bottom and at the far wall. It was concluded that, in the presence of the sand, the incompressible water solution would apply. However, as pointed out in [Hal88], in the case of resonance, the measured phase angle cannot be explained theoretically and, in the case of no resonance, the theoretical value of the incompressible water solution was not observed. The inconsistencies in the experiments are possibly due to the flexibility of the steel tank. Because of the many questions the experiments seem to be unreliable.

The second experiment is the forced vibration test of Morrow Point Dam by Hall and Duron [DH88]. Measured accelerations and water pressures were compared to finite element results. Water compressibility was either included or neglected for the theoretical results. The model was calibrated using the antisymmetric response because in this case the compressibility has little effect. Good agreement was obtained. For the symmetric response, for which compressibility is important, the compressible model showed better correlation with the measurements than the incompressible model, but the agreement was still unsatisfactory. However, the influence of compressibility effects was evident.

1.6 Objectives

Based on the literature review and on our experience, the following can be stated: In the frequency domain the problem of transmitting boundaries can be solved rigorously and efficiently for infinite domains with different geometries, including the reservoir problem, by analytical or semi-analytical methods. Frequency-domain solutions are only applicable to linear problems. If the nearfield exhibits nonlinear behavior, the analysis has to be performed in the time domain. To do so, time-domain transmitting boundaries are needed. Local, non-consistent transmitting boundaries are easy to implement but do not adequately represent the radiation condition for many problems, including the reservoir problem. On the other hand, consistent, non-local transmitting boundaries based on convolution or time-stepping boundary elements are very demanding in computer time and memory and cannot therefore be used for large three-dimensional problems.

The formulation of efficient time-domain transmitting boundaries is the main goal of this work. The method has to be applicable to dam-reservoir interaction under seismic loading. Specifically the following objectives can be stated. The work should

- Give a mathematical and intuitive insight into the problem of transmitting boundaries in the time and frequency domains.
- Show the limitations of commonly used local, non-consistent transmitting boundaries.
- Develop a numerically efficient procedure for a consistent transmitting boundary in the time domain, avoiding convolution.
- Show the efficiency and applicability of the procedure for a large model of an arch dam interacting with an infinite reservoir.

1.7 Scope

1.7.1 Model and restrictions

Besides the question of transmitting boundaries many other problems have to be dealt with when investigating the earthquake behavior of dam-reservoir systems. From the many possiblities to treat these problems one has to make some resonable, practical choises.

The fluid is considered compressible and irrotational and having small displacements and negligible viscosity effects. The incompressible case is also included for comparison.

A general geometry can be defined for the reservoir nearfield. The farfield geometry is assumed to be prismatic with an arbitrary cross-section. Reservoir bottom absorption is included by a simple dashpot model. While the farfield has to remain linear, the finite element formulation allows nonlinear behavior of the nearfield.

The dam is modeled by finite elements. Elastic solid elements are employed, but the method does not preclude any other element type. Solid elements are preferred over shell elements to reduce the programming effort and to be compatible with nonlinear joint elements [Hoh92] which will be incorporated into the computer program at a later stage.

The α-method due to Hilber, Hughes and Taylor [HHT77] is used for time integration. This algorithm is second-order accurate and permits numerical damping of higher modes. The algorithm is implemented in an implicit-explicit form [MFH89] which allows to mix implicit and explicit elements. All calculations are, however, done in a purely implicit manner. While this algorithm is standard for finite elements, any other algorithm could be used. For other algorithms see, for example, [Hug87, SB80b, Hal83]

The subject of earthquake input is not addressed in this work. For a risk analysis, a detailed study of expected peak ground acceleration, duration and frequency content of the earthquake has to be performed. The specific earthquake record used in the examples has only been chosen to demonstrate the numerical performance of the computer program. For more details on earthquake engineering, the reader is referred to the literature [Bac93].

The rigorous treatment of the scattering of incident waves and the dam-foundation interaction with a proper transmitting boundary is a subject in its own right and is not considered here. For all examples a rigid foundation was assumed. A simplified procedure using massless elastic solid elements for the foundation has been employed elsewhere [ISM93] by the author.

Although the reason for developing transmitting boundaries in the time domain is the modeling of nonlinear dam behavior, only linear examples are shown. Extension of the computer program to nonlinear analysis is a major task that is postponed to the future.

1.7.2 Related problems

The approximation of a given transfer function by a rational function is a topic also encountered in other related problems. Many of the publications referred to pertain to these related problems, but many of the techniques can, however, be adopted making the necessary adjustments. It is important to see the similarities but also the differences in these problems.

- □ Directly applied to transmitting boundaries is the *realization* of a given transfer function by a rational function where the transfer function is given numerically as discrete values.

- □ One closely related problem is the design of *recursive digital filters.* In filter design the filter coefficients are optimized such that the filter matches a given transfer function as closely as possible. The given transfer function is, however, not related to a physical problem and does therefore generally not represent a causal system. Also, the prescribed transfer function is usually not smooth but typically has jumps. These properties make the filter problem more difficult than the transmitting boundary problem. On the other hand, filters are scalar systems whereas transmitting boundaries are described by multivariable systems.

- □ Another related problem is *model reduction.* In the model reduction problem a large rational system is given and has to be reduced to a smaller system. Some methods first calculate the transfer function and then find a realization by a low-degree system. Such methods can be used directly for the transmitting boundary problem. Most methods, however, make use of the given high-degree rational system.

- ☐ System *identification* also is a problem of finding a system for a given transfer function. But the transfer function is obtained from measurements and inherently includes noise which is much larger than numerical round-off errors.

1.7.3 Presentation

The work presented includes subjects from different fields, mainly from the finite element method and from system theory. Depending on his knowledge in one or the other field, the reader might find some explanations more difficult than others. However, to give a detailed treatment of both the finite element method and the linear system theory would go beyond the extent of this work. Therefore, the finite element part is described only in general, assuming that the reader is familiar with this method. Also, the finite element method and especially the isoparametric elements are considered a standard technique which is described in many text books [Hug87]. Not only the theoretical derivations can be found there but also the practical and implementational aspects. The only less known aspect is the formulation of the fluid and interface elements. On the other hand, system theory is not familiar to civil engineers and has to be explained in more detail. Although many definitions and theorems are standard and can be found in text books [Kai80, Che84], other explanations are only found in the specialized literature and have to be collected from various sources. While there is a strong theory, numerical aspects are usually not found in text books.

Results are shown either for dimensionless parameters or for parameters in the range of relevant applications. It is beyond the scope of this work to present a conclusive parametric study. The main purpose of the results shown is to verify the numerical procedures. However, some examples are given to show the influence of parameters determining the choice of the mathematical models (compressibility, absorbing foundation).

1.8 Outline

The material is presented on two levels. On the first level, comprising Chapters 1–5, simple examples of scalar problems are treated. Neither finite elements nor system theory is employed. On the second level, including Chapters 6–12, more advanced problems are investigated. The finite element method is used for modeling the nearfield, while multivariable systems are employed for the farfield approximation.

Chapter 2 introduces simple model problems. They are intended to make the reader familiar with the problems of wave propagation, especially with the radiation condition for infinite domains. The difficulty to formulate the radiation condition in the time domain is explained. It is shown that certain problems, including the dam-reservoir-interaction problem, have a cut-off frequency which separates two qualitatively different solution regimes. Below the cut-off frequency the solution is exponentially decaying with distance, above it, the solution consists of waves propagating to infinity. In a time-domain analysis these two types of solutions coexist simultaneously because the different frequencies are not separated. The model problems are used in subsequent chapters as test problems for numerical methods.

Chapter 3 gives an overview of some commonly used numerical methods for transmitting boundaries in the time domain. The most important ones are explained and investigated in the context of the model problems, others are only covered briefly. It is shown that local, non-consistent boundaries are not appropriate for problems with a cut-off frequency. Boundary elements are shown to be basically equivalent to convolution when used in the time domain. Infinite elements assume a certain behavior with distance and cannot generally cope with both solution types below and above the cut-off frequency simultaneously.

Chapter 4 contains the theoretical background of signal processing for scalar systems. Concepts of rational transfer functions, causality and stability are covered. The Hankel matrix and singular value decomposition are introduced. All topics are explained for discrete-time and for continuous-time systems and the strong duality between the two domains is shown. Finally, the bilinear transform is used as a link between the two formulations.

Chapter 5 describes numerical methods for determining the coefficients of scalar rational transfer functions. Starting with simple least-squares methods, the transfer function is first approximated directly. Approximating the Markov parameters leads naturally to the Hankel matrix. Because the Markov parameters are much easier to determine for discrete-time systems, the continuous-time system is approximated via the mapped discrete-time system obtained by the bilinear transform. It is shown that small perturbations of the Markov parameters may lead to almost singular equations. This problem is avoided in the Carathéodory-Fejér method which is based on the singular value decomposition of the Hankel matrix. This method also yields more accurate results than all other methods.

To extend the theory to multivariable systems, the state-variable notation of systems is introduced in Chapter 6. This form is very general and allows compact matrix notation. The topics covered earlier for scalar systems are extended to multivariable systems. Additional topics which are important for the next chapter are observability, controllability and balanced realization. Again the duality between discrete-time and continuous-time systems is pointed out.

Because the extension of the Carathéodory-Fejér method to multivariable systems appears to be very involved, another method also based on singular value decomposition of the Hankel matrix is introduced. The method leads to a balanced realization, neglecting unimportant states. It is slightly less accurate than the Carathéodory-Fejér method but numerically robust and can be used for scalar as as well as for multivariable system. It is the preferred method and is implemented in the general computer program. Various examples are given to illustrate the formal theory.

Chapter 8 shows the transformation of first-order systems obtained in the last chapter into symmetric second-order systems. This is an important step towards the coupling of the farfield to the nearfield.

Chapter 9 gives the mathematical model for the fluid and analytical solutions for the two-dimensional reservoir model and for the three-dimensional models with rectangular and semi-circular cross-section. For the two-dimensional case also simplified models such as incompressible fluid and viscous boundaries are investigated. The analytical solutions are used later for verification of the numerical procedures.

Chapter 10 gives a short introduction to isoparametric finite elements. Fluid, solid and interface elements are covered. The relation between the velocity and the velocity potential at the transmitting boundary is derived leading to a finite element discretization of the cross-section of the channel. For an upstream excitation an eigenvalue problem results, while for a cross-stream excitation a boundary value problem. The size of the eigenvalue problem is reduced by approximating the frequency-dependent eigenvectors by frequency-independent Ritz vectors. Formulations for the total coupled system with solid, nearfield fluid and farfield fluid are given in the frequency domain. The formulation includes the incompressible model and the viscous boundary as special cases. These simplified models are compared to the rigorous models of general-shaped three-dimensional reservoirs with flexible arch dams.

Chapter 11 covers the practical aspects of the time-domain implementation. When using the coupled system with the Ritz vectors the the frequency-dependent part is restricted to a small submatrix. This matrix is approximated by a linear system using the methods derived in previous chapters. The total coupled system in the time domain is given. The complete procedure for constructing the rational transmitting boundary is reviewed before showing numerical solutions of the verification examples and applications of the method to large problems of arch dams.

Finally, in Chapter 12 conclusions are stated and suggestions for further work are given.

Chapter 2

Model Problems

This chapter introduces some simple model problems used to give a brief introduction to wave propagation in infinite domains and related problems. Specifically, they will be used to explain the main difficulty when formulating the radiation condition in the time domain. The model problems are as simple as possible but general enough to allow a conclusive discussion. In the following chapters they will be used to investigate some commonly used numerical solutions and some new methods which are the main contribution of this work. A similar collection of example problems for wave propagation can be found in [Gra75].

First, the simplest wave-propagation model, the semi-infinite rod, is considered. It is used to introduce the concepts of radiation condition and transmitting boundaries. It is shown that the semi-infinite rod has a simple and exact transmitting boundary.

This is not true for the next case, the spherical cavity problem. The problem has to be analyzed in the frequency domain and the transmitting boundary is formulated as a convolution integral. However, by introducing an internal variable, it is possible to formulate a transmitting boundary in differential equation form. This is generally possible if the frequency-domain boundary condition is a rational function. Differential equations can be solved numerically much more efficiently than convolution integrals.

A third example is obtained by adding an elastic or visco-elastic foundation to the semi-infinite rod which drastically changes the behavior of that model. Waves are dispersive, that is, they travel at a frequency-dependent velocity. There is a cut-off frequency below which waves do not travel at all. The frequency-domain boundary condition is no longer a rational function.

The last example is the simplest two-dimensional case, that is, a fluid channel with constant cross-section. It is shown that each mode behaves the same way as the semi-infinite rod on a elastic foundation. The extension to two dimensions also leads to some interesting geometric interpretations.

2.1 Semi-infinite rod

The simplest wave-propagation problem for an infinite domain is the semi-infinite rod as shown in Figure 2.1a. The prismatic rod has a cross-sectional area A and a mass per unit length μ. It is assumed to behave elastically with a modulus of elasticity E. The displacement in the x-direction at time t is $u(x,t)$. The governing differential equation can be found considering the momentum of an infinitesimal element as shown in Figure 2.1b.

$$EA\frac{\partial^2 u}{\partial x^2}dx = \mu \ddot{u}\, dx \tag{2.1}$$

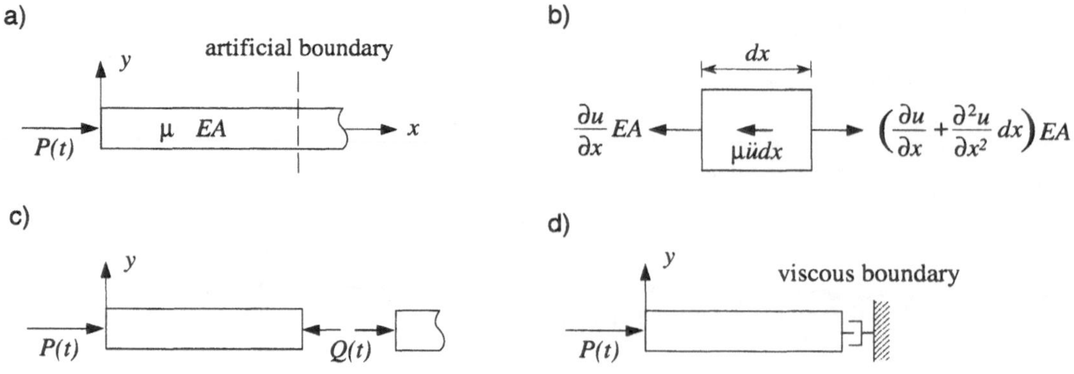

Figure 2.1: Semi-infinite rod: a) model, b) infinitesimal element, c) reaction force, d) nearfield with viscous boundary

This equation, brought into the standard form, leads to the one-dimensional wave equation:

$$\frac{\partial^2 u}{\partial x^2} = \frac{1}{c^2}\ddot{u} \tag{2.2}$$

where $c = \sqrt{EA/\mu}$ denotes the wave velocity. The general solution of this partial differential equation is given by

$$u(x,t) = f_1(t - x/c) + f_2(t + x/c) \tag{2.3}$$

where f_1 and f_2 are arbitrary (sufficiently smooth) functions. The function $f_1(t - x/c)$ represents a wave propagating in the positive x-direction with velocity c, whereas $f_2(t + x/c)$ represents a wave propagating in the opposite direction.

Radiation condition. If the semi-infinite rod is excited by a force $P(t)$ at $x = 0$, then the corresponding boundary condition is

$$P(t) = -EA\frac{\partial u}{\partial x} = c\mu\left(f_1'(t) - f_2'(t)\right) \tag{2.4}$$

where the prime denotes the derivative with respect to the argument. The second boundary condition is less obvious. Because the rod is excited on the left, we have to assume that waves can only travel to the right, otherwise they would have to come from infinity. The condition that only outgoing waves exist, is called the *radiation condition.* Since the general solution (Equation 2.3) is already given in terms of incoming and outgoing waves, the radiation condition is easily satisfied by setting $f_2 = 0$.

Transmitting boundary. In a numerical analysis it is not possible to model an infinite domain. Typically, a finite portion, the nearfield, is modeled by finite elements. The infinite portion, the farfield, is replaced by an appropriate boundary condition at the artificial boundary. An artificial boundary with a boundary condition that replaces an infinite domain is called a *transmitting boundary.* Clearly, either a fixed or a free boundary does not work, because outgoing waves would be reflected back instead of passing to the farfield. A transmitting boundary has to be formulated as a relationship between the displacement and the force at that boundary. Mathematically, a boundary with prescribed displacement (generally, prescribed field variable) is called a Dirichlet boundary, a boundary with prescribed force (generally, prescribed normal derivative) is called a Neumann boundary. Therefore, the relationship between the displacement and the force is also called a Dirichlet to Neumann (DtN) condition [HH92]. In solid mechanics this condition

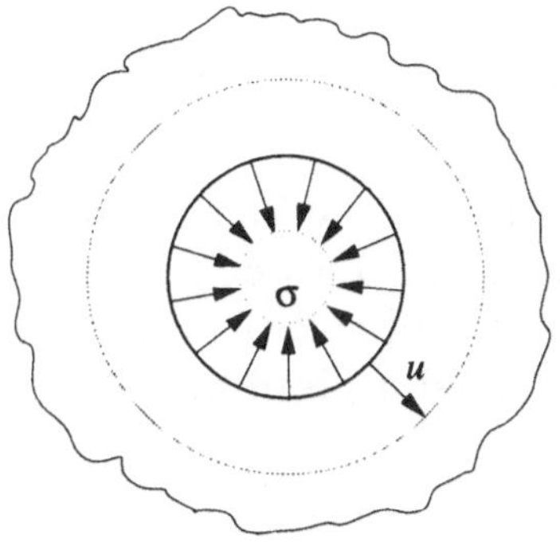

Figure 2.2: Spherical cavity

is usually expressed as a dynamic stiffness, defined as the force due to a unit displacement. For a fluid, however, the field variable is the dynamic pressure and the mathematically equivalent expression is a velocity due to a unit pressure, thus reversing displacement-like and force-like terms. Therefore, dynamic stiffness is only used when referring to a solid mechanics problem. As a more general term we prefer *impedance*.

To develop the transmitting boundary, the farfield is replaced by the reaction force $Q(t)$ as shown in Figure 2.1c. For the semi-infinite rod satisfying the radiation condition, the force Q, analogous to Equation (2.4), is

$$Q(t) = -EA\frac{\partial u}{\partial x} = c\mu f_1'(t - x/c) \tag{2.5}$$

The local derivative f_1' can be expressed in terms of the particle velocity which is given by

$$v(t) = \dot{u}(x,t) = f_1'(t - x/c) \tag{2.6}$$

The boundary condition is therefore not a force-displacement but rather a force-velocity relationship.

$$Q(t) = c\mu v(t) \tag{2.7}$$

The farfield of the semi-infinite rod can thus be replaced by a simple dashpot with a damping constant $c\mu$ which can readily be implemented in the time domain. Note, that this boundary condition is independent of the location of the transmitting boundary. It is exact for this one-dimensional case. This boundary condition is commonly referred to as viscous boundary or Lysmer-Kuhlemeyer boundary [LK69]. The nearfield with the viscous boundary is shown in Figure 2.1d.

Using $Q = -EA\partial u/\partial x = -c^2\mu\partial u/\partial x$, Equation (2.7) can also be expressed as

$$\frac{\partial u}{\partial x} + \frac{\partial u}{c\,\partial t} = 0 \tag{2.8}$$

2.2 Spherical cavity

The next problem is an infinite three-dimensional elastic solid domain with a spherical cavity. We only consider the simple case of the radially symmetric problem as shown in Figure 2.2.

It is convenient to introduce the displacement potential ψ [Fun65]. The differential equation for the displacement potential is

$$\Delta\psi = \frac{1}{c_P^2}\ddot{\psi} \tag{2.9}$$

where c_P is the compression wave (P-wave) velocity defined as

$$c_P^2 = \frac{\lambda + 2G}{\rho} \tag{2.10}$$

and ρ is the mass density and λ and G are Lame's constants. The latter are alternative parameters to Young's modulus E and Poisson's ratio ν. The relationship between the parameters is $\lambda = \nu E/[(1+\nu)(1-2\nu)]$ and $G = E/[2(1+\nu)]$. The radial displacement in terms of the displacement potential is given by

$$u = \frac{\partial \psi}{\partial r} \tag{2.11}$$

and the radial stress in terms of the displacement by

$$\sigma = \rho\, c_P^2 \frac{\partial u}{\partial r} + 2\rho\,(c_P^2 - 2\,c_S^2)\frac{u}{r} \tag{2.12}$$

where the shear-wave (S-wave) velocity is defined as

$$c_S^2 = \frac{G}{\rho} \tag{2.13}$$

Considering the radially symmetric problem, $\psi(r,t)$ is only a function of the distance r and the time t. With the Laplace operator in spherical coordinates, the wave equation (2.9) becomes

$$\Delta\psi = \frac{1}{r}\frac{\partial^2}{\partial^2 r}(r\psi) = \frac{1}{c^2}\ddot{\psi} \tag{2.14}$$

This is the one-dimensional wave equation for $r\psi$, thus

$$r\psi(r,t) = f_1(t - r/c) + f_2(t + r/c) \tag{2.15}$$

Considering only outgoing waves results in

$$\psi(r,t) = \frac{1}{r} f_1(t - r/c) \tag{2.16}$$

As for the semi-infinite rod, an exact transmitting boundary could be found for ψ. However, keeping in mind a finite element implementation, we are interested in a stress-displacement formulation. For the radial displacement we have

$$u(r,t) = -\frac{1}{c_P\, r} f_1'(t - r/c_P) - \frac{1}{r^2} f_1(t - r/c_P) \tag{2.17}$$

and for the radial stress

$$\sigma(r,t) = \rho\left(\frac{1}{r} f_1''(t - r/c_P) + \frac{4\,c_S^2}{c_P\, r^2} f_1'(t - r/c_P) + \frac{4\,c_S^2}{r^3} f_1(t - r/c_P)\right) \tag{2.18}$$

The function f_1/r can be substituted by ψ and the derivatives f_1'/r and f_1''/r by the temporal derivatives $\dot{\psi}$ and $\ddot{\psi}$, respectively. Then Equation (2.17) reduces to

$$u(r,t) = -\frac{1}{c_P}\dot{\psi} - \frac{1}{r}\psi \tag{2.19}$$

and Equation (2.18) to

$$\sigma(r,t) = \rho\left(\ddot{\psi} + \frac{4\,c_S^2}{c_P\, r}\dot{\psi} + \frac{4\,c_S^2}{r^2}\psi\right) \tag{2.20}$$

To simplify the algebra, the special case $c_P = 2\,c_S$ ($\nu = 1/3$) is used in the following. With this assumption Equation (2.20) reads

$$\sigma = \rho\left(\ddot{\psi} + \frac{c_P}{r}\dot{\psi} + \frac{c_P^2}{r^2}\psi\right) \tag{2.21}$$

The relationship between u and σ cannot, for this example, be put into a simple form as in the previous examples. It is therefore not possible to formulate a simple transmitting boundary.

Frequency-domain analysis. To proceed further, the problem is investigated in the frequency domain. This is done by assuming a steady state solution of the following form:

$$\psi(r,t) = \frac{1}{r}\hat{f}e^{i(\omega t - kr)} \tag{2.22}$$

where $\hat{f}$ is a complex frequency-dependent function, ω the circular frequency and k the wave number. This solution satisfies Equation (2.16) for

$$k = \frac{\omega}{c_P} \tag{2.23}$$

For the radial displacement we have

$$u = -\left(\frac{i\omega}{c_P} + \frac{1}{r}\right)\psi \tag{2.24}$$

and for the radial stress

$$\sigma = \rho\left((i\omega)^2 + \frac{c_P}{r}i\omega + \frac{c_P^2}{r^2}\right)\psi \tag{2.25}$$

Therefore, the stress can be expressed in terms of the displacement as

$$-\sigma = \frac{\rho\,c_P^2}{r}\left(\frac{i\omega r}{c_P} + \frac{1}{1 + i\omega r/c_P}\right)u \tag{2.26}$$

Splitting this equation into real and imaginary parts and noting that $\dot{u} = i\omega\,u$, the stress can also be written as

$$-\sigma = \frac{\rho c_P^2}{r}\frac{1}{1 + (\omega r/c_P)^2}u + \rho c_P\left(1 - \frac{1}{1 + (\omega r/c_P)^2}\right)\dot{u} \tag{2.27}$$

The first term corresponds to a frequency-dependent spring, while the second term can be interpreted as a frequency-dependent dashpot. This physical interpretation as a frequency-dependent spring and dashpot can be applied to any dynamic stiffness. The real part is equivalent to a spring, the imaginary part divided by ω is equivalent to a dashpot.

Fourier transform. The classical way to transform a function from the frequency domain to the time domain is the Fourier transform. The Fourier transform permits the representation of an arbitrary function $h(t)$ in terms of harmonic functions $e^{i\omega t}$ with a frequency-dependent amplitude $H(\omega)$. It is given by the pair of equations

$$h(t) = \frac{1}{2\pi}\int_{-\infty}^{\infty} H(\omega)\,e^{i\omega t}\,d\omega \tag{2.28}$$

$$H(\omega) = \int_{-\infty}^{\infty} h(t)\,e^{-i\omega t}\,dt \tag{2.29}$$

For the integral in Equation (2.28) to exist, the integrand has to decay to zero for high frequencies. If this is not the case, the high-frequency behavior has to be treated separately.

To transform Equation (2.26) to the time domain, consider the high-frequency behavior

$$-\sigma = \rho c_P i\omega\, u \tag{2.30}$$

This term corresponds to a dashpot, as already encountered for the semi-infinite rod. Equation (2.26) can therefore be written as

$$-\sigma = \rho\, c_P\, \dot{u} + \frac{\rho\, c_P^2}{r}\left(\frac{1}{1 + i\omega r/c_P}\right) u \tag{2.31}$$

The second term decays to zero for high frequencies and can therefore be transformed to the time domain as

$$\frac{1}{2\pi}\int_{-\infty}^{\infty}\frac{1}{1+i\omega r/c_P}e^{i\omega t}d\omega = e^{-c_P\, t/r} \tag{2.32}$$

The integral can be found in tables [CF48]. If evaluated explicitly, care has to be taken to choose the right path at the pole $i\omega = -c_P/r$.

Convolution integral. The multiplication of u with a frequency-dependent factor is transformed to a convolution integral in the time domain. For a transient motion $u(t)$ the stress is

$$-\sigma(t) = \rho\, c_P \dot{u}(t) + \frac{\rho\, c_P^2}{r}\int_0^t e^{-c_P\tau/r}u(t-\tau)\, d\tau \tag{2.33}$$

In a transient analysis $\sigma(t)$ has to be evaluated for each time step and each time the convolution integral has to be calculated. In a large model this has to be done for many degrees of freedom and can be quite time-consuming. Further, the integrand is usually not given analytically but has to be stored in memory. Thus the method might also require large memory for a large model.

Transmitting boundary in the time domain. To construct an exact transmitting boundary in the time domain, Equation (2.31) is rewritten introducing the internal variable ζ.

$$-\sigma = \rho\, c_P\, \dot{u} + \frac{\rho\, c_P^2}{r}\zeta \tag{2.34}$$

where

$$\zeta\,(1 + i\omega r/c_P) = u \tag{2.35}$$

This equation is easily transformed to the time domain as

$$\zeta + \frac{r}{c_P}\dot{\zeta} = u \tag{2.36}$$

This is a differential equation for the internal variable ζ. For a given transient motion $u(t)$ Equation (2.36) is integrated to obtain $\zeta(t)$ and then the value for σ can be calculated using Equation (2.34).

This might not look much simpler than calculating the convolution integral. Recall, however, that the differential equation (2.36) can be solved numerically using a standard time-stepping scheme involving only a few coefficients and for each time step a few multiplications. Evaluation of the convolution integral, on the other hand, requires N_T coefficients and for each time step $N_T/2$ multiplications on average, where N_T is the number of time steps, typically several thousand. Solving a differential equation by a time-stepping scheme rather than by a convolution integral is far more economical because it is a recursive algorithm, taking advantage of past results, rather than starting from zero each time. More on recursive algorithms will be presented in Chapter 4.

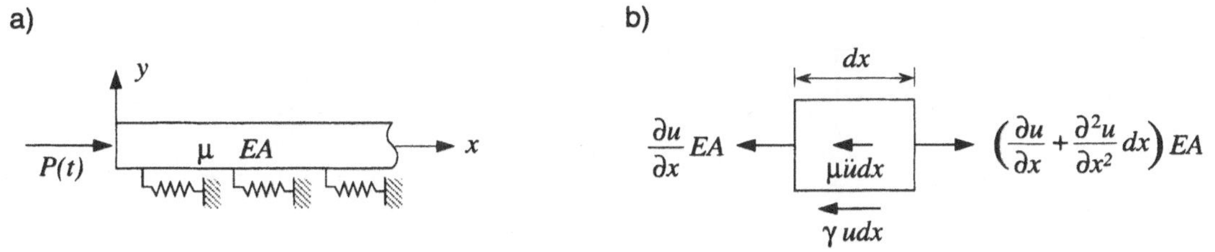

Figure 2.3: Semi-infinite rod on elastic foundation: a) model, b) infinitesimal element

If a transmitting boundary can be formulated as a differential equation rather than a convolution integral, we can drastically save computer memory and time. This is the key idea of the present work. The formulation as a differential equation is always possible if the boundary condition is a rational function in the frequency domain. In the general (non-rational) case this is only possible in an approximate way. In this work we will show a general procedure to obtain an accurate and stable approximation of a frequency-dependent boundary condition by a differential equation.

2.3 Semi-infinite rod on elastic foundation

A more complex case of a wave propagation problem is obtained by adding an elastic foundation to the semi-infinite rod considered before as shown in Figure 2.3a. Considering the infinitesimal element in Figure 2.3b, the momentum equation yields

$$EA\frac{\partial^2 u}{\partial x^2}\,dx - \gamma u\,dx = \mu\ddot{u}\,dx \tag{2.37}$$

or simplified with $\lambda^2 = \gamma/EA$

$$\frac{\partial^2 u}{\partial x^2} - \lambda^2 u = \frac{1}{c^2}\ddot{u} \tag{2.38}$$

This differential equation can no longer be solved directly in the time domain, but has to be solved in the frequency domain. This is done by assuming a steady state solution of the following form

$$u(x,t) = \hat{u}e^{i(\omega t - kx)} \tag{2.39}$$

where $\hat{u}$ is the (complex) amplitude, ω is the frequency and k the wave number in the x-direction. This equation substituted into the differential equation (2.38) yields

$$k = \pm\sqrt{\frac{\omega^2}{c^2} - \lambda^2} \tag{2.40}$$

This equation has two solutions as indicated by the $\pm$ sign. The total solution is thus

$$u(x,t) = \hat{u}_1 e^{i(\omega t - kx)} + \hat{u}_2 e^{i(\omega t + kx)} \tag{2.41}$$

Radiation condition. For $\omega^2/c^2 > \lambda^2$, the wave number k is real and positive. Waves propagating in the positive x-direction are represented by the first term and the radiation condition is met by dropping the second term in Equation (2.41). This solution is called traveling waves or *propagating waves*. If, however, $\omega^2/c^2 < \lambda^2$, then k is pure imaginary, say $k = \pm i\beta$, $\beta > 0$. We no longer have a propagating wave but an in-phase motion as can be seen from

$$e^{i(\omega t - kx)} = e^{i\omega t}e^{\pm\beta x} \tag{2.42}$$

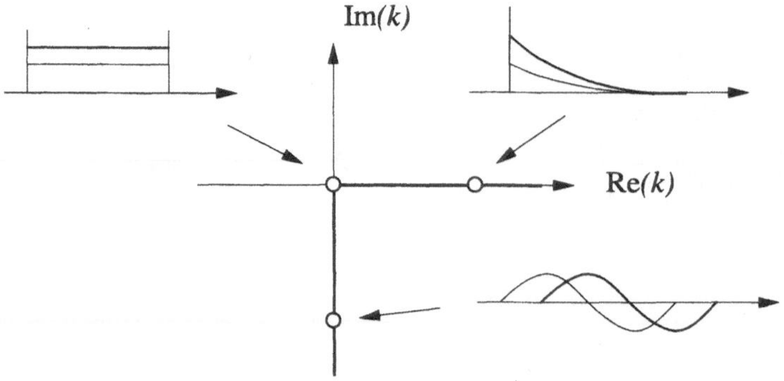

Figure 2.4: Three cases for the wave number k

In this case we have to ensure that the solution is decreasing with increasing x, that is, we have to choose $k = -i\beta$. The decaying solution is also called *evanescent waves.* The two regimes are separated by the cut-off frequency $\omega_c = c\lambda$. At the cut-off frequency the wave number is $k = 0$ and the displacement is independent of x. The three cases and the are shown in Figure 2.4 in the complex k-plane. The complex number k can be represented by polar coordinates in the complex plane as $k = \rho e^{i\theta}$. The radiation condition including both propagating waves and in-phase motion can then be stated as

$$-\frac{\pi}{2} \leq \theta \leq 0 \tag{2.43}$$

This condition will also be valid for more general cases (including damping), where k has both real and imaginary components. It should be noted that the behavior of the solution is completely different from that of the simple rod. In the low-frequency range, below the cut-off frequency, the motion is in-phase, decaying for large distances. For frequencies above the cut-off frequency, there are waves propagating towards infinity. This behavior is also typical for other problems such as a reservoir or a soil layer on a rigid foundation.

Dispersion. It is interesting to put the solution of the semi-infinite rod on an elastic foundation into the same form as the simple semi-infinite rod.

$$u(x,t) = \hat{u}_1 e^{i\omega(t-x/v_{ph})} e^{-\beta x} \tag{2.44}$$

with the phase velocity $v_{ph} = \omega/\Re(k(\omega))$ and the decay $\beta = \Im(k(\omega))$. ($\Re$ and $\Im$ denote the real and imaginary part of a complex number, respectively.) For the simple rod we have $v_{ph} = c$, that is, the waves travel with velocity c, independent of the frequency, without changing their shape. For the rod with elastic foundation, the phase velocity depends on the frequency. Different frequency contents of the wave travel with different velocities and the wave becomes distorted, that is, dispersive. The phase velocity is shown in Figure 2.5 in a dimensionless form. Below the cut-off frequency the phase velocity is infinite. Above the cut-off frequency it is

$$\frac{v_{ph}}{c} = \frac{\omega}{c\lambda\sqrt{1-(\omega/c\lambda)^2}} \tag{2.45}$$

For high frequencies, $k = \omega/c$ and the phase velocity becomes $v_{ph} = c$.

Transmitting boundary. For a transmitting boundary the farfield is replaced by the reaction force $Q(t)$ at the artificial boundary.

$$Q(t) = -EA\frac{\partial u}{\partial x} = c^2\mu\, ik(\omega)\, u(x,t) = c^2\mu\sqrt{\lambda^2 - \frac{\omega^2}{c^2}}\; u(x,t) \tag{2.46}$$

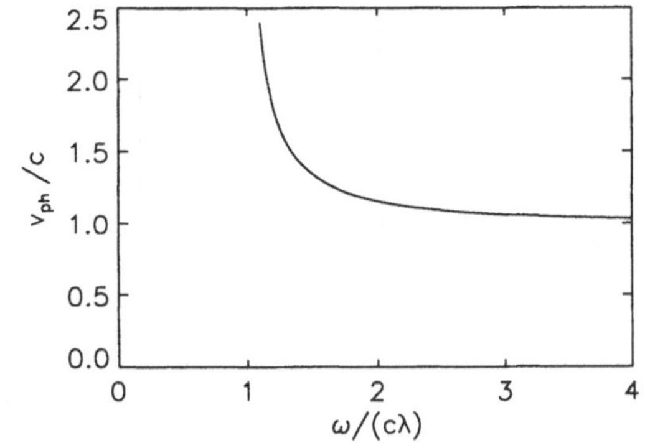

Figure 2.5: Phase velocity

Below the cut-off frequency the square root is real and the reaction force is written as

$$Q(t) = c^2 \mu \lambda \sqrt{1 - \left(\frac{\omega}{c\lambda}\right)^2}\, u(t) \qquad \omega < c\lambda \tag{2.47}$$

and the farfield can be replaced by a spring with a stiffness of $c^2\mu\lambda\sqrt{1-(\omega/c\lambda)^2}$.

For frequencies above the cut-off frequency, the square root becomes imaginary. Equation (2.47) is therefore rewritten as

$$Q(t) = \frac{c^2\mu\lambda}{\omega}\sqrt{\left(\frac{\omega}{c\lambda}\right)^2 - 1}\; i\omega\, u(t) \tag{2.48}$$

With the particle velocity $v(t)$ given by

$$v(t) = \dot{u}(x,t) = i\omega\, u(x,t) \tag{2.49}$$

the reaction force can be written as

$$Q(t) = c\mu\sqrt{1 - \left(\frac{c\lambda}{\omega}\right)^2}\, v(t) \qquad \omega > c\lambda \tag{2.50}$$

The equivalent dashpot is frequency-dependent with a value of $c\mu\sqrt{1-(c\lambda/\omega)^2}$. This dashpot cannot be directly implemented in the time domain. For $\omega \to \infty$, the value becomes $c\mu$ as for the simple rod.

The frequency-dependent spring constant k_d and damping constant c_d are plotted in Figure 2.6 in a dimensionless form.

For frequencies above the cut-off frequency, the semi-infinite rod on an elastic foundation acts like a dashpot, even if there is no material damping present. Energy is carried away with outgoing waves. Therefore, this damping is called radiation damping. Below the cut-off frequency no radiation damping exists.

To compare with other cases, the transmitting boundary condition can also be written using the spatial derivative instead of the reaction force

$$\frac{\partial u}{\partial x} + iku = \frac{\partial u}{\partial x} \pm \sqrt{\lambda^2 - \frac{\omega^2}{c^2}}\, u = 0 \tag{2.51}$$

Fourier transform. To obtain a solution of the problem in the time domain, we have to transform the dynamic stiffness Equation (2.46) by the Fourier transform. For

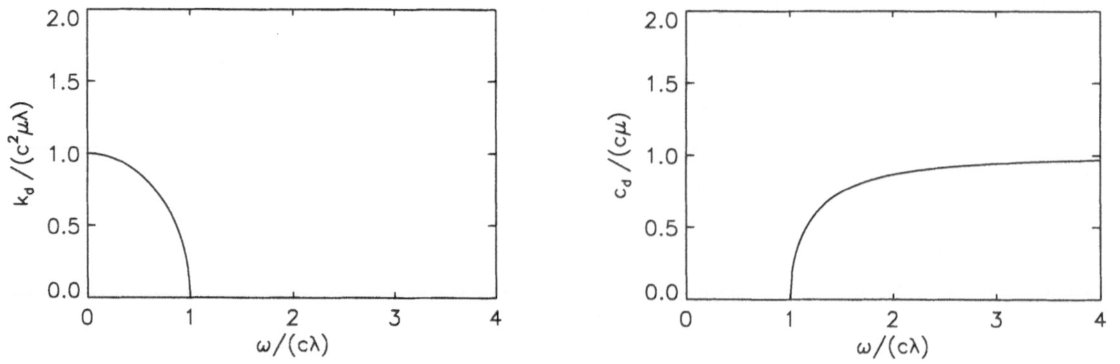

Figure 2.6: Spring and damping coefficients for semi-infinite rod on elastic foundation

the Fourier integral (Equation 2.28) to exist, $H(\omega)$ has to decrease sufficiently as $\omega \to \infty$. This is not the case for the rod since $k(\omega) \to \omega/c$ as $\omega \to \infty$. Therefore, we have to subtract the growing part and treat it separately.

$$c^2\mu(ik - \frac{i\omega}{c}) = c\mu\left(\sqrt{\lambda^2c^2 - \omega^2} - i\omega\right) \tag{2.52}$$

This function, also called the regular part of the dynamic stiffness, tends to zero for $\omega \to \infty$. It is plotted in a dimensionless form in Figure 2.7 together with the damped case in the next section. This form is mathematically more useful than the physical interpretation given earlier in Figure 2.6. The transform can be found in tables [CF48, No. 556.1] as

$$\frac{1}{2\pi}\int_{-\infty}^{\infty}\left(\sqrt{\alpha^2 + (i\omega)^2} - i\omega\right)e^{i\omega t}d\omega = \frac{\alpha}{t}J_1(\alpha t) \qquad t > 0 \tag{2.53}$$

where J_1 is the Bessel function of the first kind of order 1. Taking $\alpha = \lambda c$, the transform becomes

$$\frac{c^2\mu\lambda}{t}J_1(c\lambda t) \qquad t > 0 \tag{2.54}$$

For $t < 0$ the function vanishes. Such a function is called *causal*. The term $c\mu\, i\omega$ which was subtracted is recognized as the viscous damper introduced earlier. It can be added directly in the time domain as a dashpot.

To the multiplication of the displacement by the dynamic stiffness in the frequency domain there corresponds a convolution integral in the time domain. For a transient motion $u(t)$, the reaction force is

$$Q(t) = c^2\mu\lambda\int_0^t \frac{J_1(c\lambda\tau)}{\tau}\,u(t-\tau)\,d\tau + c\mu\dot{u}(t) \tag{2.55}$$

Since the dynamic stiffness is a causal function, the lower limit of the integral is zero. This also explains the name 'causal': For the force at time t only the displacements at times $t' = t - \tau;\ \tau > 0$ or $t' < t$ are relevant. If we want to determine the force $Q(t)$ for many different time steps, we have to evaluate the convolution integral each time. No information can be saved from one time step to the next one. This fact makes this procedure numerically very inefficient for problems with many time steps and many degrees of freedom.

Alternatively, instead of the dynamic stiffness Equation (2.46), the dynamic flexibility can be used to formulate the transmitting boundary condition. From Equation (2.46) we obtain for the displacement at the transmitting boundary

$$u(t) = \frac{Q(t)}{c\mu\sqrt{\lambda^2c^2 - \omega^2}} \tag{2.56}$$

This function decreases to zero for $\omega \to \infty$ and can therefore be transformed as [CF48, No. 557]

$$\frac{1}{2\pi}\int_{-\infty}^{\infty}\frac{e^{i\omega t}}{c\mu\sqrt{\lambda^2c^2-\omega^2}}d\omega = c\lambda J_0(c\lambda t) \tag{2.57}$$

With the convolution, the displacement is

$$u(t) = \frac{1}{c\mu}\int_0^t J_0(c\lambda\tau)\,Q(t-\tau)\,d\tau \tag{2.58}$$

2.4 Semi-infinite rod on visco-elastic foundation

An even more complex model problem is obtained by adding viscosity to the foundation. This problem will be useful as an example of a damped system. With the additional viscosity term, the differential equation is

$$\frac{\partial^2 u}{\partial x^2} - \lambda^2 u - \frac{2\epsilon}{c}\dot{u} = \frac{1}{c^2}\ddot{u} \tag{2.59}$$

Substituting the steady state solution (Equation 2.39) into this differential equation we obtain

$$k = \pm\sqrt{\frac{\omega^2}{c^2} - \lambda^2 - 2\epsilon\frac{i\omega}{c}} \tag{2.60}$$

To satisfy the radiation condition, the sign has again to be taken according to Equation (2.43).

For the Fourier transform, this problem can be brought back to the one without viscous damping by introducing $\bar{\lambda}^2 = \lambda^2 - \epsilon^2$ and $\bar{\omega} = \omega - i\epsilon c$. With this notation, Equation (2.60) becomes

$$k = \pm\sqrt{\frac{\bar{\omega}^2}{c^2} - \bar{\lambda}^2} \tag{2.61}$$

As for the undamped case, the dynamic stiffness is split up into a regular part and the high-frequency behavior.

$$c^2\mu(ik - \frac{i\bar{\omega}}{c}) = c\mu\left(\sqrt{\bar{\lambda}^2c^2 - \bar{\omega}^2} - i\bar{\omega}\right) \tag{2.62}$$

Substituting $\alpha = \bar{\lambda}c$, the Fourier transform is

$$\frac{1}{2\pi}\int_{-\infty}^{\infty}\left(\sqrt{\alpha^2 + (i\bar{\omega})^2} - i\bar{\omega}\right)e^{i\omega t}d\omega \tag{2.63}$$

Noting that $e^{i\bar{\omega}t} = e^{i\omega t}e^{\epsilon ct}$ and $d\bar{\omega} = d\omega$ we have

$$e^{-\epsilon ct}\frac{1}{2\pi}\int_{-\infty}^{\infty}\left(\sqrt{\alpha^2 + (i\bar{\omega})^2} - i\bar{\omega}\right)e^{i\bar{\omega}t}d\bar{\omega} = e^{-\epsilon ct}\frac{\alpha}{t}J_1(\alpha t) \qquad t > 0 \tag{2.64}$$

where the integrand is the same as for the undamped case. As before, this function is causal. The reaction force due to a transient motion $u(t)$ becomes

$$Q(t) = c^2\mu\bar{\lambda}\int_0^t \frac{J_1(c\bar{\lambda}\tau)}{\tau}e^{-\epsilon c\tau}u(t-\tau)\,d\tau + c\mu\,\dot{u}(t) + c^2\mu\epsilon\,u(t) \tag{2.65}$$

This corresponds to a spring, a dashpot and a convolution integral.

For later use as a model problem the regular part of the dynamic stiffness is written in a dimensionless form as

$$\kappa = \frac{ik}{\lambda} - \frac{i\omega}{c\lambda} = \pm\sqrt{1 - \left(\frac{\omega}{c\lambda}\right)^2 + 2\epsilon\frac{i\omega}{c\lambda}} - \frac{i\omega}{c\lambda} - \frac{\epsilon}{\lambda} \tag{2.66}$$

The curves for $\epsilon/\lambda = 0$ and $\epsilon/\lambda = 0.05$ are plotted in Figure 2.7.

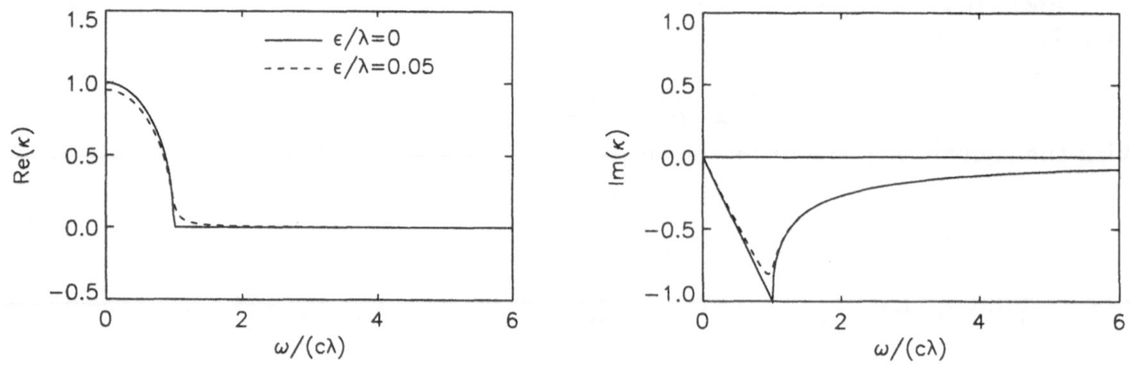

Figure 2.7: Regular part of dynamic stiffness for semi-infinite rod on visco-elastic foundation

2.5 Semi-infinite channel with constant cross-section

This section describes briefly the model for a two- or three-dimensional idealized reservoir. It will be shown that this problem leads to the same mathematical problem as the semi-infinite rod on an elastic foundation. It is therefore possible to discuss most of the theoretical and numerical aspects of dam-reservoir interaction by means of the model problems introduced earlier. A more detailed discussion of the dam-reservoir interaction problem is given in Chapter 9.

For a compressible, irrotational fluid with small velocity amplitudes, the governing differential equation is the acoustic wave equation

$$\Delta\varphi = \frac{1}{c^2}\ddot{\varphi} \tag{2.67}$$

where $\varphi = \varphi(x, y, z, t)$ is the velocity potential and c is the acoustic-wave velocity. The velocity potential describes the velocity v and the dynamic pressure p in the fluid by

$$\mathbf{v} = \mathbf{grad}\varphi \tag{2.68}$$

$$p = -\rho\dot{\varphi} \tag{2.69}$$

where ρ denotes the mass density.

We consider a semi-infinite channel with constant cross-section. The boundary conditions are the following: At the free surface, neglecting surfaces waves, the pressure vanishes

$$p = -\rho\dot{\varphi} = 0 \tag{2.70}$$

For a rigid foundation, the normal velocity is zero at the reservoir-foundation interface

$$v_n = \frac{\partial\varphi}{\partial n} = 0 \tag{2.71}$$

where n denotes the outward normal of the fluid. At the end of the channel, the normal velocity is prescribed for each point of the cross-section. Taking the upstream direction (towards ∞) as x-direction and the transmitting boundary of the channel at $x = L$, this boundary condition becomes

$$-\left.\frac{\partial\varphi}{\partial x}\right|_{x=L} = v_0(y, z, t) \tag{2.72}$$

For infinity the boundary condition is the radiation condition.

The wave equation can be solved in the frequency domain by separation of variables. The general solution is

$$\varphi = \Phi(y, z)e^{i(\omega t - kx)} \tag{2.73}$$

Φ is a function depending only on the coordinates y and z, describing the distribution of the velocity potential over the cross-section of the channel. k is again the wave number in the x-direction. Substituting the general solution into the wave equation (Equation 2.67) yields the following eigenvalue problem

$$\Delta\Phi + \lambda^2\Phi = 0 \tag{2.74}$$

where Φ is the eigenvector, subjected to the boundary conditions Equations (2.70) and (2.71). The eigenvalue λ is connected to k and ω by $\lambda^2 = \omega^2/c^2 - k^2$. This leads to the same equation for the wave number as in the model problem of the semi-infinite rod on an elastic foundation, Equation (2.40). The radiation condition is satisfied by taking the root k according to Equation (2.43) as in the model problem.

The eigenvalue problem has a different solution $\{\lambda_m, \Phi_m\}$ for each mode m, which leads to different wave numbers k_m for each mode.

$$k_m = \sqrt{\frac{\omega^2}{c^2} - \lambda_m^2} \tag{2.75}$$

The complete solution for the velocity potential is thus

$$\varphi = \sum_{m=1}^{\infty} \varphi_m = \sum_{m=1}^{\infty} A_m \Phi_m e^{i(\omega t - k_m x)} \tag{2.76}$$

and the transmitting boundary condition for each mode is

$$\frac{\partial \varphi_m}{\partial x} + ik_m\varphi = 0 \tag{2.77}$$

This shows that each mode of the cross-section of a semi-infinite channel yields the same mathematical problem as the semi-infinite rod on an elastic foundation. A more detailed analysis of dam-reservoir interaction is given in Chapter 9. There we also introduce bottom absorption, which leads to a damped problem similar to the semi-infinite rod on an visco-elastic foundation. The difference is that there the eigenvalues λ are frequency-dependent, which complicates the dam-reservoir interaction problem somewhat. Therefore, we use the semi-infinite rod on an visco-elastic foundation as a model problem in the first part of this work, leaving the analysis of dam-reservoir interaction to later chapters.

Geometric Interpretation. The radiation condition for a prismatic domain (channel) has an interesting geometric interpretation. Waves of the form

$$\varphi = \hat{\varphi}e^{i(\omega t - kx)} \tag{2.78}$$

are plane waves as shown in Figure 2.8. Plane waves are characterized by the fact that they have a constant value on planes (lines for 2D) with $\omega t - kx = const.$ The wave number k in the x-direction depends on the angle of incidence β. In the propagating direction the wave speed is c but in the x-direction the apparent speed is $c/\cos\beta$. The equation for the wave can therefore be written as

$$\varphi = \hat{\varphi}e^{i\omega(t - x\cos\beta/c)} \tag{2.79}$$

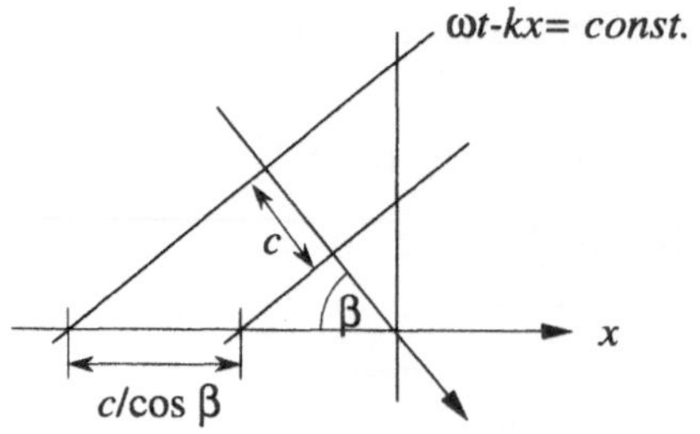

Figure 2.8: Geometric interpretation of transmitting boundary

and the wave number is

$$k = \frac{\omega}{c}\cos\beta = \frac{\omega}{c}\sqrt{1-\sin^2\beta} \tag{2.80}$$

Comparing with $k = (\omega/c)\sqrt{1-(c\lambda/\omega)^2}$ from Equation (2.75) we find

$$\frac{c\lambda}{\omega} = \sin\beta \quad \text{for} \quad c\lambda/\omega \le 1 \tag{2.81}$$

The transmitting boundary condition can be written as

$$\frac{\partial\varphi}{\partial x} + \frac{i\omega}{c}\cos\beta = 0 \tag{2.82}$$

This boundary condition can again be interpreted as a dashpot, but the damping coefficient depends on the angle of incidence.

Chapter 3

Common Numerical Procedures

Various transmitting boundaries in the time domain have been invented in the past. They all try to somehow avoid the classical convolution integral either by directly formulating a transmitting boundary condition in the time domain or by using discretization schemes specially devised for infinite domains or by transforming the frequency-domain solution to the time domain by techniques other than convolution.

Before treating the new methods proposed in this work, some commonly used numerical procedures are described in this chapter. The intention is not to give a full overview of all available methods, but rather to show different classes of methods and their restrictions. Rather than leaving the reader with general statements, the methods are analyzed using the model problems of Chapter 2. Extensive overviews can be found in the literature [ZBCE81, KT81, Kau88, Wol86, Wol88]. The first reference is out of a volume which contains several other related papers.

One classification often used in the literature [KT81] is the distinction between local and consistent boundaries. A viscous boundary is the typical case of a boundary local in space and in time, because each dashpot acts individually from others (local in space) and depends only on the present velocity (local in time). Local boundaries are efficient as far as computer resources are concerned but are usually only imperfect absorbers because they are frequency-independent. Consistent boundaries, on the other hand, are equivalent to an extension of the finite element mesh to infinity and are perfect absorbers for waves impinging with arbitrary incidence. They can be obtained by semi-analytical methods in the frequency domain. Because all degrees of freedom are coupled, they are non-local in space. In the time domain they lead to convolution integrals, which are non-local in time. Using convolution makes consistent boundaries computationally inefficient in the time domain. However, as will be shown in this work, it is possible to avoid convolution and to formulate consistent boundaries that are local in time, that is, accurate and efficient. In contrast we call the classical local boundaries 'local, non-consistent boundaries'.

The first class described in this chapter is that of local, non-consistent boundaries. These transmitting boundaries are generalizations of the viscous boundary. Although somewhat improved, they still exhibit the main drawbacks of the simple viscous boundary.

The second class treated is that of boundary elements. Boundary elements seem to be predestinated for infinite domains, because they reduce the (infinite) domain to its (finite) boundary. While this is true for the frequency domain, it will be shown that boundary elements in time are essentially the same as convolution integrals.

The third class analyzed is that of infinite elements. They are derived analogously to finite elements but use different shape functions. The aim of infinite elements is to approximate the solution of the infinite domain with a small number of degrees of freedom. This requires the use of shape functions that behave similar to the exact solution. This in turn is only possible if the solution is qualitatively known in advance and if it can be described by a few parameters. Infinite elements have been applied with some success to static and to steady state problems. For the dynamic problems considered in this work, however, they are not applicable because the solution behaves quite differently, depending on whether the frequency is below or above the cut-off frequency.

The last class includes methods that use the frequency-domain and the time-domain formulation in parallel. Gains relative to the convolution integral can only be achieved for a restricted class of problems.

As a general remark, it should be recognized that many of the analyzed methods work well for certain problems but not for others. The problem of dam-reservoir interaction seems to be one of the more challenging problems because it contains both a decaying solution and propagating waves.

3.1 Local, non-consistent boundaries

Local, non-consistent boundaries are efficient and easy to implement in any finite element code. Unfortunately, they are not appropriate for a wide class of problems, including dam-reservoir interaction and soil-structure interaction for layered soils. Nevertheless, they are often used in practice due to the lack of other methods, mostly without being aware of their drawbacks. In this section, some of the commonly used local, non-consistent boundaries are investigated. The goal is not only to give an overview of the methods but also to show the difficulties involved in developing such boundaries. The discussion in this section moves along similar lines to [Kau88] but gives some new or alternative views.

3.1.1 Engquist-Majda boundaries

The exact boundary condition in the frequency domain for the semi-infinite rod on an elastic foundation is given by (Equation 2.51)

$$\frac{\partial u}{\partial x} + iku = \frac{\partial u}{\partial x} + \frac{i\omega}{c}\sqrt{1 - \left(\frac{c\lambda}{\omega}\right)^2}\, u = 0 \tag{3.1}$$

Introducing the variable $\xi = (c\lambda/\omega)$, the exact boundary condition can be written as

$$\frac{\partial u}{\partial x} + \frac{i\omega}{c}\sqrt{1 - \xi^2}\, u = 0 \tag{3.2}$$

Engquist and Majda [EM77] suggested to approximate $\sqrt{1-\xi^2}$ by a finite continued fraction expansion, given recursively by

$$P_1 = 1 \tag{3.3}$$

$$P_2 = 1 - \frac{\xi^2}{2} \tag{3.4}$$

$$\vdots$$

$$P_{N+1} = 1 - \frac{\xi^2}{1 + P_N} \tag{3.5}$$

The approximate boundary condition of order N will thus be

$$\frac{\partial u}{\partial x} + \frac{i\omega}{c} P_N\left(\frac{c\lambda}{\omega}\right) u = 0 \tag{3.6}$$

For $N = 1$, this boundary condition becomes the viscous boundary introduced earlier:

$$\frac{\partial u}{\partial x} + \frac{i\omega}{c} u = 0 \tag{3.7}$$

For $N = 2$, the boundary condition is

$$\frac{\partial u}{\partial x} + \frac{i\omega}{c} \left(1 - \frac{1}{2}\left(\frac{c\lambda}{\omega}\right)^2\right) u = 0 \tag{3.8}$$

An alternative form is found by multiplying this equation by $2i\omega/c$.

$$\left(\frac{2i\omega}{c}\left(\frac{\partial}{\partial x} + \frac{i\omega}{c}\right) + \lambda^2\right) u = 0 \tag{3.9}$$

Taking advantage of the differential equation (2.38), λ can be expressed as

$$\lambda^2 = \left(\frac{\partial}{\partial x} - \frac{i\omega}{c}\right)\left(\frac{\partial}{\partial x} + \frac{i\omega}{c}\right) \tag{3.10}$$

which finally yields the boundary condition

$$\left(\frac{\partial}{\partial x} + \frac{i\omega}{c}\right)^2 u = 0 \tag{3.11}$$

For higher-order boundary conditions

$$\frac{\partial u}{\partial x} + \frac{i\omega}{c} P_{N+1} u = 0 \tag{3.12}$$

P_{N+1} is replaced by Equation (3.5) to obtain

$$\frac{\partial u}{\partial x} + \frac{i\omega}{c} \left(1 - \frac{(c\lambda/\omega)^2}{1 + P_N}\right) u = 0 \tag{3.13}$$

Multiplying by $(i\omega/c)(1 + P_N)$ and substituting again Equation (3.10) yields

$$\left(\frac{\partial}{\partial x} + \frac{i\omega}{c}\right)\left(\frac{\partial}{\partial x} + \frac{i\omega}{c} P_N\right) u = 0 \tag{3.14}$$

Thus the boundary condition for $N + 1$ is obtained by multiplying the one for N by $(\partial/\partial x + i\omega/c)$. The factor $i\omega$ is interpreted in the time domain as $\partial/\partial t$. The general boundary condition can therefore also be written as

$$\left(\frac{\partial}{\partial x} + \frac{1}{c}\frac{\partial}{\partial t}\right)^N u = 0 \tag{3.15}$$

Engquist and Majda also prove that this transmitting boundary is stable and they give a mathematical error analysis. They also give discrete versions for finite difference applications and an extension to the vector case of the elastic wave equation. In this overview we show plots of the reflection coefficient and the approximate stiffness.

To define the reflection coefficient, the approximative boundary condition is applied to the total solution including waves traveling in both directions.

$$u = \hat{u}_1 e^{i(\omega t - kx)} + \hat{u}_2 e^{i(\omega t + kx)} \tag{3.16}$$

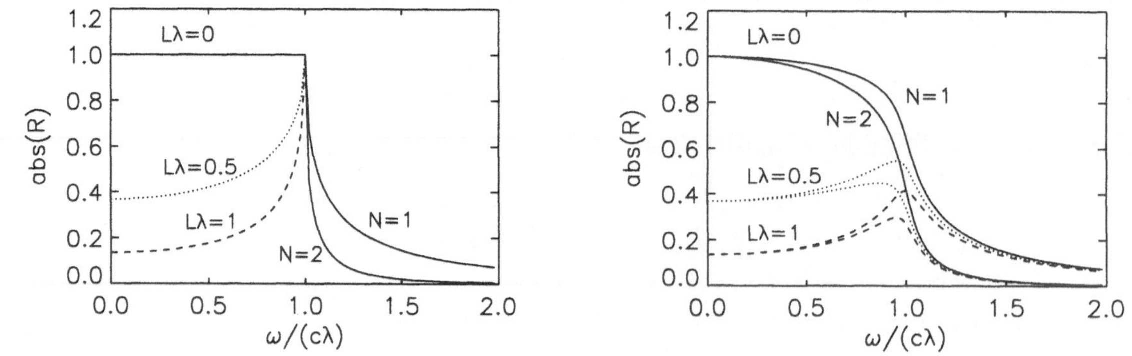

Figure 3.1: Reflection coefficient for Engquist-Majda boundaries. Left: undamped case. Right: damped case, $\epsilon/\lambda = 0.05$

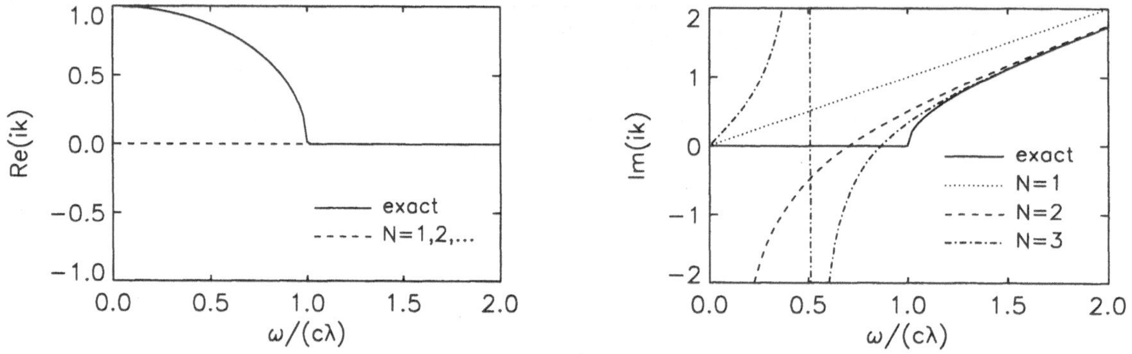

Figure 3.2: Stiffness approximation for Engquist-Majda boundaries, undamped case

The reflection coefficient is then defined as

$$R = \frac{\hat{u}_2}{\hat{u}_1} \tag{3.17}$$

Applying the general boundary condition (Equation 3.15) at a distance $x = L$ yields

$$R = -\frac{(i\omega/c - ik)^N}{(i\omega/c + ik)^N}\, e^{-2ikL} \tag{3.18}$$

Figure 3.1 shows the reflection coefficient for different orders N and distances L of the artificial boundary. All variables are given in dimensionless form. The left figure shows the undamped case. It is evident that for frequencies below the cut-off frequency only the distance L is important whereas for frequencies above the cut-off frequency the result depends only on the order N. At the cut-off frequency itself, the reflection coefficient is always 1. It should be noted that in a practical case both N and L may only be chosen up to a certain limit. High-order transmitting boundaries use high-order derivatives, which are usually not available and a large distance L increases the size of the model, making the original purpose of the transmitting boundary, to reduce the model, useless. Damping ($\epsilon/\lambda = 0.05$) reduces the reflection coefficient, especially at the cut-off frequency in the case of larger distances, as shown in the right figure.

The direct comparison of the dynamic stiffness is shown in Figure 3.2. The figure compares $(i\omega/c)P_N$ for different orders N to the exact solution ik. This comparison is made at the transmitting boundary itself ($L = 0$). As can be seen from the figures, the approximation is only valid for frequencies above the cut-off frequency. This is because the continued fraction expansion defined in Equation (3.5) approximates the square root only for $\xi < 1$, that is, for $\omega/(c\lambda) > 1$. The real part of $(i\omega/c)P_N$ is always zero because P_N is real, even for $\xi = c\lambda/\omega > 1$. To gain more insight into the problem we note that the continued fraction corresponds to a rational function because

$$P_{N+1} = 1 - \frac{\xi^2}{1 + P_N} = \frac{1 + P_N - \xi^2}{1 + P_N} \tag{3.19}$$

This rational approximation only contains even powers of ξ. To obtain real and imaginary parts we would have to consider the approximation of $\sqrt{1 + (i\xi)^2}$ and use a rational approximation including odd and even powers of $i\xi$. Generally, it will be true that approximations of ik based on even powers of $i\xi$ ($i\omega$ and λ) lead to pure imaginary values and cannot therefore fit ik for frequencies below the cut-off frequency.

A similar derivation can be made for a cylindrical or spherical boundaries [BBE84].

3.1.2 General case

A generalization of the Engquist-Majda boundaries (Equation 3.15) is

$$\prod_{n=1}^{N} \left(\frac{\partial}{\partial x} + a_n + b_n \frac{1}{c} \frac{\partial}{\partial t} \right) u = 0 \tag{3.20}$$

The reflection coefficient of the general boundary (Equation 3.20) is

$$R = -\prod_{n=1}^{N} \frac{a_n + b_n i\omega/c - ik}{a_n + b_n i\omega/c + ik} e^{-2ikL} \tag{3.21}$$

The coefficients a_n and b_n can be chosen to give a minimum reflection as explained in the next few paragraphs.

Doubly asymptotic approximation. A special case is the doubly asymptotic approximation (DAA) with N=1 and the coefficients $a_1 = \lambda$ and $b_0 = 1$

$$\frac{\partial u}{\partial x} + \frac{1}{c}\dot{u} + \lambda u = 0 \tag{3.22}$$

The doubly asymptotic approximation is a combination of the viscous boundary and a static spring. In the static case, the velocity is zero and only the spring is effective. For high frequencies, the viscous damper is exact and the spring has no influence. The boundary is thus exact for the two frequencies $\omega = 0$ and $\omega \to \infty$. The reflection coefficient for a distance $L = 0$ is shown in Figure 3.3 at the left ($a_1 = \lambda$, solid line). Compared to the purely viscous boundary ($a_1 = 0$, dotted line), the low-frequency range has improved, whereas the high-frequency range is impaired by the additional spring.

The coefficient b_n can be used to enforce an exact boundary for a certain angle of incidence. The angle of incidence β is given as $c\lambda/\omega = \sin\beta$ (Equation 2.81). If the coefficient is chosen as $b_n = \cos\beta$, the reflection coefficient is zero for the corresponding frequency because $k = (\omega/c)\cos\beta$ (Equation 2.80). The right diagram in Figure 3.3 shows the reflection coefficient for a distance $L = 0$ and coefficients $a_1 = 0$ and $b_1 = 0.5$ corresponding to an angle $\beta = 60°$ ($\omega c/\lambda = 1.16$) as a solid line. The case of the original viscous boundary ($b_1 = 1$) is shown for comparison.

A combination of the two cases is possible, but it should be noted that an improvement in one frequency range may have an adverse effect for the other range. Nevertheless, the reflection coefficient can be improved by a proper choice of the coefficients a_n and b_n. The reflection coefficient for $\omega = c\lambda$ is always 1 for the undamped model and cannot be modified.

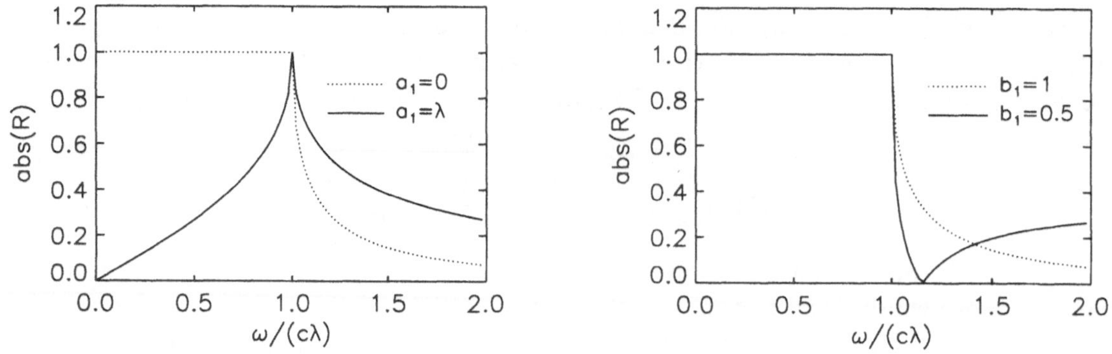

Figure 3.3: Reflection coefficient (distance L=0): Left: doubly asymptotic approximation. Right: first-order boundary with $\beta = 60°$.

3.1.3 Paraxial boundaries

Instead of formulating directly a boundary condition at the artificial boundary, paraxial boundaries are derived from a modified wave equation that includes only waves travelling in one direction. The modified wave equation is a differential equation equal or similar to the Engquist-Majda boundary conditions. Clayton and Engquist [CE77] apply the modified wave equation at the boundary resulting effectively in the same formulation as the Engquist-Majda boundaries. Other author use somewhat different formulations [Sta88, HT88]. Cohen and Jennings [CJ87] incorporate the modified wave equation into a finite element and place these elements at the artificial boundary to avoid reflections. Even though the condition is implemented in the element itself rather than at the boundary, the results are quite similar as shown by the reflection coefficients in the reference. A comparison of different paraxial boundaries applied to an elastic halfspace are given by Kausel [Kau92].

Generally speaking, paraxial boundaries are quite similar to Engquist-Majda boundaries. They are all formulated for travelling waves and do not work for evanescent waves because they contain only even powers of $i\omega$ and λ. They are thus not applicable to problems with a cut-off frequency.

3.1.4 Lindman boundary

Another boundary, similar to the general formulation, is the Lindman boundary [Lin75]. Unfortunately, the method is formulated in a finite difference context which makes it difficult to compare with other methods. A translation to differential form is shown here for the first part of the paper only, which is concerned with traveling waves (propagating waves, above the cut-off frequency). The second part which also includes evanescent waves (decaying solution, below cut-off frequency), is not analyzed.

Lindman proposed to approximate the square root in Equation (3.2) by a sum of correction functions as follows:

$$\frac{\partial u}{\partial x} + \frac{i\omega}{c}\sqrt{1-\xi^2}u = \frac{\partial u}{\partial x} + \frac{i\omega}{c}\left(1 + \sum_{n=1}^{N} h_n\right)u = 0 \tag{3.23}$$

where $\xi = (c\lambda/\omega) < 1$. The correction functions are

$$h_n = \frac{\alpha_n \xi^2}{1 - \beta_n \xi^2} u \tag{3.24}$$

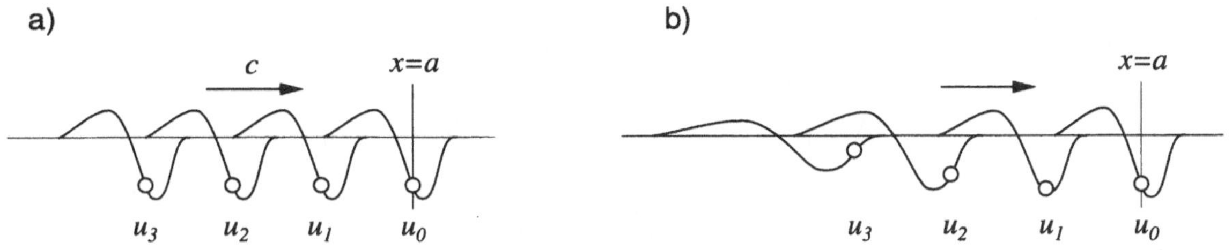

Figure 3.4: Extrapolation algorithm: a) non-dispersive wave with normal incidence, b) dispersive wave with oblique incidence

The coefficients α_n and β_n are optimized numerically for minimum reflection. Equation (3.24) can be written as a differential equation which is solved in parallel to the differential equation of the nearfield.

$$\frac{\partial^2 h_n}{c^2 \partial t^2} + \beta_n \lambda^2 h_n = -\alpha_n \lambda^2 u \tag{3.25}$$

It should be noted that in Equation (3.24) only even powers of ξ are included. This is the same as for the Engquist-Majda boundaries. Whether the Lindman boundaries for traveling waves are equivalent to the general Engquist-Majda boundaries is not obvious but seems at least likely.

Lindman also proposed a more general boundary that works for both traveling and evanescent waves. The difference scheme cannot readily be transformed to differential form because it also contains an absolute difference operator. At least it can be observed that this absolute difference operator appears in the first power thus including odd powers of ξ in the approximation. The reflection coefficients shown by Lindman [Lin75] are quite good except at the cut-off frequency where they always equal one as do the Engquist-Majda boundaries. Although very promising these boundaries are not further analyzed here.

3.1.5 Extrapolation algorithm

Liao and Wang [LW84] suggested a discrete version of a transmitting boundary. In this method a boundary value $u(a,t) = u_0(t)$ is prescribed at the artificial boundary $x = a$. The appropriate boundary value is found by extrapolating from sampling points taken at previous time steps in the interior of the nearfield. The sampling points are chosen as

$$u_m = u(t - m\Delta t, a - m\Delta x) \qquad m = 1, \ldots, N \tag{3.26}$$

Δt is the time step and Δx is the increment on the x-axis. The idea of the algorithm is to capture a wave as it passes in the x-direction. First consider the ideal case of a normal incident wave $u = f(t - x/c)$ traveling in the x-direction with a phase velocity c as depicted in Figure 3.4a. If we choose $\Delta x = c\,\Delta t$, then the values u_m all equal the value u_0 because

$$u_m = f(t - m\Delta t - (a - m\Delta x)/c) = f(t - a/c) = u_0 \tag{3.27}$$

and hence any sampling value can be used as boundary value.

In the general case the wave travels with a velocity different from c and is distorted due to dispersion as shown in Figure 3.4b. In this case the sampling values cannot be used directly but the boundary value has to be extrapolated from them. In the time domain the extrapolation is explained by the following difference scheme:

$$\begin{array}{ccccccccccc}
u_0 & & u_1 & & u_2 & & u_3 & & \cdots & & u_N \\
& \Delta u_0 & & \Delta u_1 & & \Delta u_2 & & \cdots & & \Delta u_{N-1} & \\
& & \Delta^2 u_0 & & \Delta^2 u_1 & & \cdots & & \Delta^2 u_{N-2} & & \\
& & & \cdots & & \cdots & & \cdots & & & \\
& & & & \Delta^{N-1} u_0 & & \Delta^{N-1} u_1 & & & & \\
& & & & & \Delta^N u_0 & & & & &
\end{array} \tag{3.28}$$

The differences are defined as follows

$$\begin{aligned}
\Delta u_m &= u_m - u_{m+1} \\
\Delta^2 u_m &= (u_m - u_{m+1}) - (u_{m+1} - u_{m+2}) = u_m - 2u_{m+1} + u_{m+2} \\
&\vdots \\
\Delta^N u_m &= \sum_{k=0}^{N} (-1)^k \binom{N}{k} u_{m+k}
\end{aligned} \tag{3.29}$$

where $\binom{N}{k}$ denotes the binomial coefficients. The extrapolation is performed assuming the highest-order difference $\Delta^N u_0 = 0$. Then Equation (3.29) reads

$$\Delta^N u_0 = u_0 + \sum_{k=1}^{N} (-1)^k \binom{N}{k} u_k = 0 \tag{3.30}$$

or for the boundary value

$$u_0 = -\sum_{m=1}^{N} (-1)^m \binom{N}{m} u_m \tag{3.31}$$

With this equation the prescribed boundary value u_0 can be extrapolated for each time step from the previous sampling points u_m.

The algorithm is now analyzed in the frequency domain. Consider the wave

$$u(x,t) = \hat{u}_1 e^{i(\omega t - kx)} \tag{3.32}$$

The exact value at the transmitting boundary $x = a$ is

$$u_0 = u(a,t) = \hat{u}_1 e^{i(\omega t - ka)} \tag{3.33}$$

The sampling points are

$$\begin{aligned}
u_m(t) &= \hat{u}_1 e^{i\omega(t - m\Delta t) - ik(a - m\Delta x)} \\
&= u_0(t) \left(e^{-i(\omega \Delta t - k\Delta x)}\right)^m
\end{aligned} \tag{3.34}$$

For a non-dispersive wave the wave number is $k = \omega/c$. With $\Delta x = c\Delta t$ we have $\omega\Delta t = k\Delta x$ and the last factor in Equation (3.34) is equal to one. In the general case of a dispersive wave, however, $\omega\Delta t \neq k\Delta x$ and that factor is only close to one. Then $\left(e^{-i(\omega\Delta t - k\Delta x)} - 1\right)^m$, $m = 1, \ldots$ is an infinite series converging to zero. As an approximation, the series is truncated to N terms, leading to

$$u_0 \left(e^{-i(\omega\Delta t - k\Delta x)} - 1\right)^N = 0 \tag{3.35}$$

Written as a sum with the binomial coefficients $\binom{N}{m}$ and observing Equation (3.34), we recover Equation (3.31).

The reflection coefficient is determined by substituting the total solution including the reflected wave (Equation 3.16) into the boundary condition Equation (3.35). This leads to

$$\hat{u}_1 \left(e^{-i(\omega\Delta t - k\Delta x)} - 1\right)^N e^{i(\omega t - kx)} + \hat{u}_2 \left(e^{-i(\omega\Delta t + k\Delta x)} - 1\right)^N e^{i(\omega t + kx)} = 0 \tag{3.36}$$

The reflection coefficient is defined as $R = \hat{u}_1/\hat{u}_2$ at $x = a = L$. It is

$$R = -\frac{\left(e^{-i(\omega\Delta t - k\Delta x)} - 1\right)^N}{\left(e^{-i(\omega\Delta t + k\Delta x)} - 1\right)^N} e^{-2ikL} \tag{3.37}$$

Choosing $\Delta x = c\Delta t$ and letting $\Delta t \to 0$ leads to

$$R = -\frac{(i\omega/c - ik)^N}{(i\omega/c + ik)^N} e^{-2ikL} \tag{3.38}$$

which is the same expression as for the Engquist-Majda boundaries. Therefore the extrapolation algorithm can be viewed as a discrete version of the latter. (The reflection coefficient given in [LW84] is wrong. The extrapolation there is based on the outgoing wave only. This is invalid because in a calculation it is not possible to distinguish between the incoming and the outgoing waves, which is exactly the point of the transmitting boundary problem.)

3.1.6 Superposition boundaries

Another idea to separate incoming and outgoing waves is used by the superposition boundaries. Consider a wave

$$u(x,t) = \hat{u}_1 e^{i(\omega t - kx)} + \hat{u}_2 e^{i(\omega t + kx)} \tag{3.39}$$

and impose at $x = a$ once a fixed and once a free boundary condition. For the fixed boundary

$$\hat{u}_1 e^{i(\omega t - ka)} + \hat{u}_2 e^{i(\omega t + ka)} = 0 \tag{3.40}$$

$$\hat{u}_2 = -\hat{u}_1 e^{-2ika} \tag{3.41}$$

and for the free boundary

$$-ik\hat{u}_1 e^{i(\omega t - ka)} + ik\hat{u}_2 e^{i(\omega t + ka)} = 0 \tag{3.42}$$

$$\hat{u}_2 = \hat{u}_1 e^{-2ika} \tag{3.43}$$

In the first case the reflected wave is negative while in the second case it is positive. Smith [Smi74] proposed to calculate both solutions and to average them (by superposition). In this way the reflected wave is eliminated, even for dispersive waves. The problem is that this argumentation only considers the transmitting boundary but not any other boundaries of the model. Other boundaries lead to so-called multiple reflections which vitiate the procedure as shown in Figure 3.5. The transmitting boundary is assumed to be on the right. The two models with complementary boundary conditions are drawn one above each other. The two positive waves in the first picture are reflected back at the transmitting boundary with different signs (second picture). These reflected waves, therefore, cancel when averaged. After a reflection at the left fixed boundary they travel again to the right in the third picture. Since they have now different signs, they have equal

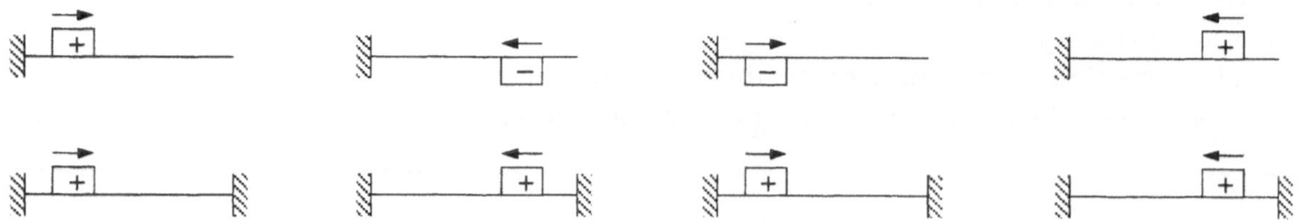

Figure 3.5: Multiple reflections for superposition boundaries

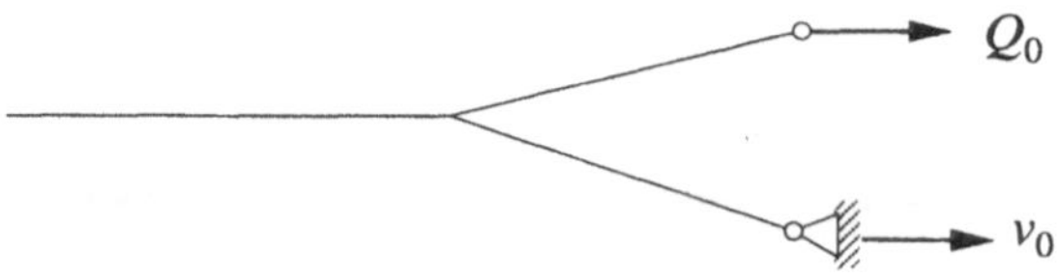

Figure 3.6: Model with complementary boundary conditions

signs after a second reflection at the transmitting boundary (picture four). This means that they no longer cancel when averaged.

The multiple reflection problem is even worse in two and three dimensions. If analyzed more carefully one sees that the procedure simply involves dividing a large problem into smaller problems taking advantage of symmetry and anti-symmetry.

Kunar and Marti [KM81] proposed a method to overcome the problem of multiple reflections by canceling the reflections "as they occur". At first glance the method seems to be simple and ingenious. The region is extended by two overlapping boundary zones with the two complementary boundary conditions as shown in Figure 3.6. After a certain time interval the solutions of the two boundary zones are averaged. The time increment has to be chosen sufficiently small in order that reflected waves do not enter the main region. To ensure equilibrium the boundary values also have to averaged. Thus the free boundary will change to a constant force boundary and the fixed boundary to a constant velocity boundary. Even though the idea is quite simple the method of analysis is not. A full analysis would have to include the time integration scheme and various parameters used in the method. For a specific discretization, an analysis of the method can be found in [Kau88]. One difficulty with the method is the choice of the interval for averaging. It has to be long enough so that waves can really develop at the boundary, but it has to short enough so that the waves do not pass into the main region. This is particularly difficult for dispersive wave with components that travel with different velocities. For frequencies below the cut-off frequency, the scheme does not work at all because there the motion does not consist of traveling waves but the whole model is excited simultaneously implying an infinite wave speed.

3.2 Boundary elements

One method that seems to be particularly suitable for problems with infinite domains is the boundary element method (BEM). In the BEM the differential equation is transformed to a boundary integral equation. Instead of having to solve the differential equation for the entire domain the problem is reduced to the boundary region. The dimensionality of the problem is reduced by one.

The BEM is well-suited for elliptic boundary value problems such as static and steady-state problems. For these problems the entire solution may be expressed in terms of the

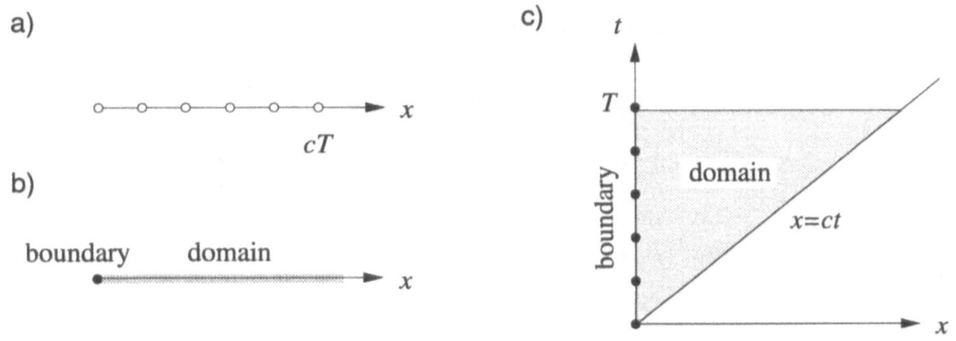

Figure 3.7: Computational space for a) finite elements, b) boundary elements in frequency domain, c) boundary elements in time domain

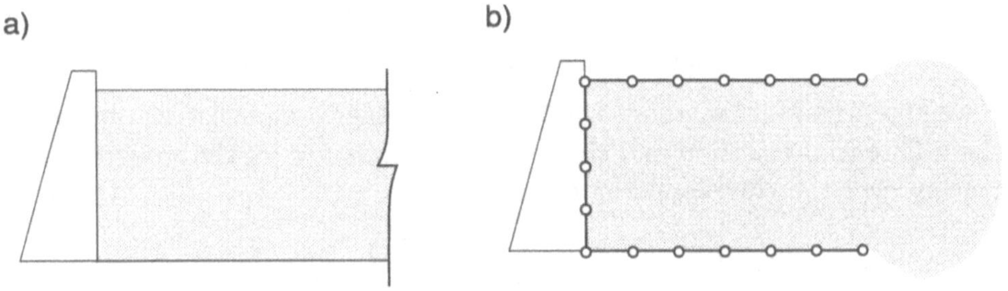

Figure 3.8: Model with infinite boundary: a) original model, b) boundary element model

current boundary values. For hyperbolic problems, however, the solution at any time depends on all previous boundary values. The reduction by one space dimension is neutralized by an additional "time dimension".

Conceptually, this is visualized in Figure 3.7 for a two-dimensional domain. Different possibilities for treating an infinite domain are shown. The first possibility is to use a large finite element mesh as shown in a). Since the waves travel at most with velocity c, the extent of the computational region does not go beyond cT for a duration time T. The second possibility is to use boundary elements in the frequency domain. The original domain is reduced to the boundary as shown in b). This solution has to be transformed to the time domain using a convolution integral. A third way is to use boundary elements in time. The original space-time domain can be reduced to the boundary consisting of the space boundary in the time interval zero to T as shown in c). The figure clearly shows that all three methods use essentially the same number of computational data points.

Another problem with BEM exists if not only the domain but also the boundaries extend to infinity. This is, for example, the case for the two-dimensional reservoir model shown in Figure 3.8. In the original model (a) the free surface and the bottom are not finite. In a BEM model (b) these boundaries have to be truncated. The calculation model differs therefore considerably from the original problem, unless the computational model is very long. A long model is feasible in this case since only the boundaries have to be modeled. This technique has been applied successfully in [HJ88, MD93, DM93] for the frequency domain. Another possibility is to use BEM only for a short nearfield together with a transmitting boundary [WWB88]. Then the advantage of the BEM, however, is lost.

The use of BEM in the frequency and time domains is demonstrated for the model problem of the semi-infinite rod on an elastic foundation. The derivation becomes somewhat lengthy, even for this simple example. The reader who is not interested in the formalism, may skip the rest of this section without loss of context.

3.2.1 Frequency-domain boundary elements

First consider the steady state case. The differential equation is given by

$$\frac{\partial^2 u}{\partial x^2} + k^2 u = 0 \qquad \text{for } x > a \tag{3.44}$$

where

$$k^2 = \frac{\omega^2}{c^2} - \lambda^2 \tag{3.45}$$

The boundary condition at $x = a$ is not specified for the moment but the second boundary condition, the radiation condition is stated as

$$\frac{\partial u}{\partial x} + iku = 0 \qquad \text{for } x \to \infty \tag{3.46}$$

Using the weighted residual method, the following integral equation is obtained by multiplying the differential equation and the radiation condition by the weighting function w and integrating over the corresponding domain.

$$\int_a^\infty \left(\frac{\partial^2 u}{\partial \xi^2} + k^2 u\right) w \, d\xi = \left(\frac{\partial u}{\partial \xi} + iku\right) w \bigg|_{\xi \to \infty} \tag{3.47}$$

where the variable x has been replaced by the dummy variable ξ for the integration. Integrating by parts twice, this equation becomes

$$\int_a^\infty u \left(\frac{\partial^2 w}{\partial \xi^2} + k^2 w\right) d\xi = \left[\frac{\partial u}{\partial \xi} w - u \frac{\partial w}{\partial \xi}\right]_{\xi = a} + \left[u \left(\frac{\partial w}{\partial \xi} + ikw\right)\right]_{\xi \to \infty} \tag{3.48}$$

For the weighting function w, the so-called *fundamental solution* is selected. The fundamental solution is the solution at ξ due to a concentrated source at x. It satisfies the differential equation

$$\frac{\partial^2 w}{\partial \xi^2} + k^2 w = \delta(x - \xi) \tag{3.49}$$

with the Dirac Delta function δ, and the radiation condition

$$\frac{\partial w}{\partial \xi} + ikw = 0 \qquad \text{for } \xi \to \infty \tag{3.50}$$

but not the boundary condition at $x = a$. It is given by

$$w(x - \xi) = -\frac{1}{2ik} e^{-ik|x-\xi|} \tag{3.51}$$

Generally, the fundamental solution is much simpler than the solution of the original problem. In this one-dimensional example this is not true. However, the important points can still be demonstrated.

Observing that the fundamental solution already satisfies the radiation condition Equation (3.50) and that $\int u(\xi)\, \delta(x - \xi)\, d\xi = u(x)$, Equation (3.48) reduces to

$$u(x) = \left[\frac{\partial u}{\partial \xi} w - u \frac{\partial w}{\partial \xi}\right]_{\xi = a} \tag{3.52}$$

that is, the displacement $u(x)$ at any point is expressed in terms of the displacement u and the derivative $\partial u/\partial\xi$ at the boundary. Generally, the boundary is a line or surface that has to be discretized by boundary elements. For this one-dimensional problem, the boundary consists only of the point $x = a$ and no discretization is necessary. To reduce the whole problem to the boundary, the source point x is moved to a. Noting the limits $x \to a$ from above ($x > \xi = a$)

$$w = -\frac{1}{2ik} \tag{3.53}$$

$$\frac{\partial w}{\partial \xi} = -\frac{1}{2} \tag{3.54}$$

we obtain

$$u(a) = -\frac{1}{2ik}\frac{\partial u}{\partial \xi} + \frac{1}{2}u(a) \tag{3.55}$$

or for the reaction force at $x = a$

$$Q = -EA\frac{\partial u}{\partial \xi} = ik\,EA\,u \tag{3.56}$$

as before. To use this equation in the time domain we have to apply the convolution integral (Equation 2.55).

3.2.2 Time-domain boundary elements

Now consider the same problem directly in the time domain. A derivation similar to the following is given in [MB82]. The differential equation is

$$\frac{\partial^2 u}{\partial t^2} - \lambda^2 u - \frac{1}{c^2}\frac{\partial^2 u}{\partial x^2} = 0 \qquad \text{for } x > a \tag{3.57}$$

The initial conditions are

$$u = \dot{u} = 0 \qquad \text{for } t = 0 \tag{3.58}$$

The boundary condition at $x = a$ will be specified later.

As in the steady state case, an integral equation is formed. The initial-value problem is considered a boundary-value problem in the space-time domain. Therefore, an additional integration over time has to be made. The upper limit of the time integrals is set to $t + \epsilon$, $\epsilon > 0$, to ensure that the singularity of the weighting function at $\tau = t$ is included.

$$\int_0^{t+\epsilon}\int_a^{\infty}\left(\frac{\partial^2 u}{\partial \xi^2} - \lambda^2 u - \frac{1}{c^2}\frac{\partial^2 u}{\partial \tau^2}\right) w\, d\xi\, d\tau = 0 \tag{3.59}$$

The weighting function w now depends on space and time. Integrating by parts the spatial derivatives with respect to ξ and the temporal derivatives with respect to τ we obtain

$$\begin{aligned}\int_0^{t+\epsilon}\int_a^{\infty} u\left(\frac{\partial^2 w}{\partial \xi^2} - \lambda^2 w - \frac{1}{c^2}\frac{\partial^2 w}{\partial \tau^2}\right) d\xi\, d\tau = \\ -\int_0^{t+\epsilon}\left[\frac{\partial u}{\partial \xi}w - u\frac{\partial w}{\partial \xi}\right]_{\xi=a}^{\xi\to\infty} d\tau + \frac{1}{c^2}\int_a^{\infty}\left[\frac{\partial u}{\partial \tau}w - u\frac{\partial w}{\partial \tau}\right]_{\tau=0}^{t+\epsilon} d\xi\end{aligned} \tag{3.60}$$

Note the convention $[\cdot]_a^b = [\cdot]_b - [\cdot]_a$. Again, the fundamental solution is selected as the weighting function. It is the response at time τ at location ξ due to an impulse at time t at location x and has to satisfy the differential equation

$$\frac{\partial^2 w}{\partial \xi^2} - \lambda^2 w - \frac{1}{c^2}\frac{\partial^2 w}{\partial \tau^2} = \delta(x-\xi)\,\delta(t-\tau) \tag{3.61}$$

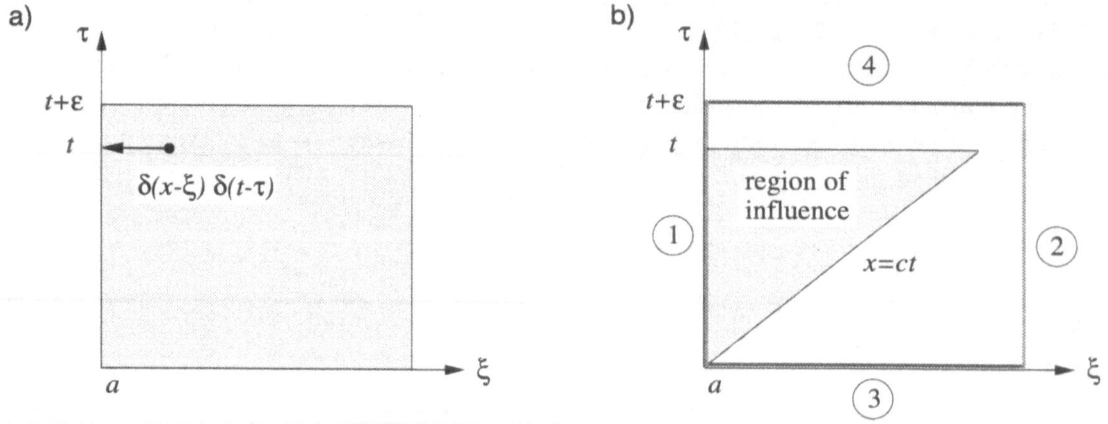

Figure 3.9: Integration regions for BEM in time: a) interior domain, b) integration paths 1–4 on boundary

Substituting this equation into the integral equation (3.60), the left-hand side becomes

$$\int_0^{t+\epsilon} \int_a^{\infty} u(\xi,\tau)\,\delta(x-\xi)\,\delta(t-\tau)\,d\xi\,d\tau = u(x,t) \tag{3.62}$$

The integration domain is depicted in Figure 3.9a. Because the upper bound of the time integration is $\tau = t + \epsilon$, the singularity $\delta(t - \tau)$ is fully included.

For the right-hand side of the integral equation (3.60) there are four integration paths as shown in Figure 3.7b. Path 1 along $\xi = a$ will give the main contribution. Path 2 along $\xi \to \infty$ and path 4 along $\tau = t + \epsilon$ give no contributions because they are beyond the region of influence. The integral along $\tau = 0$, path 3, vanishes because of the zero initial conditions. The integral Equation (3.60) therefore reduces to

$$u(x,t) = \int_0^{t+\epsilon} \left[\frac{\partial u}{\partial \xi} w - u \frac{\partial w}{\partial \xi}\right]_{\xi=a} d\tau \tag{3.63}$$

As for the steady state case, the displacement $u(x,t)$ is expressed in terms of the values at the boundary which now consists of the line $\xi = a$, $0 < \tau < t + \epsilon$. To have only the boundary involved in the calculations, the source point x is subsequently moved to $x = a$.

The fundamental solution is found by applying the Fourier transform to the steady state solution Equation (3.51) (see [CF48, No. 866])

$$w(x-\xi, t-\tau) = -\frac{c}{2} J_0\left(\lambda\sqrt{c^2(t-\tau)^2 - (x-\xi)^2}\right) H\left((t-\tau) - \frac{|x-\xi|}{c}\right) \tag{3.64}$$

where H denotes the Heaviside function.

To write Equation (3.63) at the boundary, w and $\partial w/\partial \xi$ have to be evaluated at $\xi = a$ and then the limit $x \to a$ from above ($x > a$) has to be taken. Note, one has to be careful taking the limits when the Heaviside function is involved. The final results are

$$w = -\frac{c}{2} J_0\big(c\lambda(t-\tau)\big) H(t-\tau) \tag{3.65}$$

$$\frac{\partial w}{\partial \xi} = -\frac{1}{2}\,\delta(t-\tau) \tag{3.66}$$

With the above results, Equation (3.63) becomes

$$u(a,t) = -\frac{c}{2}\int_0^t J_0\big(c\lambda(t-\tau)\big)\frac{\partial u}{\partial \xi}\,d\tau + \frac{1}{2}u(a,t) \tag{3.67}$$

The ϵ has been dropped because there are no more singularities in the integrand. Noting that $Q(t) = -c^2\mu\, \partial u/\partial\xi$, the following expression is obtained:

$$u(a,t) = \frac{1}{c\mu} \int_0^t J_0\Big(c\lambda(t-\tau)\Big) Q(\tau)\, d\tau \tag{3.68}$$

This is the same as the convolution integral of the dynamic flexibility formulation (Equation 2.58).

The example shows the typical behavior of the boundary element method in the time domain, stated already at the beginning of this section: Although the space dimension of the problem is reduced by one, an additional time dimension appears. The formulation is basically equivalent to a convolution integral and hence offers no advantage over conventional methods.

3.3 Infinite elements and mapping

As an introduction, the first part of this section is devoted to finite elements. It is shown that the spatial discretization should not be too coarse in order to avoid excessive dispersion. Thus the technique of modeling the farfield by large-size elements is not applicable to wave problems.

Inspired by the finite element method different authors have developed infinite elements. Infinite elements can be derived either by using special shape functions, usually functions that decay for large distances such as e^{-x} or $1/x$, or by mapping the infinite domain into a finite one. A good overview is given by Bettes and Bettes [BB84]. A prerequisite for the formulation of infinite elements is some knowledge of the solution, that is, the shape functions have to be close to the actual solution of the problem and also the boundary conditions at infinity have to be known, a least approximately. As pointed out in [BB84], infinite elements work well for static and steady state problems. For static problems a shape function decaying uniformly to zero at infinity is usually appropriate. For steady state wave problems the shape functions have to be oscillatory with the correct period [CC87]. For transient wave problems the solution cannot be described by a few degrees of freedom with appropriate shape functions but instead a large number of degrees of freedom is essential.

Infinite elements for fluid problems have been proposed among others by Olson and Bathe [OB85b]. They use a DDA approach, that is, viscous dampers for the high-frequency behavior and a static infinite-element solution for the low-frequency behavior. Needless to say such elements are not appropriate for the dam-reservoir problem.

Another method using similar concepts of mapping is proposed by Nath [Nat81]. Instead of a single infinite element, a large reservoir consisting of many elements is mapped into a smaller region. Regularly spaced nodes in the mapped domain correspond to nodes with increasing spacing towards the far end, at which a fixed boundary condition is applied. Neither the increasing spacing nor the fixed boundary seem to be appropriate assumptions for the dam-reservoir problem.

3.3.1 Spatial discretization, finite elements

Before discussing infinite elements, the finite element method is demonstrated by the example of a rod on an elastic foundation. The example is used as an introduction to finite and infinite elements. A more general description of finite elements applicable to fluid and solid elements is given in a separate chapter (Chapter 10). The finite elements solution is also used to show the effect of spatial discretization.

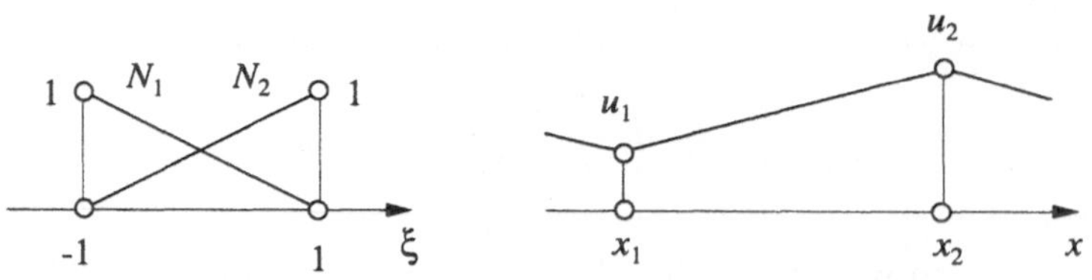

Figure 3.10: Shape functions and mapping to parent element

Consider a finite rod with the differential equation

$$\frac{\partial^2 u}{\partial x^2} + k^2 u = 0 \qquad \text{for } a < x < b \tag{3.69}$$

with

$$k^2 = \frac{\omega^2}{c^2} - \lambda^2 \tag{3.70}$$

The boundary conditions are

$$Q_a = -EA\,\frac{\partial u}{\partial x} \qquad \text{at } x = a \tag{3.71}$$

and

$$Q_b = EA\,\frac{\partial u}{\partial x} \qquad \text{at } x = b \tag{3.72}$$

The starting point is again, as for the BEM, an integral equation obtained by a weighted residual method. The differential equation (multiplied by EA) and the boundary conditions are multiplied by a weighting function w and integrated over the domain.

$$\int_a^b EA\left(\frac{\partial^2 u}{\partial x^2} + k^2 u\right) w\,dx = \left[\left(-EA\frac{\partial u}{\partial x} - Q\right) w\right]_{x=a}^{b} \tag{3.73}$$

The first term of the integrand is integrated by parts to obtain

$$\int_a^b EA\,\frac{\partial u}{\partial x}\frac{\partial w}{\partial x}\,dx - \int_a^b c^2 \mu\, k^2 u w\,dx = [Qw]_{x=a}^{b} \tag{3.74}$$

where $EA = c^2\mu$ has been substituted in the second integral. The domain $a < x < b$ is now divided into elements. For each element the functions u and w are interpolated between the nodal values by the so-called shape functions. Continuity is enforced because the elements share their common nodal values. The method is shown for a set of elements with linear shape functions. Consider the function $u(x)$ depicted in Figure 3.10. For the element between nodes 1 and 2 it is described by

$$u(x) = N_1(\xi)\,u_1 + N_2(\xi)\,u_2 \tag{3.75}$$

where $-1 \leq \xi \leq 1$ is the local non-dimensional coordinate of the element and N_1 and N_2 are the shape functions given by

$$N_1(\xi) = (1-\xi)/2 \tag{3.76}$$

$$N_2(\xi) = (1+\xi)/2 \tag{3.77}$$

Note that $N_1(-1) = 1$ and $N_1(1) = 0$ and analogously for N_2. Further $N_1(\xi) + N_2(\xi) = 1$. The discretization is over space only, while the nodal values are still functions of time. This is also called semi-discretization.

The coordinate x is interpolated using the same shape functions (isoparametric elements)

$$x(\xi) = N_1(\xi)\, x_1 + N_2(\xi)\, x_2 \tag{3.78}$$

In the further development the values for $\partial x/\partial\xi$ and $\partial\xi/\partial x$ are needed. They are

$$\frac{\partial x}{\partial \xi} = \frac{\partial N_1}{\partial \xi} x_1 + \frac{\partial N_2}{\partial \xi} x_2 = \frac{x_2 - x_1}{2} = \frac{L}{2} \tag{3.79}$$

and

$$\frac{\partial \xi}{\partial x} = \left(\frac{\partial x}{\partial \xi}\right)^{-1} = \frac{2}{L} \tag{3.80}$$

The derivative of the displacement u with respect to x is

$$\frac{\partial u}{\partial x} = \left(\frac{\partial N_1}{\partial x} u_1 + \frac{\partial N_2}{\partial x} u_2\right) = B_1(\xi)\, u_1 + B_2(\xi)\, u_2 \tag{3.81}$$

where by the chain rule

$$B_1(\xi) = \frac{\partial N_1}{\partial \xi}\frac{\partial \xi}{\partial x} = -\frac{1}{L} \tag{3.82}$$

$$B_2(\xi) = \frac{\partial N_2}{\partial \xi}\frac{\partial \xi}{\partial x} = \frac{1}{L} \tag{3.83}$$

The displacement and its derivative for an element with nodes at x_1 and x_2 are written as

$$u(\xi) = \begin{bmatrix} N_1(\xi) & N_2(\xi) \end{bmatrix} \begin{bmatrix} u_1 \\ u_2 \end{bmatrix} = \mathbf{N}^e \mathbf{u}^e \tag{3.84}$$

$$\frac{\partial u}{\partial x} = \begin{bmatrix} B_1(\xi) & B_2(\xi) \end{bmatrix} \begin{bmatrix} u_1 \\ u_2 \end{bmatrix} = \mathbf{B}^e \mathbf{u}^e \tag{3.85}$$

and for the whole domain as

$$u = \mathbf{N}\mathbf{u} \quad \text{and} \quad \frac{\partial u}{\partial x} = \mathbf{B}\mathbf{u} \tag{3.86}$$

Using the same interpolation the weighting function is

$$w = \mathbf{N}\mathbf{w} \quad \text{and} \quad \frac{\partial w}{\partial x} = \mathbf{B}\mathbf{w} \tag{3.87}$$

The integral Equation (3.74) evaluated over the domain yields

$$\mathbf{w}^{\mathrm{T}} \int_a^b EA\,\mathbf{B}^{\mathrm{T}}\mathbf{B}\, dx\, \mathbf{u} - \mathbf{w}^{\mathrm{T}} \int_a^b k^2 c^2 \mu\, \mathbf{N}^{\mathrm{T}}\mathbf{N}\, dx\, \mathbf{u} = \mathbf{w}^{\mathrm{T}}\mathbf{Q} \tag{3.88}$$

The vector $\mathbf{Q}$ contains the reaction forces Q_a and Q_b for the nodes at both ends and zero for the internal nodes. The equation has to be satisfied independently of the arbitrary weighting function $\mathbf{w}$. Therefore

$$(\mathbf{K}\mathbf{u} + \lambda^2 c^2 \mathbf{M} - \omega^2 \mathbf{M})\mathbf{u} = \mathbf{Q} \tag{3.89}$$

with

$$\mathbf{K} = \int_a^b EA\,\mathbf{B}^{\mathrm{T}}\mathbf{B}\, dx \tag{3.90}$$

$$\mathbf{M} = \int_a^b \mu\, \mathbf{N}^{\mathrm{T}}\mathbf{N}\, dx \tag{3.91}$$

These are the classical stiffness and mass matrices for the rod. The integrals over the domain $a < x < b$ are carried out for each element individually and the matrices $\mathbf{K}$ and $\mathbf{M}$ are assembled afterwards. The integral for the stiffness matrix of an individual element is, after changing the integration variable from x to ξ,

$$\mathbf{K}^e = \int_{x_1}^{x_2} EA\,\mathbf{B}^{eT}\mathbf{B}^e\,dx = \int_{-1}^{1} EA\,\mathbf{B}^{eT}\mathbf{B}^e \left|\frac{\partial x}{\partial \xi}\right| d\xi = \int_{-1}^{1} EA\,\mathbf{B}^{eT}\mathbf{B}^e (L/2)\,d\xi \tag{3.92}$$

Using the previously derived functions for $\mathbf{B}$, the element stiffness matrix is

$$\mathbf{K}^e = EA\int_{-1}^{1} \begin{bmatrix} -1/L \\ 1/L \end{bmatrix} \begin{bmatrix} -1/L & 1/L \end{bmatrix} \frac{L}{2}\,d\xi = \frac{EA}{L}\begin{bmatrix} 1 & -1 \\ -1 & 1 \end{bmatrix} \tag{3.93}$$

Similarly the consistent mass matrix is derived

$$\mathbf{M}^e = \mu\int_{-1}^{1} \begin{bmatrix} (1-\xi)/2 \\ (1+\xi)/2 \end{bmatrix} \begin{bmatrix} (1-\xi)/2 & (1+\xi)/2 \end{bmatrix} \frac{L}{2}\,d\xi = \frac{\mu L}{6}\begin{bmatrix} 2 & 1 \\ 1 & 2 \end{bmatrix} \tag{3.94}$$

Dispersion due to discretization. Consider a node j and its neighboring nodes $j-1$ and $j+1$ in a regularly spaced finite element model. The dynamic equilibrium for node j is (with $EA = c^2\mu$)

$$\frac{c^2\mu}{L}(2u_j - u_{j-1} - u_{j+1}) - k^2c^2\frac{\mu L}{6}(4u_j + u_{j-1} + u_{j+1}) = 0 \tag{3.95}$$

Now assume a steady state motion

$$u(x,t) = u_j e^{i(\omega - k_0 x)} \tag{3.96}$$

where k_0 is the wave number of the discrete model. The wave number k of the continuous problem plays now only the role of a material property defined as $k^2 = \omega^2/c^2 - \lambda^2$. The displacements are

$$u_{j-1} = u_j e^{ik_0L} \quad \text{and} \quad u_{j+1} = u_j e^{-ik_0L} \tag{3.97}$$

Assembling the proper terms from the stiffness and mass matrix, the dynamic equilibrium at node j becomes

$$\frac{c^2\mu}{L}(2 - e^{ik_0L} - e^{-ik_0L}) = \frac{k^2c^2\mu L}{6}(4 + e^{ik_0L} + e^{-ik_0L}) \tag{3.98}$$

or

$$(kL)^2 = 6\,\frac{1-\cos k_0L}{2+\cos k_0L} \tag{3.99}$$

Equation (3.99) is an implicit formulation for k_0. It is solved by calculating kL for different values of k_0L. If the values are chosen along the path indicated in Figure 3.11, the values for kL become real valued. The wave number k_0 of the discrete model is plotted in Figure 3.12. For small values of k, the limit is $k_0 = k$. For higher values of k, up to $kL = \sqrt{12}$ (corresponding to $k_0L = \pi$), the discrete model is a reasonable approximation. Because k_0 is real, the wave travels as indicated in Figure 3.11. At $kL = \sqrt{12}$ the wave is standing and no energy is transmitted in the rod. For $kL > \sqrt{12}$, the wave is also standing but decays with distance. The limiting value of $kL = \sqrt{12}$ turns out to be the largest eigenvalue of a single rod element.

This example shows clearly that, at least for simple shape functions, the spacing of the nodes should not be too large, because otherwise even a long finite element model does not behave properly. For $\lambda = 0$ (no elastic foundation), the limit is about half the wave length. Similar effects show up due to the time discretization. These are, however, not investigated here. Although there are beneficial counter-effects for certain combinations of spatial and temporal discretization schemes, the discretization should generally be accurate independently for space and for time.

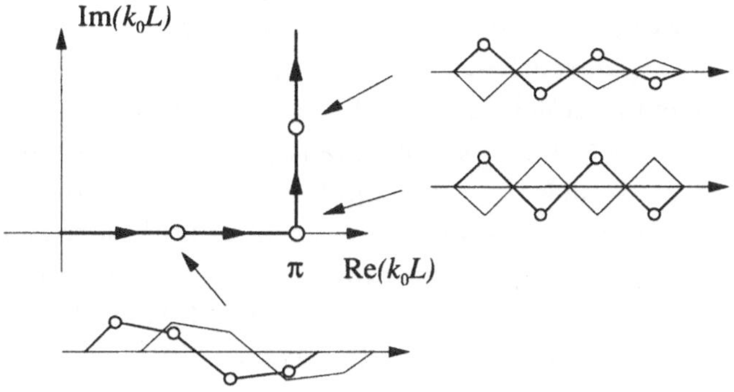

Figure 3.11: Dispersion due to discretization: typical cases

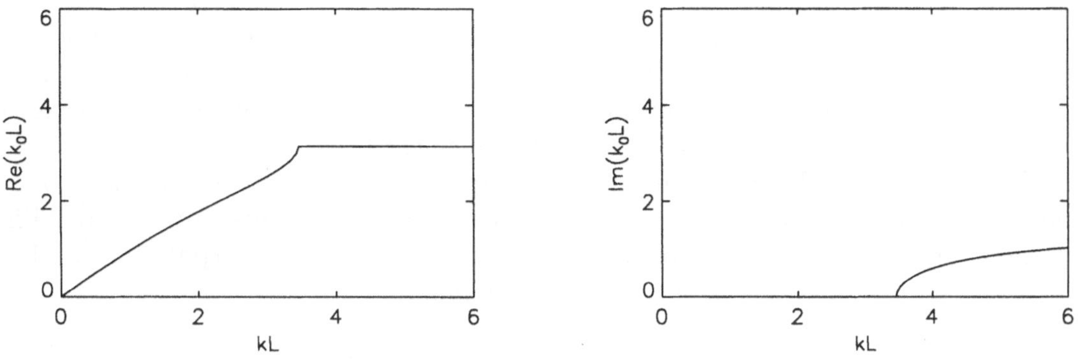

Figure 3.12: Dispersion due to discretization: wave numbers

3.3.2 Infinite elements

Infinite shape function. The construction of a single infinite element $a \leq x < \infty$ starts with the integral equation (3.74) as in the finite element formulation. Because the infinite element is supposed to simulate the transmitting boundary condition, not the exact boundary condition is imposed at the far end $x = b$ but rather a simple condition, that is, zero force or zero displacement. For the shape function, an obvious choice is the exact solution. However, to be more flexible, a free parameter $\tilde{k}$ is introduced instead of the wave number. Then the displacement is

$$u = u_a \, e^{i\tilde{k}(a-x)} \tag{3.100}$$

and similarly the weighting function

$$w = w_a \, e^{i\tilde{k}(a-x)} \tag{3.101}$$

Since with this choice the displacement at $x = b$ is specified, the boundary condition there is chosen as $Q_b = 0$. With the substitution $EA = c^2\mu$ the integral equation (3.74) is

$$u_a w_a c^2 \mu \left(- \int_a^b \tilde{k}^2 \, e^{2i\tilde{k}(a-x)} dx - \int_a^b k^2 e^{2i\tilde{k}(a-x)} dx \right) = Q w_a \tag{3.102}$$

The integrals can be evaluated analytically. This leads to the relationship

$$Q = \frac{ik + i\tilde{k}}{2} c^2 \mu \left(1 - e^{2i\tilde{k}(a-b)} \right) u_a \tag{3.103}$$

For choosing a value $\tilde{k}$, first consider the case $\tilde{k} = k$. For $\omega < c\lambda$, the exponential function in Equation (3.103) decays to zero for large values of b because $i\tilde{k} = \sqrt{\lambda^2 - \omega^2/c^2}$ is a real number. The expression in parentheses converges to 1 and the exact answer is obtained. For $\omega > c\lambda$, the exponential function is oscillating and so is the stiffness of the infinite element depending on the position b of the right boundary. To assure convergence, a real part must be included in $i\tilde{k}$ such that the exponential function decays for large values of b. Then, however, the stiffness of the infinite element is no longer exact.

Generally, a reasonable answer can be expected if $\tilde{k} \approx k$ is chosen. Since usually the exact value is not known, it has to be guessed. The solution obtained from the infinite element lies between the guessed value and the exact solution. For problems where k is not a fixed value, such as for a reservoir with different modes or any wave problem in the time domain, infinite elements are not expected to give dependable results. For static problems, however, they seem to give good results.

Mapping. Another interesting infinite element has been proposed by Bettes [BB84] for statics. It has also been applied to dynamic fluid problems by Olson and Bathe [OB85b] for the low-frequency behavior combined with a viscous boundary for high frequencies. Although these elements cannot be used for the dam-reservoir problem, they nevertheless show the principles and the limitations of such elements.

The derivation is based on geometric mapping. The geometric mapping by shape functions has already been explained for isoparametric finite elements. Here, different functions are chosen for the geometric mapping and for the interpolation of the field variables. The shape functions for geometric mapping are

$$N_1 = \frac{-2\xi}{1-\xi} \qquad N_2 = \frac{1+\xi}{1-\xi} \tag{3.104}$$

The coordinate mapping is

$$x = N_1\, x_1 + N_2\, x_2 \tag{3.105}$$

Note that $N_1 + N_2 = 1$, which guarantees that the mapping is invariant for translation. To simplify the algebra consider the special case

$$x_1 = a \qquad x_2 = 2a \tag{3.106}$$

Then the x-coordinate is

$$x = \frac{2a}{1-\xi} \tag{3.107}$$

The values $\xi = -1,\ 0,\ 1$ correspond to $x = x_1,\ x_2,\ \infty$, respectively. Later on, the derivatives are also needed. They are

$$\frac{\partial x}{\partial \xi} = \frac{2a}{(1-\xi)^2} \tag{3.108}$$

and

$$\frac{\partial \xi}{\partial x} = \left(\frac{\partial x}{\partial \xi}\right)^{-1} = \frac{(1-\xi)^2}{2a} \tag{3.109}$$

Different shape functions are used for the interpolation of field variables. They are

$$H_1 = -\frac{\xi(1-\xi)}{2} \qquad H_2 = 1 - \xi^2 \tag{3.110}$$

The displacement

$$u = H_1\, u_1 + H_2\, u_2 \tag{3.111}$$

interpolates the nodal values at the nodes 1 and 2 and vanishes at $\xi = 1$ (corresponding to $x \to \infty$). The derivatives of the shape functions are

$$B_1(\xi) = \frac{\partial H_1}{\partial \xi}\frac{\partial \xi}{\partial x} = -\frac{(1-2\xi)(1-\xi)^2}{4a} \tag{3.112}$$

$$B_2(\xi) = \frac{\partial H_2}{\partial \xi}\frac{\partial \xi}{\partial x} = -\frac{\xi(1-\xi)^2}{a} \tag{3.113}$$

The stiffness matrix is given by the integral

$$\mathbf{K} = c^2\mu \int_{-1}^{1} \begin{bmatrix} B_1^2 & B_1\,B_2 \\ B_1\,B_2 & B_2^2 \end{bmatrix} \frac{2a}{(1-\xi)^2}\,d\xi = \frac{c^2\mu}{15a}\begin{bmatrix} 23 & -26 \\ -26 & 32 \end{bmatrix} \tag{3.114}$$

and the mass by

$$\mathbf{M} = \mu \int_{-1}^{1} \begin{bmatrix} H_1^2 & H_1\,H_2 \\ H_1\,H_2 & H_2^2 \end{bmatrix} \frac{2a}{(1-\xi)^2}\,d\xi = \frac{a\mu}{3}\begin{bmatrix} 1 & -2 \\ -2 & 16 \end{bmatrix} \tag{3.115}$$

The integrals have been evaluated analytically, but special numerical integration schemes are also available [BB84] for more general cases. For this example, the analytical calculations have been performed using the program MAPLE, a system for symbolic mathematical computation [CGG+90]. The element is now used to predict the stiffness of the semi-infinite rod on an elastic foundation in the static case. It is found by solving the equation

$$(\mathbf{K} + c^2\lambda^2\mathbf{M})\begin{bmatrix} u_1 \\ u_2 \end{bmatrix} = \begin{bmatrix} Q \\ 0 \end{bmatrix} \tag{3.116}$$

The solution

$$u_1 = \frac{Qa}{c^2\mu}12\frac{2+5\lambda^2a^2}{3+74\lambda^2a^2+15\lambda^4a^4} \tag{3.117}$$

is compared to the exact solution

$$u_{ex} = \frac{Q}{\lambda c^2\mu} \tag{3.118}$$

in Figure 3.13. The value u_1/u_{ex} is plotted for different values of $a\lambda$. Note, that the leading term in the approximate solution can be written as $Q\lambda a/(\lambda c^2\mu)$ and the function plotted is

$$\frac{u_1}{u_{ex}} = \frac{12\,\lambda a\,(2+5\lambda^2a^2)}{3+74\lambda^2a^2+15\lambda^4a^4} \tag{3.119}$$

It is close to one for a wide range of $a\lambda$, but it clearly depends on that parameter and thus on the element size. The method is therefore not reliable if we have no idea of what this parameter should be.

In the references [BB84, OB85b] there are many problems shown that work very well, but for the problems addressed here, infinite elements are not useful and they are therefore not investigated further.

3.4 Mixed time-frequency domain methods

This section reviews some methods that switch between time-domain and frequency-domain during the calculation.

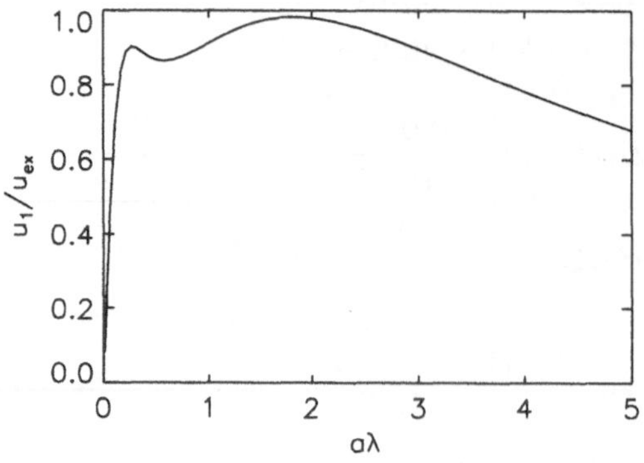

Figure 3.13: displacement of infinite element for semi-infinite rod on elastic foundation

3.4.1 Hybrid method

One approach is the hybrid frequency-time-domain procedure suggested by Darbre [DW88]. The method works iteratively. Starting from a linear analysis in the frequency domain, out-of-balance forces due to nonlinear effects are calculated. These forces are used as a correction for a subsequent linear analysis.

One problem is, that adding correction forces at one time step, may alter the response at later times. The typical effect is that after a few iterations the results have converged up to a certain time step. Therefore, it is suggested to calculate the complete time history not at once but to use several segments. Even then one has typically to perform several hundred iterations making the analysis very costly. The second problem is that convergence is not always achieved. Even for simple problems convergence is only guaranteed for mild nonlinearities [DW88]. An application related to dams is presented in [FC90].

3.4.2 "Recursive" method

Mohasseb and Wolf [MW89] suggested a method where the nearfield is calculated in the time domain but the interaction forces are calculated after each time step in the frequency domain. The reaction forces are stored in the frequency domain and updated after each time step, rather than calculated afresh. Because the reaction forces are reused, the authors call their method "recursive". However, the update is not a true recursion. For certain applications only a few frequencies have to be considered and the number of operations can therefore be reduced considerably compared to a convolution.

Chapter 4

Scalar Systems, Signal Processing

In this chapter a new class of methods, the rational methods are introduced. These methods are mathematically described by linear time-invariant systems having a rational transfer function. A system is a mathematical abstraction of a physical model (mechanical, electrical). It has an input-output relationship described by a differential or difference equation, an impulse response function or a transfer function. The internal structure is usually of minor interest and has generally no physical significance but general attributes such as stability, and minimal realization have to be investigated.

One important aspect of systems is the duality between discrete-time systems and continuous-time systems. Discrete-time systems, expressed as difference equations, are often more intuitive and are useful for numerical implementation. Continuous-time systems, expressed as differential equations are closer to mathematical analysis. The connection between the two formulations is the bilinear transform. Besides its practical interest, the duality between discrete-time and continuous-time systems is also theoretically appealing.

After describing briefly the motivation for using rational methods, the theoretical background for discrete-time and continuous-time systems is given. Although some topics are standard textbook material, others, such as the Hankel matrix and the singular value decomposition, are a specialized introduction to the following derivations. Only linear time-invariant systems are considered. The chapter is restricted to scalar systems, that is, systems with only one input and one output variable. These systems can be treated by the signal processing formulation. The more general state-variable description is used for multivariable systems (multi-input, multi-output) and is given later in Chapter 6.

4.1 Rational systems

Consider first a one degree-of-freedom system given by

$$\ddot{u} + \omega_n^2 u = p(t) \tag{4.1}$$

with zero initial conditions. This differential equation can be solved in two different ways. The first is by the convolution integral (also called Duhamel integral)

$$u(t) = \int_0^t h(t-\tau)\, p(\tau)\, d\tau \tag{4.2}$$

where $h(t-\tau) = \sin\omega(t-\tau)$ is the impulse response. For numerical evaluation the functions are discretized at equidistant time steps $t_n = n\Delta t$. The discrete form of the convolution integral is then

$$u(t_n) = \Delta t \left(\frac{h(t_0)}{2}\, p(t_0) + \frac{h(t_n)}{2}\, p(t_n) + \sum_{k=1}^{n-1} h(t_{n-k})\, p(t_k) \right) \tag{4.3}$$

The term $h(t_{n-k})\,p(t_k)$ is the response at t_n due to an impulse $p(t_k)$ at an earlier time step t_k. The superposition of the response due to all past impulses yields the present response. The factor 1/2 for $h(t_0)$ and for $h(t_n)$ counts for the half width of the contributing area at both ends of the integration interval. The computational effort to calculate the convolution integral is investigated next. For each desired time step t_n, a sum with n terms has to be evaluated. For N_T time steps the number of coefficients $h(t_k)$ to be stored is N_T, the number of multiplications is $1 + 2 + 3 + \cdots + (N_T + 1) \approx N_T^2/2$. The memory demand is thus proportional to the number of time steps, the computational effort proportional to the square of the number of time steps.

The second method to solve Equation (4.1) is by a time-stepping scheme such as the Newmark method. These schemes are based on the discretization of the differential equation itself rather than the impulse response. The commonly used algorithms can be written in the displacement difference-equation form [Hug87, p. 527] as

$$u(t_n) = -a_1\,u(t_{n-1}) - a_2\,u(t_{n-2}) + b_0\,p(t_n) + b_1\,p(t_{n-1}) + b_2\,p(t_{n-2}) \tag{4.4}$$

The method proceeds from one time step to the next. Regarding the computational effort, there are only a few coefficients and the number of multiplications is the same for each time step. In the above formulation there are 4 coefficients. The number of multiplications for each time step is also 4, leading to $4N_T$ multiplications for N_T time steps. Hence the memory demand is independent of the number of time steps and the computational effort is proportional to the number of time steps. The second method is thus generally much more efficient.

The transmitting boundary is not given as differential equation but rather as frequency-domain solution $F(\omega)$ (usually dynamic stiffness) defined by

$$\hat{u}\,e^{i\omega t} = F(\omega)\,\hat{p}\,e^{i\omega t} \tag{4.5}$$

where $\hat{u}$ and $\hat{p}$ are the complex frequency-dependent amplitudes. Instead of transforming the transfer function to the time domain and applying a convolution integral, the idea is now to use a time stepping scheme. Two different ways can be taken. One way is to find a time stepping scheme analogous to Equation (4.4) which has the same transfer function as the original frequency-domain solution. Because the output at a given time step depends not only on the previous input as for the convolution but also on the previous output, the algorithm is recursive. The other way is to first find a linear differential equation with constant coefficients which is then discretized using a standard time integration algorithm. This leads again to a recursive scheme.

4.1.1 Recursive realization

To find directly a recursive realization, Equation (4.4) is generalized to

$$u(t_n) = -\sum_{k=1}^{N} a_k\,u(t_{n-k}) + \sum_{k=0}^{M} b_k\,p(t_{n-k}) \tag{4.6}$$

This equation appears in various fields of engineering. In signal processing and filter theory it is called recursive digital filter or infinite impulse response (IIR) filter. In system identification it is called ARMA model. AR = 'auto regressive' is the first part and MR = 'moving average' is the second part of the equation.

Equation (4.6) can be transformed to the frequency domain to obtain its transfer function $H(\omega)$. By definition, $H(\omega)\,e^{i\omega t} = u$ for the input $p = e^{i\omega t}$. Considering only discrete time steps $t_n = n\Delta t$, Equation (4.6) becomes

$$H(\omega)\, e^{i\omega n\Delta t} = -H(\omega) \sum_{k=1}^{N} a_k\, e^{i\omega(n-k)\Delta t} + \sum_{k=0}^{M} b_k\, e^{i\omega(n-k)\Delta t} \tag{4.7}$$

Solved for $H(\omega)$ we obtain

$$H(\omega) = \frac{\sum_{k=0}^{M} b_k\, e^{-i\omega k\Delta t}}{\sum_{k=0}^{N} a_k\, e^{-i\omega k\Delta t}} = \frac{b_0 + b_1\, e^{-i\omega\,\Delta t} + \cdots + b_M\, e^{-i\omega\, M\,\Delta t}}{a_0 + a_1\, e^{-i\omega\,\Delta t} + \cdots + a_N\, e^{-i\omega\, N\,\Delta t}} \tag{4.8}$$

where the constant 1 has been written as a_0 to get a more general expression. This is a rational transfer function. The recursive algorithm, Equation (4.6), will solve the problem if $H(\omega)$ is an approximation of $F(\omega)$. The transfer function of a difference equation is periodic with a period that depends on the time increment Δt. The original frequency-domain solution has to be restricted to a frequency interval which coincides with this period, that is, it has to be band-limited.

4.1.2 Differential equation formulation

Instead of directly looking for a recursive realization, we can first find a differential equation. This differential equation can be coupled with the finite element differential equations to one system which is then discretized in time. The advantage of this procedure is that the problem of approximation and the discretization in time are uncoupled and the same time integration scheme can be used for both the finite and the infinite domain. This strategy also allows for variable time steps in a nonlinear analysis.

The counterpart of the difference equation, Equation (4.4), is the differential equation

$$a_0\, u^{(N)} + \cdots + a_{N-1}\, \dot{u} + a_N\, u = b_{M-N}\, p^{(M)} + \cdots + b_{N-1}\, \dot{p} + b_N\, p \tag{4.9}$$

The transfer function is $(M \leq N)$

$$H(\omega) = \frac{\sum_{k=N-M}^{N} b_k\, (i\omega)^{N-k}}{\sum_{k=0}^{N} a_k\, (i\omega)^{N-k}} = \frac{b_N + b_{N-1}\, i\omega + \cdots + b_{N-M}\, (i\omega)^M}{a_N + a_{N-1}\, i\omega + \cdots + a_0\, (i\omega)^N} \tag{4.10}$$

Because this transfer function is a rational function, the differential equation is the analogue to the recursive realization. If the parameters of this transfer function are selected such that $H(\omega)$ is an approximation of the exact transfer function $F(\omega)$, then the differential equation, Equation (4.9), can be used to model the infinite domain. Contrary to the discrete-time case, this transfer function is not periodic and hence not restricted to band-limited transfer functions.

4.2 Discrete-time systems

The recursive realization introduced in the last section can be viewed as a discrete-time system. This section gives the necessary theoretical background on discrete-time systems for the next chapter where different methods of determining the coefficients for the recursive realization scheme are explained. The theory of digital signal processing may be found in many text books, for example in the one by Rabiner and Gold [RG75]. A compact overview is given in the article [RCH+72].

4.2.1 z-transform

For discrete-time systems it is convenient to introduce the z-transform. The z-transform is the discrete-time counterpart of the Laplace transform. The starting point for the derivation is the discrete-time convolution. It is written in the general form

$$u(n) = \sum_{k=-\infty}^{\infty} h_k\, p(n-k) \tag{4.11}$$

where the discrete-time impulse response h_k, $k = -\infty, \cdots, \infty$ has been used instead of the continuous-time impulse response $h(t)$. The variables $u(t_n)$ and $p(t_n)$ have been changed to the customary notation $u(n)$ and $p(n)$. Setting $p = e^{i\omega t}$ and $u = H(e^{i\omega\Delta t})e^{i\omega t}$, the transfer function is found to be

$$H(e^{i\omega\Delta t}) = \sum_{k=-\infty}^{\infty} h_k\, e^{-i\omega k\Delta t} \tag{4.12}$$

The transfer function is written here as a function of $e^{i\omega\Delta t}$ rather than a funcion of ω to be consistent with subsequent definitions.

$H(e^{i\omega\Delta t})$ is a periodic function with period $2\pi/\Delta t$. Therefore, Equation (4.12) may be considered as a Fourier series. The coefficients of the Fourier series are

$$h_k = \frac{\Delta t}{2\pi} \int_{-\pi/\Delta t}^{\pi/\Delta t} H(e^{i\omega\Delta t})\, e^{i\omega k\Delta t}\, d\omega \tag{4.13}$$

With the change of variable $z = e^{i\omega\Delta t}$ and $d\omega = dz/(iz\Delta t)$, Equations (4.12) and (4.13) yield the z-transform pair

$$H(z) = \sum_{k=-\infty}^{\infty} h_k\, z^{-k} \tag{4.14}$$

$$h_k = \frac{1}{2\pi i} \oint_C H(z)\, z^{k-1}\, dz \tag{4.15}$$

The integral in the last equation (4.15) has to be evaluated counterclockwise along the path C, the unit circle in the complex z-plane. Note that Equation (4.14) is a periodic function as it is the transform of a discrete function. Equation (4.15), on the other hand, is not periodic because it is a transform of a continuous function. The function $H(z)$ is defined on the whole complex plane. The theory of complex analysis and especially the theory of analytic functions is applicable. Although the transform pair Equations (4.14) and (4.15) has been derived here using $z = e^{i\omega\Delta t}$, it does not depend on this particular mapping between z and ω. It can be derived directly using complex analysis (use Equation 4.30). The z-transform is thus much more than just a convenient abbreviation $z = e^{i\omega\Delta t}$.

Rational transfer function. The z-transform can now be used to write the transfer function Equation (4.8) of the recursive realization. Instead of writing the transfer function for polynomials of arbitrary degrees M and N we show only the two cases we are interested in, that is, the proper and the strictly proper transfer functions. The most general proper transfer function is

$$H(z) = \frac{\sum_{k=0}^{N} b_k z^{-k}}{\sum_{k=0}^{N} a_k z^{-k}} = \frac{b_0 + b_1 z^{-1} + \cdots + b_N z^{-N}}{a_0 + a_1 z^{-1} + \cdots + a_N z^{-N}} = \frac{b_0 z^N + \cdots + b_N}{a_0 z^N + \cdots + a_N} \tag{4.16}$$

where some of the b_k can be zero. The most general strictly proper transfer function is

$$H(z) = \frac{\sum_{k=1}^{N} b_k z^{-k}}{\sum_{k=0}^{N} a_k z^{-k}} = \frac{b_1 z^{-1} + \cdots + b_N z^{-N}}{a_0 + a_1 z^{-1} + \cdots + a_N z^{-N}} = \frac{b_1 z^{N-1} + \cdots + b_N}{a_0 z^N + \cdots + a_N} \tag{4.17}$$

The terms proper and strictly proper relate to the relative order of the numerator and denominator polynomial. A proper rational transfer function has a numerator polynomial of order equal or less than the degree of the denominator polynomial. For a strictly proper transfer function the order of the numerator has to be less than the order of the denominator.

Partial fractions realization. Since $H(z)$ is a rational function of z it can be expanded into partial fractions as

$$H(z) = \frac{b_0\, z^N + \cdots + b_N}{a_0\, z^N + \cdots + a_N} = c_0 + \sum_{k=1}^{N} \frac{c_k}{z - z_k} \tag{4.18}$$

where the constant c_0 is only present for proper transfer functions and vanishes for strictly proper transfer functions. The constants z_k are the poles. They are the roots of the polynomial

$$\sum_{k=0}^{N} a_k\, z^{N-k} = 0 \tag{4.19}$$

The form Equation (4.18) is only valid if all poles z_k are distinct. This restriction is usually true in practical applications and it is always true for the rational approximations derived in Chapter 7. (For an input-output symmetric system the matrix $\mathbf{A}$ obtained by the balanced realization method can always be transformed into diagonal form.) To avoid unnecessary complications only this case is therefore considered here.

The partial fractions expansion is usually written in a slightly different form relating more closely to the difference equations of the recursive realization. Using powers of z^{-1} instead of z yields

$$H(z) = \hat{c}_0 + \sum_{k=1}^{N} \frac{\hat{c}_k}{1 - z_k\, z^{-1}} \tag{4.20}$$

The relation between the coefficients c_k and $\hat{c}_k$ is established in the following. We expand

$$\begin{aligned} \frac{c_k}{z - z_k} = \frac{c_k\, z^{-1}}{1 - z_k\, z^{-1}} &= c_k\, z^{-1} \left(1 + z_k\, z^{-1} + (z_k\, z^{-1})^2 + \cdots\right) \\ &= c_k\, z_k^{-1} \left(z_k\, z^{-1} + (z_k\, z^{-1})^2 + \cdots\right) \end{aligned} \tag{4.21}$$

using the geometric-series formula $1/(1-x) = 1 + x + x^2 + \cdots$. Analogously we expand

$$\begin{aligned} \frac{\hat{c}_k}{1 - z_k\, z^{-1}} &= \hat{c}_k \left(1 + z_k\, z^{-1} + (z_k\, z^{-1})^2 + \cdots\right) \\ &= \hat{c}_k + \hat{c}_k \left(z_k\, z^{-1} + (z_k\, z^{-1})^2 + \cdots\right) \end{aligned} \tag{4.22}$$

Comparing the two expansions, we see that

$$\hat{c}_k = c_k\, z_k^{-1} \tag{4.23}$$

$$\hat{c}_0 = c_0 - \sum_{k=1}^{N} \hat{c}_k \tag{4.24}$$

For the expansions to converge on the unit circle $|z| = 1$ we have to assume that $|z_k| < 1$. For $|z_k| > 1$ we can use the denominator $1 - z z_k^{-1}$ instead of $1 - z_k\, z^{-1}$ and expand it into a series. This leads to identical relations as Equations (4.23) and (4.24).

Convolution. A neat application of the z-transform is the derivation of the convolution for discrete-time systems. The multiplication of two transfer functions corresponds to the multiplication of two polynomials in z:

$$H(z)\,G(z) = \sum_{k=-\infty}^{\infty} h_k\, z^{-k} \sum_{l=-\infty}^{\infty} g_l\, z^{-l} = \sum_{k=-\infty}^{\infty} \sum_{l=-\infty}^{\infty} h_k\, g_l\, z^{-(k+l)} \tag{4.25}$$

With a change of variable $m = k + l$, this equation becomes the expected convolution:

$$H(z)\,G(z) = \sum_{m=-\infty}^{\infty} \sum_{l=-\infty}^{\infty} h_{m-l}\, g_l\, z^{-m} \tag{4.26}$$

4.2.2 Stability and causality of rational systems

Stability. To investigate stability, the partial fractions form of the transfer function (Equation 4.20) is considered. Transformed back to the time domain, it leads to the difference equations

$$x_k(n) = z_k\, x_k(n-1) + \hat{c}_k\, p(n) \tag{4.27}$$

$$u(n) = \hat{c}_0\, p(n) + \sum_{k=1}^{N} x_k(n) \tag{4.28}$$

where the $x_k(n)$ are temporary variables (also called state variables or internal variables). From this realization a stability criterion is derived. For each time step x_k is multiplied by z_k. To ensure that x_k does not grow to infinity it is necessary that $|z_k| < 1$. The realization is thus stable if all poles z_k lie within the unit circle.

Causality. For the derivation of the z-transform, the coefficients h_k have not been restricted to the positive time axis. On the other hand it is clear that for physical systems the response depends only on the past input whereas the future input has no influence. This is referred to as the causality condition. In terms of the impulse response h_k a system is causal if $h_k = 0$ for $k < 0$.

In the following, the rational transfer function (Equation 4.16) is analyzed regarding causality. The impulse response h_n can be found by applying the transformation Equation (4.15) to the partial fraction expansion Equation (4.18). This leads to

$$h_n = \frac{c_0}{2\pi i} \oint_C z^{n-1}\, dz + \sum_{k=1}^{N} \frac{c_k}{2\pi i} \oint_C \frac{z^{n-1}}{z - z_k}\, dz \tag{4.29}$$

The following integrals are calculated using the Cauchy theorem and the calculus of residues [Pap62, pp. 292, 296].

$$\frac{1}{2\pi i} \oint_C z^{n-1}\, dz = \begin{cases} 1, & n = 0 \\ 0, & \text{otherwise} \end{cases} \tag{4.30}$$

$$\frac{1}{2\pi i} \oint_C \frac{z^{n-1}}{z - z_k}\, dz = \begin{cases} z_k^{n-1}, & n > 0 \text{ and } |z_k| < 1 \\ -z_k^{n-1}, & n \le 0 \text{ and } |z_k| > 1 \\ 0, & \text{otherwise} \end{cases} \tag{4.31}$$

The second case in Equation (4.31) can be derived from the first one by a change of variables $z = 1/y$ and $z_k = 1/y_k$ and $dz = -1/y^2$. The unit circle is mapped into itself with the direction reversed, leading to a sign change of the integral. Thus we have

$$\frac{1}{2\pi i} \oint_C \frac{z^{n-1}}{z - z_k}\, dz = \frac{1}{2\pi i} \oint_C \frac{y^{1-n}}{y^{-1} - y_k^{-1}} \frac{1}{y^2}\, dy = -\frac{1}{2\pi i} \oint_C \frac{y^{-n}\, y_k}{y - y_k} = -y_k^{-n}\, y_k = -z_k^{n-1} \tag{4.32}$$

Poles z_k inside the unit circle are mapped to y_k outside and vice versa.

With these integrals the impulse response coefficients are

$$h_0 = c_0 - \sum_{\{k:\ |z_k|>1\}} c_k z_k^{-1} = \hat{c}_0 + \sum_{\{k:\ |z_k|<1\}} \hat{c}_k \tag{4.33}$$

and for $n \geq 1$

$$h_n = \sum_{\{k:\ |z_k|<1\}} c_k z_k^{n-1} = \sum_{\{k:\ |z_k|<1\}} \hat{c}_k z_k^n \tag{4.34}$$

and

$$h_{-n} = - \sum_{\{k:\ |z_k|>1\}} c_k z_k^{-n-1} = - \sum_{\{k:\ |z_k|>1\}} \hat{c}_k z_k^{-n} \tag{4.35}$$

Thus h_n can be separated into a causal part, $n \geq 0$ and an anti-causal part, $n < 0$. The coefficient h_0 is included in the causal part. The causal part without h_0 is called the strictly causal part. The strictly causal part is determined by the poles inside the unit circle, the anti-causal part by the poles outside. The value h_0 is the mean value of the periodic transfer function as can be seen by Equation (4.13). For a strictly proper transfer function the value c_0 vanishes and if, additionally, the system is stable, we have $h_0 = 0$. Thus a causal system can be approximated by a proper stable transfer function, a strictly causal system by a strictly proper stable transfer function.

An equivalent statement results when considering the convergence of series [CBV76, p. 150]. The geometric series

$$z^{-1} + z_k\, z^{-2} + \cdots + z_k^{n-1}\, z^{-n} + \cdots = \frac{1}{z - z_k} \qquad |z_k\, z^{-1}| < 1 \tag{4.36}$$

converges for z on the unit circle if $|z_k| < 1$. On the other hand, the series

$$z_k^{-1} + z_k^{-2}\, z + \cdots + z_k^{-n-1}\, z^n + \cdots = -\frac{1}{z - z_k} \qquad |z_k^{-1}\, z| < 1 \tag{4.37}$$

converges on the unit circle if $|z_k| > 1$. More generally, using Laurent series, a function $H(z)$ analytic in $|z| \leq 1$ can be represented on the unit by a series $h_0 + h_1\, z^{-1} + h_2\, z^{-2} + \cdots$ and likewise a function $H(z)$ analytic in $|z| \geq 1$ by a series $h_{-1}\, z + h_{-2}\, z^2 + \cdots$.

Causality versus stability. It should be noted that the concepts of causality and stability are different. For a proper (or strictly proper) rational transfer function the causality criterion and the stability criterion are the same. Therefore the terms are sometimes used interchangeably. Note, however, that an improper transfer function, even with stable poles, is non-causal since

$$\frac{b_0\, z^{N+1} + \cdots + b_{N+1}}{z^N + \cdots + a_N} = b_0\, z + h_0 + h_1\, z^{-1} + \cdots \tag{4.38}$$

Another observation, which to some extent explains the close relationship between causality and stability, is that the forward recursion, Equation (4.27), can be changed to a backward recursion as

$$x_k(n-1) = \frac{1}{z_k}\Big(x_k(n) - \hat{c}_k\, p(n)\Big) \tag{4.39}$$

This shows that the transfer function of a causal system can also be used to describe a anti-causal system. If the forward recursion is stable, that is, if $|z_k| < 1$, then the backward recursion is unstable because $|1/z_k| > 1$ and vice versa. The same observation will be made for continuous-time systems. As we will see there, the choice between the two cases depends on the integration path of the inversion integral. For the z-transform the integration path is the unit circle and the causal part of a system is defined by the singularities inside this circle. A larger or smaller circle, however, would include or exclude more or less sigularities and lead to different results.

4.2.3 Hankel matrix

One important question is how to choose the degree N of a recursive realization. The answer is given by the Hankel matrix which plays a key role for the method developed in this work.

The z-transform of a recursive realization can be written either as a rational function (Equation 4.16) or as a polynomial (Equation 4.14). As an example we take $N = 2$ and assume a strictly causal system.

$$\frac{b_1 z^{-1} + b_2 z^{-2}}{1 + a_1 z^{-1} + a_2 z^{-2}} = h_1 z^{-1} + h_2 z^{-2} + \cdots \tag{4.40}$$

Multiplied by the denominator, the relation becomes

$$b_1 z^{-1} + b_2 z^{-2} = h_1 z^{-1} + (a_1 h_1 + h_2) z^{-2} + (a_2 h_1 + a_1 h_2 + h_3) z^{-3} + \cdots \tag{4.41}$$

Comparing coefficients of z^{-k} the following sequence is constructed

$$h_1 = b_1 \tag{4.42}$$
$$h_2 = b_2 - a_1 h_1 \tag{4.43}$$
$$h_3 = -a_2 h_1 - a_1 h_2 \tag{4.44}$$
$$h_4 = -a_2 h_2 - a_1 h_3 \tag{4.45}$$
$$\vdots$$

The sequence reflects the fact that the impulse response is the response due a unit input at time 0. Disregarding the first two equations, the third and later equations can be written in vector form as

$$\begin{bmatrix} h_3 \\ h_4 \\ \vdots \end{bmatrix} = -a_2 \begin{bmatrix} h_1 \\ h_2 \\ \vdots \end{bmatrix} - a_1 \begin{bmatrix} h_2 \\ h_3 \\ \vdots \end{bmatrix} \tag{4.46}$$

Thus the the vector on the left-hand side is linear combination of the two right-hand-side vectors. The vectors can be put in a matrix as

$$\mathbf{\Gamma} = \begin{bmatrix} h_1 & h_2 & h_3 & \cdots \\ h_2 & h_3 & h_4 & \cdots \\ h_3 & h_4 & h_5 & \cdots \\ \vdots & \vdots & \vdots & \ddots \end{bmatrix} \tag{4.47}$$

This matrix is called the (infinite-dimensional) Hankel matrix. (According to the general definition of the Hankel matrix the entries are the same along anti-diagonals, that is, $h_{i,j}$ depends only on $i + j$). In the above example, the Hankel matrix has only two linearly independent columns and is therefore of rank 2. A general theorem of system theory states that the degree N of a system (the degree of the denominator of a rational transfer function) is equal to the rank of the Hankel matrix. For the further development, the Hankel matrix is assumed to be bounded, that is, $\sum h_k^2 < \infty$.

Example: The transfer function

$$\frac{z^{-1}}{1 - 1/2\, z^{-1}} = z^{-1} + \frac{1}{2} z^{-2} + \frac{1}{4} z^{-3} + \cdots \tag{4.48}$$

has the rank-one Hankel matrix

$$\mathbf{\Gamma} = \begin{bmatrix} 1 & 1/2 & 1/4 & \cdots \\ 1/2 & 1/4 & 1/8 & \cdots \\ 1/4 & 1/8 & 1/16 & \cdots \\ \vdots & \vdots & \vdots & \ddots \end{bmatrix} \tag{4.49}$$

4.2.4 Singular value decomposition

Regarding the rank of a Hankel matrix, two questions arise. Firstly, how can this rank be determined and, secondly, is there a simpler form that takes advantage of the low rank of the matrix. The central key to these questions is the singular value decomposition (SVD). Any finite-rank matrix, singular or non-singular, can be decomposed into

$$\mathbf{\Gamma} = \mathbf{U\Sigma V}^{\mathrm{T}} = \begin{bmatrix} u_{11} & u_{12} & \cdots & u_{1,N} \\ u_{21} & u_{22} & \cdots & u_{2,N} \\ \vdots & \vdots & & \vdots \end{bmatrix} \begin{bmatrix} \sigma_1 & & & \\ & \sigma_2 & & \\ & & \ddots & \\ & & & \sigma_N \end{bmatrix} \begin{bmatrix} v_{11} & v_{21} & \cdots \\ v_{12} & v_{22} & \cdots \\ \vdots & \vdots & \\ v_{1,N} & v_{2,N} & \cdots \end{bmatrix} \tag{4.50}$$

The diagonal matrix $\mathbf{\Sigma}$ contains the singular values σ_i. The number of singular values N equals the rank of the Hankel matrix and thus the degree of the system. The matrices $\mathbf{U}$ and $\mathbf{V}$ are orthonormal, that is

$$\mathbf{U}^{\mathrm{T}}\mathbf{U} = \mathbf{I} \qquad \mathbf{V}^{\mathrm{T}}\mathbf{V} = \mathbf{I} \tag{4.51}$$

The Hankel matrix (provided it is bounded) can be viewed as a linear operator. In this context the singular value decomposition can be interpreted differently as follows:

$$\mathbf{\Gamma x} = \sum_{i=1}^{N} \sigma_i(\mathbf{v}_i^{\mathrm{T}}\mathbf{x})\mathbf{u}_i \tag{4.52}$$

where $\mathbf{u}_i = [u_{1,i}, u_{2,i}, \cdots]^{\mathrm{T}}$ and $\mathbf{v}_i^{\mathrm{T}} = [v_{1,i}, v_{2,i}, \cdots]$ are the singular vectors. Equation (4.52) states that $\mathbf{\Gamma x}$ can be expressed as a linear combination of the singular vectors $\mathbf{u}_i$ with a factor of $\sigma_i(\mathbf{v}_i^{\mathrm{T}}\mathbf{x})$.

Example: The Hankel matrix in the previous examples, Equation (4.49), has a singular value decomposition

$$\mathbf{\Gamma} = \frac{\sqrt{3}}{2}\begin{bmatrix} 1 \\ 1/2 \\ 1/4 \\ \vdots \end{bmatrix} \begin{bmatrix} \frac{4}{3} \end{bmatrix} \frac{\sqrt{3}}{2} \begin{bmatrix} 1 & 1/2 & 1/4 & \cdots \end{bmatrix} \tag{4.53}$$

For the scaling of the singular vectors the sum $1 + (1/2)^2 + (1/4)^2 + \cdots = 4/3$ has been used.

The infinite-dimensional case is, of course, only of theoretical interest. For practical purposes it has to be approximated by a finite number of dimensions. There are robust numerical methods to perform the singular value decomposition. As a matter of fact, the singular value decomposition is the only dependable numerical method for determining the rank of a matrix.

4.2.5 Discrete Fourier transform (DFT)

For numerical integration, Equation (4.15) is first transformed by the change of variable $z = e^{i\Omega}$.

$$h_k = \frac{1}{2\pi i}\int_{-\pi}^{\pi} H(e^{i\Omega})\, e^{i(k-1)\Omega}\, ie^{i\Omega} d\Omega = \frac{1}{2\pi}\int_{-\pi}^{\pi} H(e^{i\Omega})\, e^{ik\Omega}\, d\Omega \tag{4.54}$$

This integral is evaluated approximating the integral by a sum. The integration interval is divided into N_F subintervals of length $\Delta\Omega = 2\pi/N_F$ and the integration variable is discretized as $\Omega_n = n\Delta\Omega = 2\pi n/N_F$. Substituting the integral by a sum and $d\Omega$ by $\Delta\Omega$ leads to

$$h_k = \frac{1}{N_F}\sum_{n=0}^{N_F-1} H(e^{i\Omega_n})\, e^{2\pi ikn/N_F} \tag{4.55}$$

For the inverse transformation (Equation 4.14) the sum can be approximated by taking only N_F terms as

$$H(e^{i\Omega_n}) = \sum_{n=0}^{N_F-1} h_k\, e^{-2\pi ikn/N_F} \tag{4.56}$$

Note that, because both the time domain and the frequency domain are discrete, both h_k and $H(e^{i\Omega_n})$ are periodic. Therefore, also the starting point of the summation index Equation (4.56) may be chosen arbitrarily.

Equations (4.55) and (4.56) constitute a discrete Fourier transform (DFT) pair. Even if they have been determined only as approximations to pair Equations (4.14) and (4.15), it can be shown that they are exact inverses of each other. For a real-valued impuls response h_k the Fourier spectrum has the property $H(e^{i\Omega_n}) = \bar{H}(e^{-i\Omega_n})$ where the bar denotes the complex conjugate. The usual procedure is thus to calculate $H(e^{in\Delta\Omega})$ for $n = 0, \ldots, N_F/2$ and to construct the values for $n = N_F/2 - 1, \ldots, N_F$ from $H(e^{in\Delta\Omega}) = H(e^{i(n-N_F)\Delta\Omega}) = \bar{H}(e^{i(N_F-n)\Delta\Omega})$. The value at $n = N_F/2$ has to be real, because it has to be its own complex conjugate. The DFT is implemented efficiently by the Fast Fourier Transform (FFT).

4.2.6 Sampling

The way we introduced the z-transform was as the transfer function of the convolution operator (Equation 4.11). This operator consists of sampling the input function $p(t)$ at time steps $n\Delta t$ as $p(n)$, multiplying the sampled values by the coefficients h_k and taking the sum. The sampling can be written using the Dirac Delta function $\delta(t)$ as

$$p(n) = \int_{\infty}^{\infty} p(\tau)\, \delta(n\Delta t)\, d\tau \tag{4.57}$$

With this definition the discrete-time convolution Equation (4.11) can be written as a continuous-time convolution integral $u(n) = \int_{-\infty}^{\infty} h_\delta(\tau)\, p(t-\tau)\, d\tau$, where

$$h_\delta(t) = \sum_{k=-\infty}^{\infty} h_k\, \delta(t - k\Delta t) \tag{4.58}$$

The discrete-time transfer function Equation (4.12) is therefore effectively the continuous-time transfer function of h_δ. The function h_δ represents a sequence of Dirac Delta functions with values h_k.

One important question is, of course, whether the sampled input $p(n)$ and, even more importantly, the sampled output $u(n)$ adequately describe the continuous-time functions $p(t)$ and $u(t)$. An example of the discretization problem is shown in Figure 4.1. The curves

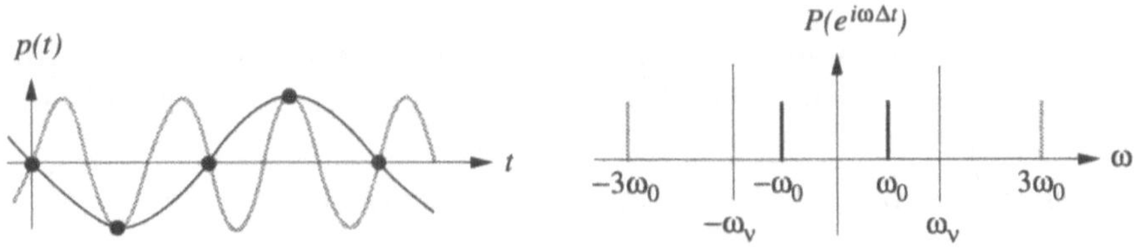

Figure 4.1: Sampling theorem: aliasing

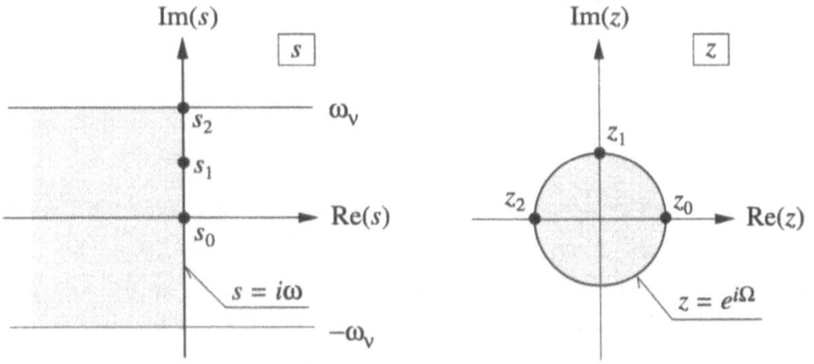

Figure 4.2: Sampling theorem: mapping

$p(t) = \sin \omega_0 t$ and $p(t) = \sin 3\omega_0 t$ both pass through the same sampling points. The two curves correspond to the peaks in the Fourier spectrum at $\pm\omega_0$ and $\pm 3\omega_0$, respectively. The ambiguity of the functions $p(t)$ is in fact the reason for the periodic behaviour of $P(e^{i\omega\Delta t})$. To make the function $p(t)$ unique, the Fourier spectrum should be zero outside of the window $\pm\omega_\nu$, that is, it has to be band-limited. The frequency

$$\omega_\nu = \pi/\Delta t \tag{4.59}$$

is called the *Nyquist frequency*. This is the content of the sampling theorem by Shanon [Pap62, p. 50]. In an experimental setup, it is important to have band-limited signals when sampling and it is also important to use an appropriate sampling rate. In a numerical context only the discretization interval is of concern. For a band-limited function $p(t)$ the continuous function can be uniquely recovered from the sampling points by [Pap62]

$$p(t) = \sum_{k=-\infty}^{\infty} p(k) \frac{\sin(\omega_\nu t - k\pi)}{\omega_\nu t - k\pi} \tag{4.60}$$

The relation between the continuous-time s-domain and the discrete-time z-domain can also be visualized as a mapping between the two complex planes as shown in Figure 4.2. The imaginary axis $s = i\omega$ is mapped into the unit circle $z = e^{i\omega\Delta t} = e^{i\Omega}$. To make this mapping one-to-one, Ω has to be restriced to the interval $-\pi < \Omega < \pi$. In terms of ω this is $-\omega_\nu < \omega < \omega_\nu$, with the Nyquist frequency ω_ν defined in Equation (4.59). This means again that the transfer function has to be band-limited. The mapping between the z-domain and the s-domain is

$$s = \ln z/\Delta t \qquad z = e^{s\Delta t} \qquad -\omega_\nu < \Im(s) < \omega_\nu \tag{4.61}$$

The unit disk is mapped to the strip $-\omega_\nu < \Im(s) < \omega_\nu$ and $\Re(s) < 0$ as indicated in the figure. In Section 4.4 we will introduce a different mapping, the bilinear transform, which maps the unit circle one-to-one into the entire imaginary axis.

4.2.7 Norms

To determine the quality of an approximation $H(z)$ with respect to an exact function $F(z)$ we have to apply a norm to the error function $E(z) = F(z) - H(z)$. We introduce two commonly used norms.

The Euclidean norm, or mean-square norm is defined by

$$\|E(z)\|_2 = \left(\frac{1}{2\pi}\int_0^{2\pi} |E(e^{i\Omega})|^2\, d\Omega\right)^{1/2} = \left(\sum_{k=-\infty}^{\infty} e_k^2\right)^{1/2} = \|e\|_2 \tag{4.62}$$

The equality is due to Parseval's formula [Pap62, p. 27].

The maximum norm or Chebyshev norm is the maximum value of the error on the unit circle C.

$$\|E(z)\|_\infty = \max_\Omega |E(e^{i\Omega})| \tag{4.63}$$

The choice of a norm depends on the minimization method used as well as the kind of approximation we require. A common problem in filter design is to find a filter with a stop band, that is, a range of frequencies where the transfer function is as close to zero as possible. In this case the maximum norm is appropriate. On the other hand in model reduction, that is, approximating a model with one of lower degree, the mean-square norm is preferable.

Here we just mention another norm which is used in the context of rational approximation, namely the Hankel norm. For details the reader is referred to the literature [Glo84].

4.3 Continuous-time systems

Analogously to Section 4.2, this section gives the theoretical background for continuous-time systems.

4.3.1 Laplace transform

The continuous-time counterpart of the z-transform is the Laplace transform. The Laplace transform is a generalization of the Fourier transform for ω not restricted to the real axis but extended to the complex plane. The variable $s = \alpha + i\omega$ is used to denote values on the complex plane. Only the definitions and some explanations are given here. For a detailed treatment see Papoulis [Pap62, pp. 2,169] or Siebert [Sie86]. We first introduce the more popular unilatral Laplace transform, and then the more general bilateral Laplace transform.

Unilateral Laplace transform. Formally, the Laplace transformation is obtained by substituting s for $i\omega$ in the Fourier transform (Equation 2.28). If we assume that $h(t) = 0$ for $t < 0$, that is, if we consider causal functions, the integral starts at zero and we have

$$H(s) = \int_0^\infty h(t)\, e^{-st}\, dt \tag{4.64}$$

Writing $s = \alpha + i\omega$ in the integrand, the transform becomes

$$H(s) = \int_0^\infty h(t)\, e^{-\alpha t} e^{-i\omega t}\, dt \tag{4.65}$$

This can also be viewed as the Fourier transform of $h(t)\, e^{-\alpha t}$.

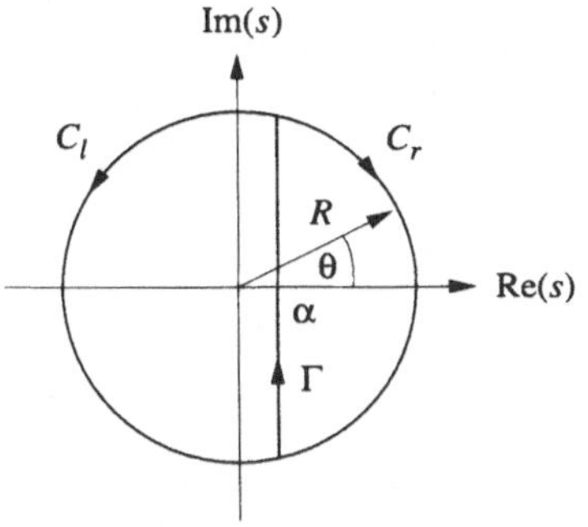

Figure 4.3: Integration paths for inverse Laplace transform

The integral only exists if the integrand is decaying. If the function $h(t)$ is increasing, the factor $e^{-\alpha t}$ can be used to make the integrand decaying by choosing the parameter α large enough, say $\alpha > \gamma$. The Laplace transform is thus defined on the right half-plane $\Re(s) > \gamma$. As an example consider the function $h(t) = e^{bt}$. The factor $e^{b}\, e^{-\alpha t}$ decays for $\alpha > \gamma \geq b$. The transform is

$$H(s) = \frac{1}{s-b} \tag{4.66}$$

defined on the right half-plane $\Re(s) > b$.

The inverse transformation is (Equation 2.29)

$$h(t)\, e^{-\alpha t} = \frac{1}{2\pi} \int_{-\infty}^{\infty} H(s)\, e^{i\omega t}\, d\omega \tag{4.67}$$

Multiplying this equation by $e^{\alpha t}$ and changing the integration variable to $s = \alpha + i\omega$ yields

$$h(t) = \frac{1}{2\pi i} \int_{\alpha - i\infty}^{\alpha + i\infty} H(s)\, e^{st}\, ds \tag{4.68}$$

Equations (4.64) and (4.68) constitute a Laplace transform pair. The path of integration is the Bromwich path defined as the vertical line $\Re(s) = \alpha$ in the region of definition, that is in the right half-plane $\Re(s) > \gamma$. For the example the path of integration is $\Re(s) = \alpha > b$, that is a vertical line to the right of the singularity $s = b$. Generally, for the unilateral Laplace transform the region of definition is some half-plane to the right of all singularites and is not usually specified explicitely. The advantage of the unilateral Laplace transform over the Fourier transform is that it can also handle growing functions, such as the above example $h(t) = e^{bt}$ with $b > 0$. Such functions, however, are not used in this work.

We now want to show that the inverse transformation indeed recovers a causal function. Consider the contour in the analytic region $\Re(s) < \gamma$ as depicted in Figure 4.3. It consists of the part Γ on the vertical line $\Re(s) = \alpha$ and the segment C_r on the circle with radius R. The integration along the path C_r defined by $s = R\, e^{i\theta}$ is

$$\frac{1}{2\pi} \int_{C_r} H(s)\, e^{st}\, ds = \lim_{R \to \infty} \frac{1}{2\pi i} \int_{-\pi/2}^{\pi/2} H(s)\, e^{Rt\cos\theta}\, e^{iRt\sin\theta}\, R\, d\theta \tag{4.69}$$

Note that $|e^{iRt\sin\theta}| = 1$ and that $\cos\theta > 0$. Further we assume that $H(s) \to 0$ for $R \to \infty$. Then, for $t < 0$ and $R \to \infty$, the expression $H(s)\, R\, e^{Rt\cos\theta} \to 0$ and the integral vanishes. This is the essence of Jordan's lemma [Pap62, p. 300]. Because the integral along a closed contour in an analytic region is zero, the integral along Γ is also zero and $h(t) = 0$ for $t < 0$.

For $t > 0$ we consider the closed contour consisting of the line Γ and the segement C_l to the left of α. Because singularities are included in this contour, the integral is not zero. Using the same reasoning as before one can show that the integral along C_l is zero in the limit as $R \to \infty$. The integral along Γ is therefore equal to the integral of the closed contour $C_l \cup \Gamma$:

$$h(t) = \frac{1}{2\pi} \int_{\alpha - i\infty}^{\alpha + i\infty} H(s)\, e^{st}\, ds = \frac{1}{2\pi} \int_{C_l \cup \Gamma} H(s)\, e^{st}\, ds \tag{4.70}$$

For example, the inversion integral of the transfer function $H(s) = 1/(s - b)$ used before can be evaluated using the Cauchy formula [Pap62, pp. 295] which reads

$$\frac{1}{2\pi i} \oint_C \frac{g(s)}{s - s_k} ds = g(s_0) \tag{4.71}$$

where C is a closed contour counterclockwise around the pole $s = s_k$. Therefore the function $h(t)$ is

$$h(t) = \int_\Gamma \frac{e^{st}}{s - b} ds = \begin{cases} e^{bt} & \text{for } t > 0 \\ 0 & \text{for } t < 0 \end{cases} \tag{4.72}$$

for b positive or negative.

Bilateral Laplace transform. Although the unilateral Laplace transform is the one mostly used in engineering applications we prefer the more general bilateral Laplace transform in this work. The bilateral transform is not restricted to causal functions. Therefore, the integral in Equation (4.64) has to be taken over the whole time axis as

$$H(s) = \int_{-\infty}^{\infty} h(t)\, e^{-st}\, dt \tag{4.73}$$

The integrand has to decay in both the positive and the negative time direction. The factor $h(t)\, e^{-\alpha t}$ has to decay for positive times and for some $\alpha > \gamma_1$. It also has to decay for negative times and $\alpha < \gamma_2$. If $\gamma_1 \leq \gamma_2$ the transform is defined in the vertical strip $\gamma_1 < \alpha < \gamma_2$. This excludes functions growing in the positive or negative time direction.

To define a Laplace transform the region of definition has to be specified along with the function $H(s)$. The same function $H(s)$ may lead to different functions $h(t)$ for different regions of definition. Consider again the example $H(s) = 1/(s - b)$. Different cases are shown in Figure 4.4. Taking the path of integration to the right of the singularity $s = b$ yields the causal function

$$h(t) = \begin{cases} e^{bt} & \text{for } t > 0 \\ 0 & \text{for } t < 0 \end{cases} \tag{4.74}$$

as already shown for the unilateral transform. This case is depicted in the figure on the left. For $b < 0$ this function is decaying and describes a causal stable system. For $b > 0$ the corresponding system is causal but unstable. On the other hand, taking the region of definition as the half-plane to the left of the singularity leads to the anti-causal function

$$h(t) = \begin{cases} 0 & \text{for } t > 0 \\ -e^{bt} & \text{for } t < 0 \end{cases} \tag{4.75}$$

This result can be found similarly to the first case. The negative sign is due to the fact that the intergration on the right contour C_r is evaluated clockwise. The anti-causal case is shown in the figure on the right. For $b < 0$ this function is blowing up and describes an anti-causal and unstable system whereas for $b > 0$ the system is anti-causal and stable. If we take the region of definition always such that it includes the imaginary axis, the

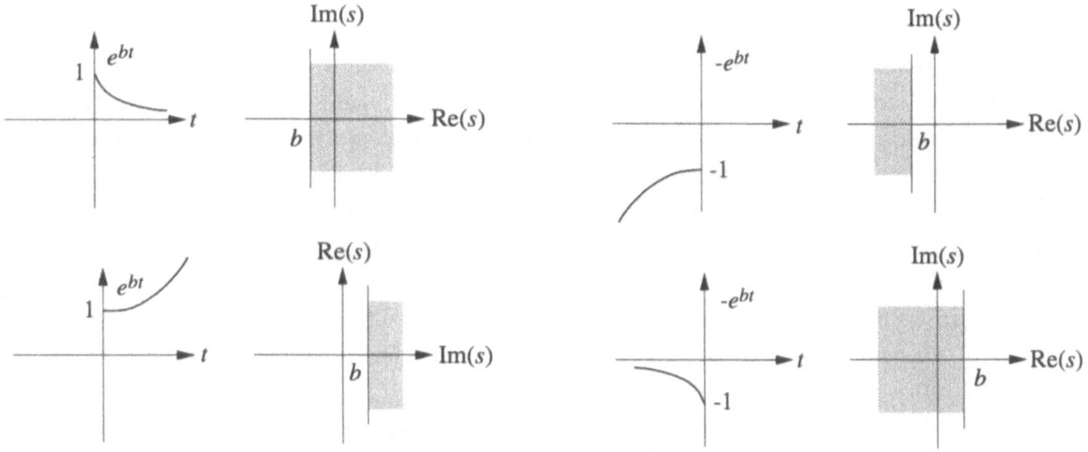

Figure 4.4: Laplace transforms and regions of definition

integration path is to the right for $b < 0$ and to the left for $b > 0$. Then we get the desired result that $b < 0$ leads to a causal stable system and $b > 0$ to a anti-causal unstable system.

Laplace versus Fourier transform. The Laplace transform has been determined as a generalization of the Fourier transform. The question is whether the Fourier transform can also be considered as a special case of the Laplace transform. A rigorous answer can be found in [Pap62, p. 172]. Here we only give an intuitive explanation. If all singularities are to the left of the imaginary axis, the integration path for the inversion integral can be taken along the imaginary axis and the Fourier transform coincides with the Laplace transform. If, however, some singularities are to the right of the imaginary axis, the corresponding Fourier transform does not exist. Existence of the Fourier transform means physically that the steady-state response due to an input $e^{i\omega t}$ is $H(i\omega)\,e^{i\omega t}$, that is proportional to the input. Consider again the example $H(s) = 1/(s-b)$. The response due to an input e^{pt} is described by the differential equation

$$\dot{y} = by + e^{pt} \tag{4.76}$$

Taking the causal case with $y(0) = 0$, the solution is

$$y(t) = \frac{e^{ibt}}{b} + \frac{e^{pt}}{p-b} \tag{4.77}$$

First consider the Fourier transform. For $p = i\omega$ the solution is

$$y(t) = \frac{e^{bt}}{b} + \frac{e^{i\omega t}}{i\omega - b} \tag{4.78}$$

The steady-state solution tends to $e^{i\omega t}$ only if the first term dampes out after some time, that is for $b < 0$. For $b > 0$ the first term grows faster than the second one and the response tends to $y = e^{bt}/b$, which is not proportional to the input. Hence the Fourier transform does not exist in that case.

Now consider the Laplace transform with $p = \alpha + i\omega$. We have

$$y(t) = \frac{e^{bt}}{b} + \frac{e^{\alpha t}\,e^{i\omega t}}{\alpha + i\omega - b} \tag{4.79}$$

For $\alpha > b > 0$ both terms grow with time but the second faster that the first one and the steady-state solution tends to $y = (e^{\alpha t} e^{i\omega t})/(\alpha + i\omega - b)$, which is proportional to the input. A good discussion of the subject can be found in [Sie86, Ch. 12 and 13].

Convention used in this work. The physical systems considered in this work have transfer functions that are analytic on the right half-plane. These functions have a decaying impulse response whose Fourier transform exist. The bilateral Laplace transform is used as an analytic continuation of the $i\omega$-axis to some region in the complex plane. Therefore, we always take the region of definition such that it includes the imaginary axis. The analytic properties of the Laplace transform allow using contour integration and employing Cauchy's theorem [Pap62, p. 292], which often considerably simplifies the evaluation of the inversion integral. For rational transfer functions this can simply be accomplished with the calculus of residues.

With the above convention we have the following general causality condition: If a Laplace transform $H(s)$ is analytic on the right half-plane and tends to zero for large s, then it is the transform of a causal function [Pap62, p. 214]. In the case of rational transfer functions the condition is that all poles have to lie on the left half-plane.

Hilbert transform. If a transfer function is analytic on the right half-plane and tends to zero for large s, then the corresponding system is causal. From that fact one can derive a relation between the real and imaginary parts of a transfer function along the imaginary axis $s = i\omega$. The real and imaginary parts are related to each other by the Hilbert transform [Pap62, p. 198] (also known as Kramers-Kronig relationship). The transfer function of a causal system is defined by either its real or its imaginary part and one can be calculated from the other. This is a theoretically remarkable fact. In this work, however, we do not employ it because the analysis yields directly a complex transfer function.

Analytic functions. In the derivation of the Laplace transform and its inverse analytic functions play an important role. Here we discuss the subject of analyticity in a pratical context. More rigorous definitions can be found in textbooks [Pap62, App. II], [CBV76, Ch. 2].

Functions are analytic if they are continuous and satisfy the Cauchy-Riemann equations. As a practical rule, we can state that all elementary functions and sums and products of them are analytic. Polynomials, exponential and trigonometric functions are analytic on the entire complex plane. Rational functions are analytic except at the poles. Multivalued functions such as roots, inverse trigonometric and logarithmic functions are analytic if we only consider a single branch. They fail to be analytic at the branch points and branch-cuts where they are not continuous.

Transfer functions derived from physical models are generally analytic except at singularities. They are typically combinations of elementary functions in s and they are real for $s = 0$. The latter is necessary to yield a real-valued impulse response. The condition for a real impulse response is $H(\bar{s}) = \bar{H}(s)$. This means that the imaginary part has to pass through zero on the real axis $\Im(s) = 0$. This was already discussed in the context of the Fourier transform where the condition is $H(-i\omega) = \bar{H}(i\omega)$. An intersting case is the damping usually introduced for soil by choosing a complex modulus of elasticity. Depending on the model used, the transfer function might not be analytical on the right half-plane yielding a non-causal system. The damping based on physical models such as the one used in the example of a rod on a visco-elastic foundation (Section 2.4) or the Kelvin models used in [WO85] lead to causal systems. On the other hand, the most popular model of using $E^* = E(1 + 2\xi i)$ [Wol88] is not analytical. This can be seen in the simplest case for the single-degree-of-freedom system $H(s) = 1/[K(1 + 2\xi i) + s^2 M]$. This function is only valid for $\Im(s) > 0$. For $\Im(s) < 0$ it must be $H(s) = 1/[K(1 - 2\xi i) + s^2]$ to yield a real-valued impulse response. Thus the function has a discontinuitiy along the real axis and is therefore not analytic.

Rational transfer function. The rational transfer function (Equation 4.10) is now written using the Laplace transform. As in discrete-time case we are only interested in proper transfer functions

$$H(s) = \frac{\sum_{k=0}^{N} b_k\, s^{N-k}}{\sum_{k=0}^{N} a_k\, s^{N-k}} = \frac{b_N + b_{N-1}\, s + \cdots + b_0\, s^N}{a_N + a_{N-1}\, s + \cdots + a_0\, s^N} \tag{4.80}$$

and in strictly proper transfer functions

$$H(s) = \frac{\sum_{k=1}^{N} b_k\, s^{N-k}}{\sum_{k=0}^{N} a_k\, s^{N-k}} = \frac{b_N + b_{N-1}\, s + \cdots + b_1\, s^{N-1}}{a_N + a_{N-1}\, s + \cdots + a_0\, s^N} \tag{4.81}$$

There is a difference between the discrete-time and the continuous-time formulation that is somewhat puzzling at first sight. Discrete-time systems are usually written in negative powers of z, but continuous-time system use positive powers of s. This customary notation takes into consideration the realization in the time domain. The natural formulation for the recursion is the backward difference with the corresponding variable z^{-1}. For continuous-time systems we usually use a differential equation corresponding to the variable s. However, to get an analogy between the discrete-time and the continuous-time case, we have to formulate the discrete-time transfer function using positive powers of z. Therefore, the discrete-time transfer function (Equation 4.16) is written in both forms, with negative and with positive powers of z. The relation between the two formulations in partial fractions form is given in Equations (4.23) and (4.24).

Example for non-rationtional transfer function. As an example for a non-rational Laplace transform we use the transfer function of a rod on a visco-elastic foundation. This function will be used in later chapters to demonstrate different approximation methods. The function is (Equation 2.66, with $i\omega = s$)

$$\kappa(s) = \frac{ik}{\lambda} - \frac{s}{c\lambda} = \sqrt{1 + \left(\frac{s}{c\lambda}\right)^2 + 2\epsilon\frac{s}{c\lambda}} - \frac{s}{c\lambda} - \frac{\epsilon}{\lambda} \tag{4.82}$$

with a viscosity coefficient $\epsilon/\lambda = 0.05$. This is a non-rational transfer function and it is defined on the on the whole complex plane. Because the square root is a two-valued function, it is analytic except on a branch-cut [Pap62, p. 288]. The branch-cut is shown in Figure 4.5 as a line from $(-1-\epsilon/\lambda i)$ to $(+1-\epsilon/\lambda i)$. The shift to $-\epsilon/\lambda$ is due to the viscous term. The function itself, real and imaginary part, are shown by a gray scale. Positive values are shown white, negative values black. The same function is shown in Figure 4.6 as a three-dimensional plot. From this figure we see that the function is smooth except at the the branch-cut.

4.3.2 Stability and causality of rational systems

Stability. A rational transfer function, Equation (4.10), can be expanded into partial fractions as

$$H(s) = c_0 + \sum_{k=1}^{N} \frac{c_k}{s - s_k} \tag{4.83}$$

The poles s_k are defined as the roots of the polynomial

$$\sum_{k=0}^{N} a_k\, s^{N-k} = 0 \tag{4.84}$$

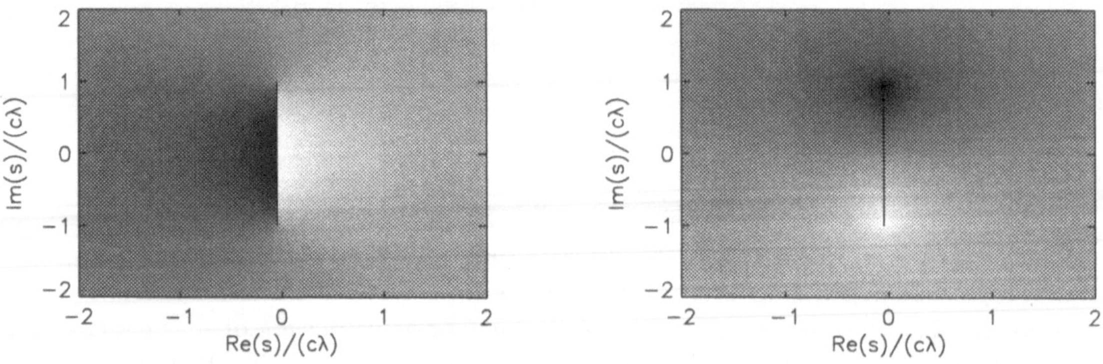

Figure 4.5: Transfer function in complex plane: branch cut. Real part on the left, imaginary part on the right.

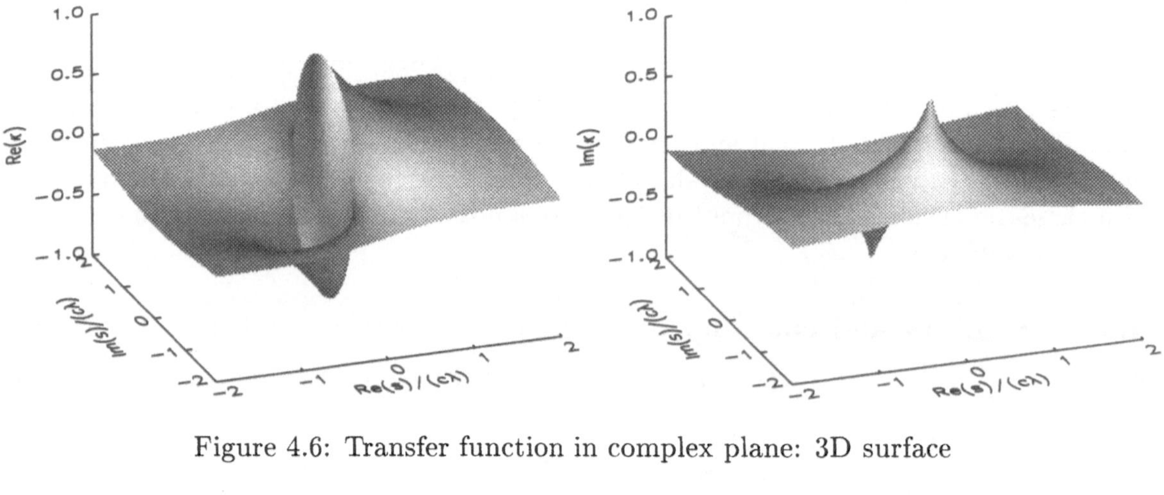

Figure 4.6: Transfer function in complex plane: 3D surface

The partial fraction expansion is equivalent to the following time domain equations

$$\dot{x}_k(t) = s_k\, x_k(t) + c_k\, p(t) \tag{4.85}$$

$$u(t) = c_0\, p(t) + \sum_{k=1}^{N} x_k(t) \tag{4.86}$$

where the x_k are temporary variables (internal variables, state variables). The stability condition is derived from the differential equation for the state variables $x_k(t)$, Equation (4.85). The solution of the homogenous differential equation

$$\dot{x}_k = s_k\, x_k \tag{4.87}$$

is

$$x_k = e^{s_k t} = e^{\Re(s_k)t} e^{i\Im(s_k)t} \tag{4.88}$$

This solution is stable for

$$\Re(s_k) < 0 \tag{4.89}$$

Thus a realization is stable if all poles lie on the left half-plane.

Causality. Causality has been discussed already in the dervation of the Laplace transform. Poles are singularities where the function is not analytic. Hence all poles on the left half-plane lead to the causal part of the function and all poles on the right half-plane yield the anti-causal part. As in the discrete-time case, the stability and the causality criterion are the same provided the transfer function decays at infinity.

4.3.3 Markov parameters

Analogously to the discrete-time power series Equation (4.14) we can write for a strictly proper transfer function

$$H(s) = \sum_{k=1}^{\infty} h_k\, s^{-k} \tag{4.90}$$

The coefficients h_k are called the Markov parameters. Note that for the discrete-time case the Markov parameters are identical to the impulse response. In the continuous-time case the Markov parameters are related to the impulse response as follows [Kai80, p. 70]: A causal impulse response can be expanded into a Taylor series

$$h(t) = h(0) + \dot{h}(0)\, t + \ddot{h}(0)\, \frac{t^2}{2} + \cdots + h^{(k-1)}(0)\, \frac{t^{k-1}}{(k-1)!} + \cdots \qquad t > 0 \tag{4.91}$$

Observing the Laplace transform pair ($h(t) = 0$ for $t < 0$)

$$\frac{t^{k-1}}{(k-1)!} \leftrightarrow \frac{1}{s^k} \qquad k \geq 1 \tag{4.92}$$

the transfer function is given by

$$H(s) = h(0)\, \frac{1}{s} + \dot{h}(0)\, \frac{1}{s^2} + \cdots + h^{(k-1)}(0)\, \frac{1}{s^k} + \cdots \tag{4.93}$$

Comparing the last equation with the definition of the Markov parameters (Equation 4.90) yields

$$h_k = \left. \frac{d^{k-1}}{dt^{k-1}} h(t) \right|_{t=0} \qquad k \geq 1 \tag{4.94}$$

For a (not strictly) proper transfer function there is an additional term h_0. This coefficient leads to a constant value in the transfer function and corresponds to a Dirac Delta function $\delta(t)$ in the impulse response, which can be implemented as a static spring. Positive powers of s lead to derivatives of the Dirac Delta function and can be implemented as dampers. They do not correspond to the anti-causal part of the system as do positive powers of z in the discrete-time case.

The anti-causal part of a system cannot be described by the continuous-time Markov parameters because the Laplace transform pair in Equation (4.92) describes only causal functions. This is illustrated by the following example. Consider the transfer function

$$H(s) = \frac{1}{s-b} = \frac{1}{s} + \frac{b}{s^2} + \frac{b^2}{s^3} + \cdots \qquad |s| > |a| \tag{4.95}$$

The geometric series converges outside the circle with radius $|b|$ around the origin $s = 0$. The region of definition is either to the left or to the right of that circle but does not include the imaginary axis. If we choose the right half-plane, the function $h(t)$ is causal independently of whether b is positive or negative. There is no way to choose an integration path which results in a causal function for a negative value of b and in a anti-causal function for a positive value of b.

The practical importance of the Markov parameters appears to be somewhat limited in the continuous-time case but they are useful in some theoretical derivations.

4.3.4 Integral Hankel operator

A similar difficulty as for the Markov parameters arises for the analogue of the discrete-time Hankel matrix. Although the continuous-time Hankel matrix can be formed using the Markov parameters, there is a fundamental mathematical problem with this procedure. The discrete-time (infinite dimensional) Hankel matrix is bounded because the Markov parameters are identical to the impulse response. The (infinite dimensional) Hankel matrix containing the continuous-time Markov parameters is, however, not bounded. As a simple example consider the continuous-time impulse response e^{-t}. According to Equation (4.94) the Markov parameters are all $h_k = 1$ and the sequence is hence unbounded. The key for finding the continuous-time analogue of the Hankel matrix, is to consider the Hankel matrix as a linear operator

$$\mathbf{\Gamma x} = \begin{bmatrix} h_1 & h_2 & h_3 & \cdots \\ h_2 & h_3 & h_4 & \cdots \\ h_3 & h_4 & h_5 & \cdots \\ \vdots & \vdots & \vdots & \ddots \end{bmatrix} \begin{bmatrix} x_0 \\ x_1 \\ x_2 \\ \vdots \end{bmatrix} \tag{4.96}$$

$\mathbf{\Gamma x}$ is a vector with the kth component

$$(\mathbf{\Gamma x})_k = h_k\, x_0 + h_{k+1}\, x_1 + \cdots = \sum_{i=0}^{\infty} h_{k+i}\, x_i \tag{4.97}$$

This sum exists for a bounded discrete-time impulse response but would not exist for an unbounded sequence of continuous-time Markov parameters. Analogously, the Hankel integral operator is defined as

$$(\Gamma x)(t) = \int_0^{\infty} h(t+\tau)\, x(\tau)\, d\tau \tag{4.98}$$

which is a similar form as the convolution integral, except for the sign in $h(t+\tau)$. An in-depth treatment of the continuous-time Hankel operator and singular value decomposition (see next section) requires a knowledge of functional analysis and is beyond the scope of this work. They are not used directly for the further development. Nevertheless, this brief description should help to complete the picture. A more involved discussion of continuous-time Hankel operators can be found in [Par88, LK82].

4.3.5 Singular value decomposition of integral Hankel operator

Analogously to the singular value decomposition of the discrete-time Hankel matrix (Equation (4.50)

$$\mathbf{\Gamma x} = \mathbf{U\Sigma V}^{\mathrm{T}}\mathbf{x} = \sum_{i=1}^{N} \sigma_i(\mathbf{v}_i^{\mathrm{T}}\mathbf{x})\mathbf{u}_i \tag{4.99}$$

The singular value decomposition of the continuous-time Hankel integral operator is

$$(\mathbf{\Gamma x})(t) = \sum_{i=1}^{N} \sigma_i \left(\int_0^\infty v_i(\tau)\, x(\tau)\, d\tau \right) u_i(t) \tag{4.100}$$

with $u_i(t)$ and $v_i(t)$ orthonormal, that is

$$\int_0^\infty u_i(t)\, u_j(t)\, dt = \begin{cases} 1 & i = j \\ 0 & i \neq j \end{cases} \tag{4.101}$$

and likewise for $v(t)$.

Example: Find the singular value decomposition of the Hankel integral operator associated with $h(t) = e^{-t}$

$$(\Gamma x)(t) = \int_0^\infty e^{-(t+\tau)} x(\tau)\, d\tau = \frac{1}{2} \left(\int_0^\infty \sqrt{2}\, e^{-\tau} x(\tau)\, d\tau \right) \sqrt{2}\, e^{-t} \tag{4.102}$$

therefore $\sigma_1 = 1/2$ and $u(t) = v(t) = \sqrt{2}\, e^{-t}$ and

$$\|u(t)\|_2^2 = \|v(t)\|_2^2 = \int_0^\infty |u(t)|^2\, dt = \int_0^\infty 2e^{-2t} dt = 1 \tag{4.103}$$

4.3.6 Numerical Fourier transform

In the section on discrete-time systems (Subsection 4.2.5), the DFT has been used to approximate the z-transform. It can, however, also be used to approximate the Fourier transform or the Laplace transform along the imaginary s-axis. We consider the Laplace transform pair Equations (4.68) and (4.73) for $s = i\omega$ and discretize t and ω as $t = k\Delta t$ and $\omega = n\Delta\omega$, where $\Delta t = T/N_F$ and $\Delta\omega = 2\pi/T$ for a time period T and N_F time steps. The largest frequency which can be represented is the Nyquist frequency $\omega_\nu = \pi/\Delta t$. Thus for a fine frequency discretization we need a large time interval and a fine time discretization leads to a large maximum frequency. The product ωt becomes $\omega t = 2\pi kn/N_F$. Substituting integrals by sums and differentials by deltas leads to

$$H(in\Delta\omega) = \Delta t \sum_{k=0}^{N_F-1} h(k\Delta t)\, e^{-2\pi ikn/N_F} \tag{4.104}$$

$$h(k\Delta t) = \frac{1}{N_F \Delta t} \sum_{n=0}^{N_F-1} H(in\Delta\omega)\, e^{2\pi ikn/N_F} \tag{4.105}$$

This is again the DFT pair, except for the factor Δt which is necessary to scale the numerical approximation to the values of the continuous Fourier transform. Since both transforms are periodic, the summation indices may start at arbitrary values, as long as they cover a complete period. As mentioned earlier (Subsection 4.2.5), $H(in\Delta\omega)$ is only computed for $n = 0, \ldots, N_F/2$ corresponding to $\omega \leq \omega_\nu$. The rest is constructed using complex conjugate values. The numerical computation is performed by the Fast Fourier Transform.

One subtlety has to be pointed out in this context. The values $h(k\Delta t)$ in Equation (4.105) are numerical approximations to the continuous-time impulse response $h(t)$. The values in Equation (4.55) are approximations to the discrete-time impulse response h_k. The two values are related by $h(k\Delta t) = h_k\,\Delta t$. This is the same relation as has been used implicitly in Equation (4.3) for the numerical approximation of the convolution integral. A special case arises for the values h_0 of a causal system. The value h_0 in Equation (4.105) will, according to the theory of Fourier series, be the mean value of $h(t^-)$ and $h(t^+)$. For a causal impulse response with $h(t^-) = 0$ this value will effectively be $h(0)/2$. The relationship $h_0 = h(0)/2$ for a causal system also shows up when comparing the continuous-time convolution Equation (4.3) with the discrete-time convolution Equation (4.11). As a consequence, when interested in the causal, continuous-time impulse response $h(t)$, the value of the discrete-time impulse response h_0 should be doubled as $h(0) = 2h_0$. However, to evaluate a convolution the value h_0 has to be left unchanged.

4.3.7 Norms

Analogously to the discrete-time case we consider two norms. The Euclidean or mean-square norm is defined as

$$\|E(s)\|_2 = \left(\frac{1}{2\pi}\int_{-\infty}^{\infty} |E(i\omega)|^2\, d\omega\right)^{1/2} = \left(\int_{-\infty}^{\infty} |e(t)|^2\, dt\right)^{1/2} = \|e(t)\|_2 \qquad (4.106)$$

The equality of the mean-square norm in the frequency and in the time domain is due to Parseval's formula [Pap62, p. 27].

The maximum norm or Chebyshev norm is defined as the maximum value along the axis $s = i\omega$.

$$\|E(s)\|_\infty = \max_\omega |E(i\omega)| \qquad (4.107)$$

4.4 Bilinear transform

Having found analogies between discrete-time and continuous-time systems for all important aspects of systems, it would be appealing to have a general connection between the two domains. Simply interchanging s and z as described for the sampling in Subsection 4.2.6 obviously is not the solution. The proper link is the bilinear transform. The bilinear transform is a powerful tool for both theoretical and numerical applications. It can be used to state theorems in one domain and prove them in the other domain. Numerical computations can be performed in one domain and transformed to the other domain.

4.4.1 Definition of mapping

The bilinear transform is defined by the following mapping

$$s = \alpha\frac{z-1}{z+1} \qquad z = \frac{\alpha+s}{\alpha-s} \qquad \alpha > 0 \qquad (4.108)$$

This mapping transforms the imaginary axis $s = i\omega$ into the unit circle $|z| = 1$ and vice versa. The arbitrary parameter α has to be restricted to $\alpha > 0$ to ensure that the left

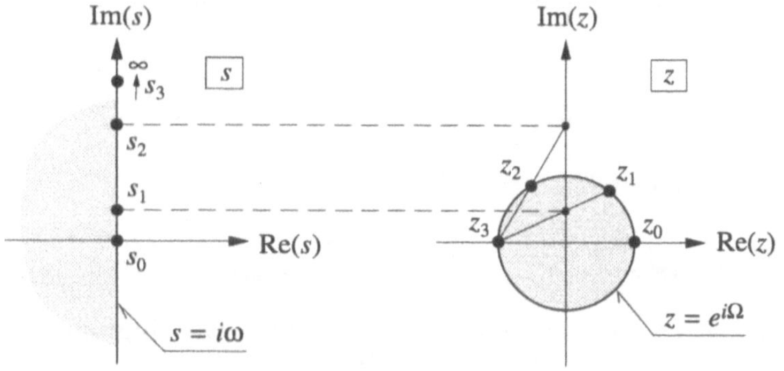

Figure 4.7: Bilinear transformation

half-plane $\Re(s) < 0$ is transformed to the unit disk $|z| < 1$. With this restriction, the corresponding regions of stability and causality are transformed into each other. The mapping is illustrated in Figure 4.7. Of special interest is the the mapping between the values on the imaginary axis $s = i\omega$ and the values on the unit circle $z = e^{i\Omega}$. Noting that $z\bar{z} = 1$ on the unit circle, this mapping is given by

$$i\omega = s = \alpha\frac{z-1}{z+1}\left(\frac{\bar{z}+1}{\bar{z}+1}\right) = \alpha\frac{z-\bar{z}}{2+z+\bar{z}} = i\alpha\frac{\sin\Omega}{1+\cos\Omega} = i\alpha\tan\frac{\Omega}{2} \tag{4.109}$$

The relationship

$$\omega = \alpha\tan\frac{\Omega}{2} \tag{4.110}$$

has a simple geometric interpretation for $\alpha = 1$, as shown in the figure. The point $z = e^{\pm\pi} = -1$ maps to $s \to \pm\infty$ depending on by which direction it is reached.

Using the bilinear transform a continuous-time transfer function $H(s)$ is related to the discrete-time transfer function $\tilde{H}(z)$ by

$$H(s) = H\left(\alpha\frac{z-1}{z+1}\right) = \tilde{H}(z) \tag{4.111}$$

For $s = i\omega$ we can also use the simpler relation

$$H(i\omega) = H(i\alpha\tan\Omega/2) = \tilde{H}(e^{i\Omega}) \tag{4.112}$$

A stricly proper transfer function $H(s)$ is generally transformed to a proper transfer function $\tilde{H}(z)$. This transfer function can be split up into a constant plus a strictly proper transfer function $\tilde{H}(z) = H(\alpha)+\tilde{H}_s(z)$. Using the relationship $(z+1)/(z-1) = 1+2/(z-1)$ we can write a strictly proper transfer function ($h_0 = 0$) as

$$\begin{aligned} H(s) &= h_1\, s^{-1} + h_2\, s^{-2} + \cdots \\ &= h_1\frac{z+1}{\alpha(z-1)} + h_2\left(\frac{z+1}{\alpha(z-1)}\right)^2 + \cdots \\ &= h_1\,\alpha^{-1} + h_2\,\alpha^{-2} + \cdots + h_1\frac{2}{\alpha(z-1)} + h_2\left(\frac{2}{\alpha(z-1)}\right)^2 + \cdots \\ &= H(\alpha) + \tilde{H}_s(z) \end{aligned} \tag{4.113}$$

Example: The spherical cavity problem introduced in Section 2.2 has the dynamic stress-displacement relation Equation (2.31). The first term, the dashpot can be implemented directly in the time domain. The second part has, up to normalizing factors, the form $H(s) = 1/(1+s)$. This transfer function is transformed with $\alpha = 3$ as

$$H(s) = \frac{1}{1+s} = \frac{1}{3\frac{z-1}{z+1}} = \frac{z+1}{4z-2} = \frac{1}{4} + \frac{3}{8}\frac{z^{-1}}{1-1/2\,z^{-1}} = \tilde{H}(z) \tag{4.114}$$

The constant term is indeed $H(\alpha) = H(3) = 1/4$.

4.4.2 Singular values

The singular values of the Hankel matrix of $\tilde{H}(z)$ equal the singular values of the Hankel integral operator of $H(s)$. A proof is given in [LK82]. Another proof, given in [Glo84] will be repeated in Chapter 6. Here it is shown only for the following example.

Example: The transfer function (Equation 4.114)

$$\tilde{H}(z) = \frac{1}{4} + \frac{3}{8}\frac{z^{-1}}{1-1/2\,z^{-1}} \tag{4.115}$$

is almost identical to example Equation (4.48). The first term does not enter the Hankel matrix. The factor 3/8 of the second term carries through the calculation, leading to a singular value of $\sigma_1 = (3/8)(4/3) = 1/2$ (Equation 4.53). This is the same value as for $H(s)$ found in example Equation (4.102).

4.4.3 Discrete Fourier transform

If the transfer function $H(s)$ is not given in analytical but in numerical form along the imaginary axis $s = i\omega$, the inverse Fourier transform can numerically be computed by the DFT. Difficulties arise because the non-periodic transfer function $H(i\omega)$ is approximated by a periodic function. Also, to include high frequencies, the number of evenly spaced sampling points N_F has to be large. These problems can be avoided by using the integral $\oint \tilde{H}(z)\,dz$ on the unit circle instead of calculating $\int H(s)\,ds$ along $s = i\omega$. The function $\tilde{H}(z)$ is a periodic function and the distribution of the discretization points is much more appropriate for many transfer functions. The transfer function is evaluated at frequency points which correspond to equidistant intervals in the discrete-time domain given by

$$\omega_k = \alpha \tan\frac{\Omega_k}{2} = \alpha \tan\frac{k\pi}{N_F} \tag{4.116}$$

where N_F is the number of frequency intervals on the unit circle. The regular spacing on the unit circle is distorted on the imaginary axis of the s-plane. The distribution of the continuous-time frequencies is shown in Figure 4.8 for the parameter $\alpha = 1$. The parameter α can be used to adjust the values ω_k such that the dense region, corresponding to $\Omega = \pi/2$, is where the transfer function is important. The discrete-time impulse response can be calculated numerically by the DFT Equation (4.55) as

$$h_n = \frac{1}{N_F}\sum_{k=0}^{N_F-1} H(i\alpha \tan k\pi/N_F)\, e^{2\pi i k n/N_F} \tag{4.117}$$

The discrete-time impulse response h_n is used to find a rational transfer function $\tilde{H}(z)$ which is then transformed to the continuous-time case by the bilinear transformation. The derivation of a rational transfer function from its impulse response is described in the next chapter, the back-transformation of a discrete-time transfer function to the continuous-time domain is explained in the next section.

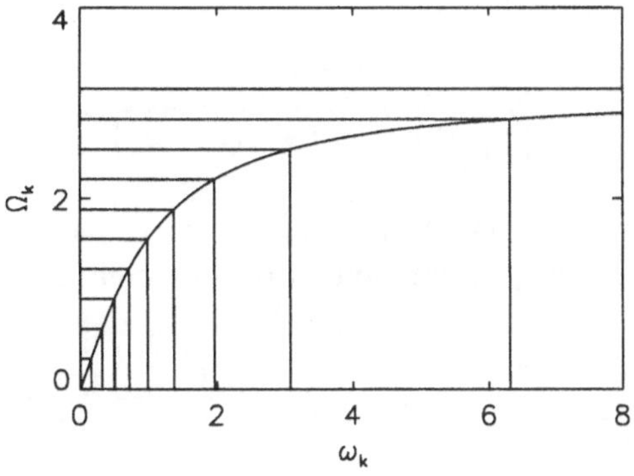

Figure 4.8: Distribution of continuous-time frequencies ω for regularly spaced discrete-time frequencies Ω

An interesting side topic we want to touch briefly here is the numerical inverse Laplace transform. The problem is to find the continuous-time impulse response from the function

$$\tilde{H}(z) = \sum_{k=0}^{\infty} h_k\, z^{-k} \tag{4.118}$$

Using the bilinear transform yields

$$H(s) = \sum_{k=0}^{\infty} h_k \left(\frac{\alpha - s}{\alpha + s}\right)^k \tag{4.119}$$

Unfortunately, the terms $(\alpha - s)^k/(\alpha + s)^k$ do not have an inverse Laplace transform. Following Henrici [Hen78, p. 44], a modified transfer function $\tilde{G}(z) = (\alpha + s)\,\tilde{H}(z) = 2\alpha(1 + 1/(z+1))\,\tilde{H}(z)$ is considered. The DFT of $\tilde{G}(z)$ yields

$$H(s) = \frac{1}{\alpha + s} G(s) = \frac{g_0}{\alpha + s} + g_1 \frac{\alpha - s}{(\alpha + s)^2} + \cdots \tag{4.120}$$

The terms $(\alpha - s)^k/(\alpha + s)^{k+1}$ are transformed to the time domain by Laguerre polynomials. An almost identical procedure is given in [Pap62, p. 210]. The method does not seem to have found its way into the civil engineering community. Here it is mentioned only as a side remark and is not analyzed any further. For more details see the references.

We conclude with the transformation of a rational discrete-time transfer function to a continuous-time transfer function.

4.4.4 Back-transformation for rational transfer function

The back-transformation of the discrete-time rational transfer function into the continuous-time domain is most easily performed in the partial fraction form (Equations 4.20 and 4.83). The relation

$$H(s) = \tilde{H}\left(\frac{\alpha + s}{\alpha - s}\right) \tag{4.121}$$

is written in partial fractions form as

$$c_0 + \sum_{k=1}^{N} \frac{c_k}{s - s_k} = \tilde{c}_0 + \sum_{k=1}^{N} \frac{\tilde{c}_k}{\frac{\alpha+s}{\alpha-s} - z_k} \tag{4.122}$$

The right-hand-side expression has to be brought into the form of the left-hand side.

$$\tilde{c}_0 + \sum_{k=1}^{N} \frac{\tilde{c}_k}{\frac{\alpha+s}{\alpha-s} - z_k} = \tilde{c}_0 + \sum_{k=1}^{N} \frac{\tilde{c}_k(\alpha - s)}{\alpha(1 - z_k) + s(1 + z_k)} = \tilde{c}_0 + \sum_{k=1}^{N} \frac{\tilde{c}_k \frac{\alpha-s}{z_k+1}}{s - \alpha\frac{z_k-1}{z_k+1}} \tag{4.123}$$

This expression already has the right denominator. Writing each term in the summation as a constant plus a stricly proper rational function yields

$$c_0 + \sum_{k=1}^{N} \frac{c_k}{s - s_k} = \tilde{c}_0 - \sum_{k=1}^{N} \frac{\tilde{c}_k}{z_k + 1} + \sum_{k=1}^{N} \frac{\frac{2\alpha\tilde{c}_k}{(z_k+1)^2}}{s - \alpha\frac{z_k-1}{z_k+1}} \tag{4.124}$$

By a comparison of the individual terms in the last equation, we see that

$$s_k = \alpha\frac{z_k - 1}{z_k + 1} \tag{4.125}$$

$$c_0 = \tilde{c}_0 - \sum_{k=1}^{N} \frac{\tilde{c}_k}{1 + z_k} \tag{4.126}$$

$$c_k = \frac{2\alpha\tilde{c}_k}{(1 + z_k)^2} \tag{4.127}$$

Note that the constant term c_0 in the continuous-time case relates to the constant value of the transfer function for $s \to \infty$ whereas $\tilde{c}_0$ in the discrete-time case relates to the mean value of the periodic transfer function. Therefore $\tilde{c}_0 = 0$ does not imply $c_0 = 0$ and vice versa.

Chapter 5

Approximation by Scalar Systems

In the last chapter we discussed the motivation and the theoretical background for employing the rational formulation. In this chapter, several methods for determining the coefficients of a rational transfer function are derived and illustrated by examples.

The first two sections describe simple approximation methods using the least-squares method. The approximation is either performed directly for the transfer function considering a number of sampling points or for the Markov parameters of the discrete-time system. Best results are obtained when using the Markov parameters of the discrete-time system obtained by the bilinear transform and then transforming the result back to the continuous-time domain. The third section describes the important numerical and theoretical tool of the singular value decomposition. The singular value decomposition can be used to solve the least squares problem but the main application is the methods based on the singular value decomposition of the Hankel matrix which are described in the fourth section. These methods play the key role of this work. In this chapter we give one such method applicable to scalar problems. Another method which can also be used for multivariable problems will be discussed later (Chapter 7).

5.1 Approximation of transfer function (frequency sampling)

5.1.1 Discrete-time systems

An obvious choice for finding the coefficients of an approximant is to enforce its transfer function at selected sampling points, thus the name frequency sampling. In this section we use a discrete-time system (difference scheme) to approximate the original (exact) solution. Before analyzing the recursive formulation we first consider the non-recursive case. The approximant Equation (4.8) with $N = 0$ has to match the original transfer function $F(i\omega)$ at the sampling points ω_r.

$$\sum_{k=0}^{M} b_k \, z_r^{-k} = F(i\omega_r) \tag{5.1}$$

with $z_r = e^{i\omega_r \Delta t}$. Writing this for L sampling points in matrix form as $\mathbf{A}\mathbf{x} = \mathbf{b}$ leads to

$$\begin{bmatrix} z_1^{-M} & \cdots & z_1^{-1} & 1 \\ \vdots & & & \vdots \\ z_L^{-M} & \cdots & z_L^{-1} & 1 \end{bmatrix} \begin{bmatrix} b_M \\ \vdots \\ b_1 \\ b_0 \end{bmatrix} = \begin{bmatrix} F(i\omega_1) \\ \vdots \\ F(i\omega_L) \end{bmatrix} \tag{5.2}$$

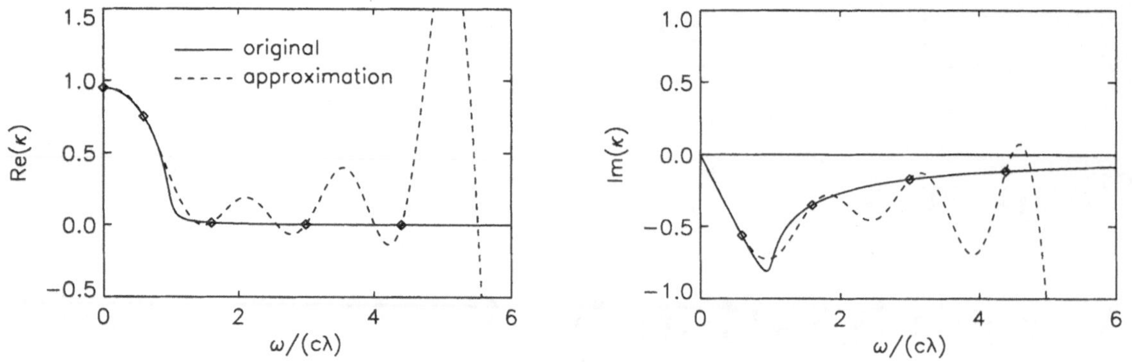

Figure 5.1: Non-recursive approximation by linear equations ($M = 8$)

Since this is an equation with complex coefficients but real unknowns, it is necessary to write it for the real and the imaginary parts separately. In this way one obtains for each sampling point two real equations. For $\omega = 0$ only one equation is obtained since the imaginary part is satisfied automatically. Denoting the real part by $\Re$ and the imaginary part by $\Im$ and taking the appropriate number of sampling points we obtain

$$\begin{bmatrix} \Re(\mathbf{A}) \\ \Im(\mathbf{A}) \end{bmatrix} \mathbf{x} = \begin{bmatrix} \Re(\mathbf{b}) \\ \Im(\mathbf{b}) \end{bmatrix} \tag{5.3}$$

which is a set of linear equations for the unknowns b_k. The transfer function of the approximant is

$$H(e^{i\omega\Delta t}) = \sum_{k=0}^{M} b_k \, e^{-i\omega\Delta t} \tag{5.4}$$

The formulation Equation (5.1) corresponds to a short convolution with a few but optimized parameters. The standard convolution has a large number of parameters that are not usually obtained by linear equations but by the Fourier transform.

The technique is applied to the transfer function of a rod on a visco-elastic foundation (Equation 2.66),

$$F(i\omega) = \kappa(i\omega) = \frac{ik}{\lambda} - \frac{i\omega}{c\lambda} = \sqrt{1 - \left(\frac{\omega}{c\lambda}\right)^2 + 2\epsilon\frac{i\omega}{c\lambda}} - \frac{i\omega}{c\lambda} - \frac{\epsilon}{\lambda} \tag{5.5}$$

with a viscosity coefficient $\epsilon/\lambda = 0.05$. The parameter Δt in the approximant is taken as $c\lambda\,\Delta t = 0.5$. Figure 5.1 shows the original transfer function and its approximation. The sampling points are also indicated, five for the real part and four for the imaginary part. The approximant exhibits large oscillations between the sampling points, which is typical for an interpolating polynomial. The approximation can be improved by selecting optimal sampling points, but their location can only be determined for specific examples[Wer82]. In ordinary interpolation problems, polynomials are usually replaced by spline functions. Spline functions, however, cannot be used here because they are not easily implemented in the time domain. We will see that suitable functions are rational transfer functions. They are good approximants and are readily transformed to the time domain.

For the recursive case we match the transfer function

$$\frac{\sum_{k=0}^{N} b_k \, z_r^{-k}}{\sum_{k=0}^{N} a_k \, z_r^{-k}} = F(i\omega_r) \tag{5.6}$$

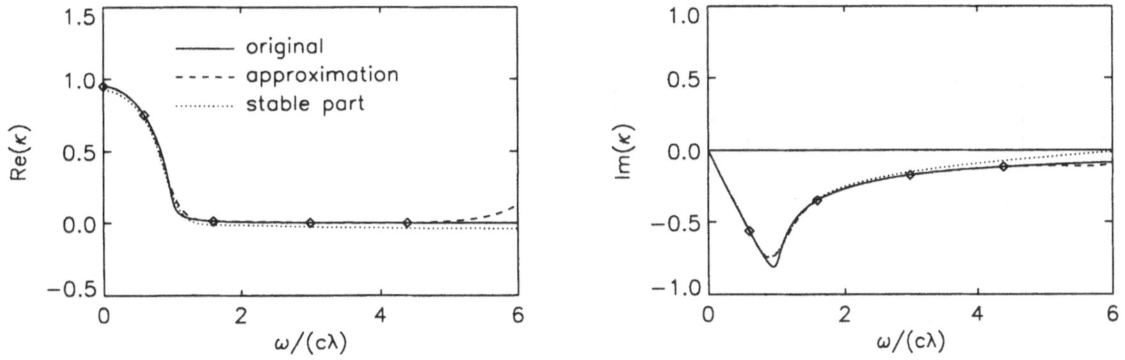

Figure 5.2: Recursive approximation by linear equations ($N = 4$)

at the sampling points ω_r. The proper transfer function with $M = N$ has been chosen because a constant representing the mean is expected. The equation can be rewritten as (recall that $a_0 = 1$)

$$\sum_{k=0}^{N} b_k\, z_r^{-k} - F(i\omega_r) \sum_{k=1}^{N} a_k\, z_r^{-k} = F(i\omega_r) \tag{5.7}$$

or, in matrix form as

$$\begin{bmatrix} -F(i\omega_1)\, z_1^{-N} & \cdots & -F(i\omega_1)\, z_1^{-1} & z_1^{-N} & \cdots & z_1^{-1} & 1 \\ \vdots & & \vdots & \vdots & & \vdots & \vdots \\ -F(i\omega_L)\, z_L^{-N} & \cdots & -F(i\omega_L)\, z_L^{-1} & z_L^{-N} & \cdots & z_L^{-1} & 1 \end{bmatrix} \begin{bmatrix} a_N \\ \vdots \\ a_1 \\ b_N \\ \vdots \\ b_1 \\ b_0 \end{bmatrix} = \begin{bmatrix} F(i\omega_1) \\ \vdots \\ F(i\omega_L) \end{bmatrix} \tag{5.8}$$

Again, each equation is written for the real and for the imaginary part. Taking $L = N$ sampling points plus the equation for $\omega = 0$ yields $2N + 1$ equations for the same number of unknowns. Solving the linear equations determines the coefficients a_k, b_k. The transfer function of the approximant is

$$H(e^{i\omega\Delta t}) = \frac{\sum_{k=0}^{N} b_k\, e^{-i\omega\Delta t}}{\sum_{k=0}^{N} a_k\, e^{-i\omega\Delta t}} \tag{5.9}$$

The same example as before is used for demonstration. The degree of the approximation is chosen as $N = 4$ which turns out to be appropriate for this case. Figure 5.2 shows the approximation with the sampling points. Compared to the non-recursive formulation with the same number of coefficients, the approximation is much better. The main reason is that the approximant is a rational function.

One problem is that a periodic (discrete-time) function is used to approximate the non-periodic (continuous-time) transfer function. Implicitely, one is approximating a periodic transfer function which coincides with the original function only up to the Nyquist frequency ω_ν as shown in Figure 5.3. Above the Nyquist frequency the original transfer function is treated as zero, according to the sampling theorem by Shanon [Pap62, p. 50]. The Nyquist frequency depends on the value Δt and is $\omega_\nu = \pi/\Delta t$, which is half the period. In this example we have $c\lambda\,\Delta t = 0.5$ and the Nyquist frequency is given by $\omega_\nu/(\lambda c) = 2\pi$. The frequency range shown in the figure is just below that value. One has to be careful

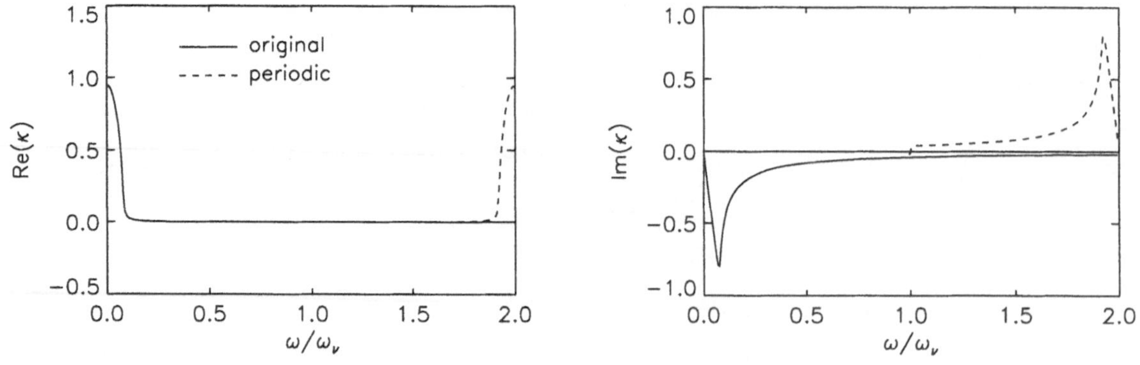

Figure 5.3: Continuous-time and discrete-time transfer function

not to specify sampling points above the Nyquist frequency. Sampling points above the Nyquist frequency would effectively correspond to points below the Nyquist frequency with a complex conjugate value because $H(e^{i(\omega_\nu-\omega)\Delta t}) = \bar{H}(e^{i\omega\Delta t})$.

Due to symmetry, the imaginary part of the periodic transfer function has to pass through zero at the Nyquist frequency. If the original transfer function is not zero there, the periodic transfer function has a jump and the tail of the original transfer function is cut off according to the sampling theorem. This leads generally to a non-causal system and hence to an unstable approximant. A useful approximation is only possible if the Nyquist frequency is high enough meaning that the the time step has to be small enough.

The problems just addressed is exactely what happend in the last example. The four poles are at $z_1 = -1.290, z_2 = 0.604, z_{3,4} = 0.736 \pm 0.356i$. Clearly, the first pole is outside the unit circle. As a remedy, which may seem somewhat arbitrary at this point, we try to remove that unstable pole. The approximant is expanded into partial fractions. For the example we have

$$H(z) = 0.0942 + \frac{0.0596}{z+1.290} + \frac{0.197}{z-0.604} + \frac{0.0244+0.0739i}{z-0.736+0.356i} + \frac{0.0244-0.0739i}{z-0.736-0.356i} \tag{5.10}$$

It turns out that neglecting the unstable term does not change the approximation much as demonstrated in Figure 5.2. The figure clearly shows that good results can be obtained with this procedure.

In the above procedure, the quality of the approximation depends much on the choice of the sampling points. A good approximation is only achieved after a number of trials. This disadvantage is removed by specifying a larger number of sampling points ($L > N$) and solving a least-squares problem. Figure 5.4 shows the results for the same example as before using $L = 20$ sampling points in the range from zero to $\omega/(c\lambda) = 4$. As seen from the figure, the results are essentially the same.

A theoretical problem is that the least-squares procedure does not minimize the error

$$E(i\omega) = H(e^{i\omega\Delta t}) - F(i\omega) = \frac{P(e^{i\omega\Delta t})}{Q(e^{i\omega\Delta t})} - F(i\omega) \tag{5.11}$$

but the weighted error

$$E'(i\omega) = P(e^{i\omega\Delta t}) - Q(e^{i\omega\Delta t})\,F(i\omega) \tag{5.12}$$

The difficulty is that Equation (5.11) leads to a nonlinear optimization problem for the coefficients a_k, b_k which is much more complicated to solve than the linear problem related to Equation (5.12). From a practical point of view, the linear optimization leads to

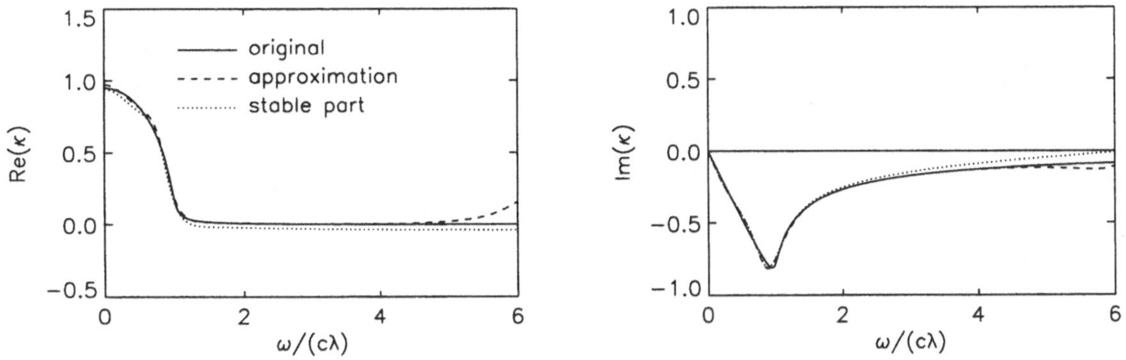

Figure 5.4: Recursive approximation by least squares ($N = 4$, $L = 20$)

good results. This is mostly due to the fact that the original transfer function is well behaved. In other applications, such as filter design, the transfer function is not smooth and the mathematical problems are much more involved. A typical transfer function of a filter, for example, is unity in a pass band and zero in a stop band. From the large number of publications relating to digital filter design we reference just a few articles [Ste70, Dec72, Dec74, Dud74].

If one wants to improve the results obtained from the linear optimization, this can be done by iteratively solving a sequence of linear problems. This method is a variant of the differential correction algorithm [Dud74].

$$E_{n+1}(i\omega) = \frac{P_{n+1}(e^{i\omega\Delta t})}{Q_n(e^{i\omega\Delta t})} - F(i\omega)\frac{Q_{n+1}(e^{i\omega\Delta t})}{Q_n(e^{i\omega\Delta t})} \tag{5.13}$$

5.1.2 Continuous-time systems

The problem encountered with the periodic behavior of a discrete-time approximant can be avoided by using a continuous-time system. The continuous-time rational transfer function is given by

$$H(s) = \frac{\sum_{k=1}^{N} b_k\, s^{N-k}}{\sum_{k=0}^{N} a_k\, s^{N-k}} \tag{5.14}$$

The strictly proper transfer function is used because the original transfer function tends to zero for large frequencies. The equation can be rewritten as (recall that $a_0 = 1$)

$$\sum_{k=1}^{N} b_k\, s_r^{N-k} - F(i\omega_r) \sum_{k=1}^{N} a_k\, s_r^{N-k} = F(i\omega_r)\, s_r^N \tag{5.15}$$

with $s_r = i\omega_r$. In matrix form this is

$$\begin{bmatrix} -F(i\omega_1)\, s_1^0 & \cdots & -F(i\omega_1)\, s_1^{N-1} & s_1^0 & \cdots & s_1^{N-1} \\ \vdots & & \vdots & \vdots & & \vdots \\ -F(i\omega_L)\, s_L^0 & \cdots & -F(i\omega_L)\, s_L^{N-1} & s_L^0 & \cdots & s_L^{N-1} \end{bmatrix} \begin{bmatrix} a_N \\ \vdots \\ a_1 \\ b_N \\ \vdots \\ b_1 \end{bmatrix} = \begin{bmatrix} F(i\omega_1)\, s_1^N \\ \vdots \\ F(i\omega_L)\, s_L^N \end{bmatrix} \tag{5.16}$$

Because both the original transfer function and the approximant are now non-periodic, the approximation is better than for the discrete-time approximation. Figure 5.5 shows

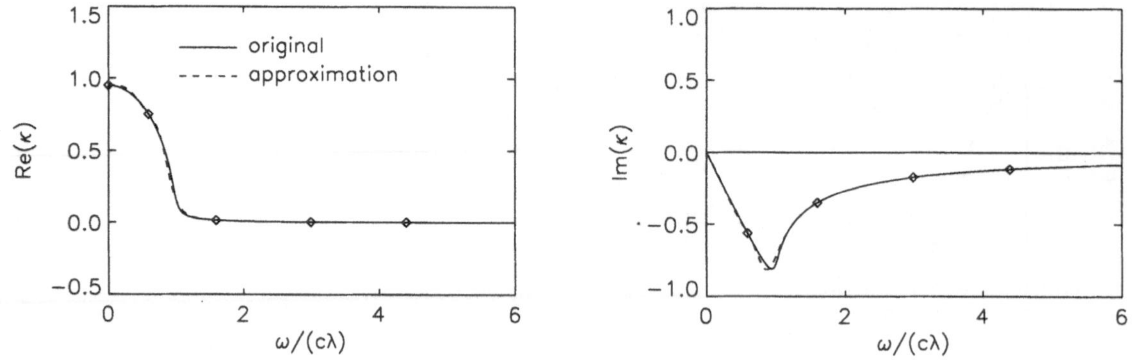

Figure 5.5: Continuous-time approximation by linear equations ($N = 4$)

the comparison. There is also no problem with stability in this example. Poles are at $s_{1,2} = -0.493 \pm 0.354i$ and $s_{3,4} = -0.225 \pm 0.841i$. Clearly, all poles have negative real parts and are hence stable.

5.2 Approximation of Markov parameters

5.2.1 Markov parameters of discrete-time system

Another possibility for determining the coefficients of a rational transfer function is to use the Markov parameters f_k. For the discrete-time case the Markov parameters equal the impulse response and are calculated by the discrete Fourier transform Equation (4.105). We approximate the strictly causal part of the transfer function $F(z) = \sum_{k=0}^{\infty} f_k\, z^{-k}$ by the rational approximant with $M = N - 1$ as

$$F(z) - f_0 = \frac{\sum_{k=1}^{N} b_k\, z^{-k}}{\sum_{k=0}^{N} a_k\, z^{-k}} = \sum_{k=1}^{\infty} f_k\, z^{-k} \tag{5.17}$$

Multiplying with the denominator one gets

$$\begin{aligned} \sum_{k=1}^{N} b_k\, z^{-k} &= \left(\sum_{k=0}^{N} a_k\, z^{-k}\right)\left(\sum_{k=1}^{\infty} f_k\, z^{-k}\right) \\ b_1\, z^{-1} + \cdots + b_N\, z^{-N} &= f_1\, a_0\, z^{-1} + (f_2\, a_0 + f_1\, a_1) z^{-2} + \cdots \end{aligned} \tag{5.18}$$

The coefficients a_k, b_k are found by comparing powers of z. Thus

$$b_k = \sum_{j=0}^{k-1} f_{k-j}\, a_j \qquad k = 1, 2, \ldots, N \tag{5.19}$$

$$0 = \sum_{j=0}^{N} f_{k-j}\, a_j \qquad k = N+1, N+2, \ldots \tag{5.20}$$

For $1 \le k \le N$ we get

$$\begin{bmatrix} b_N \\ b_{N-1} \\ \vdots \\ b_1 \end{bmatrix} = \begin{bmatrix} f_1 & f_2 & \cdots & f_N \\ 0 & f_1 & \cdots & f_{N-1} \\ \vdots & \vdots & & \vdots \\ 0 & 0 & \cdots & f_1 \end{bmatrix} \begin{bmatrix} a_{N-1} \\ a_{N-2} \\ \vdots \\ a_0 \end{bmatrix} \tag{5.21}$$

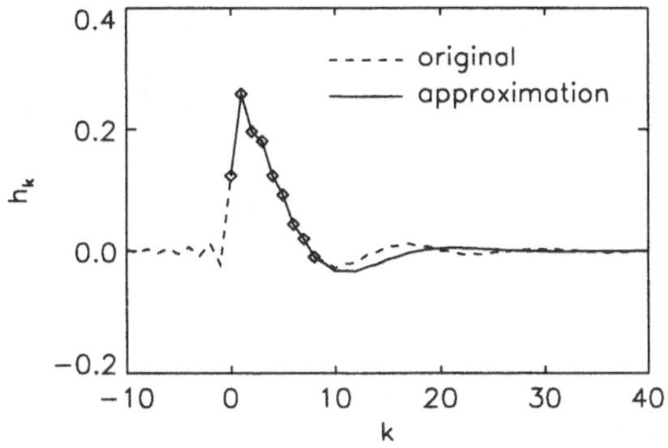

Figure 5.6: Padé approximation ($N = 4$): Markov parameters

For $k > N$, Equation (5.20) is rearraged to ($a_0 = 1$)

$$-f_k = \sum_{j=1}^{N} f_{k-j}\, a_j \tag{5.22}$$

which is written in matrix form as

$$\begin{bmatrix} f_1 & f_2 & \cdots & f_N \\ f_2 & f_3 & \cdots & f_{N+1} \\ \vdots & \vdots & & \vdots \\ f_L & f_{L+1} & \cdots & f_{L+N} \end{bmatrix} \begin{bmatrix} a_N \\ a_{N-1} \\ \vdots \\ a_1 \end{bmatrix} = \begin{bmatrix} -f_{N+1} \\ -f_{N+2} \\ \vdots \\ -f_{L+N+1} \end{bmatrix} \tag{5.23}$$

The last matrix is the (finite dimensional) Hankel matrix $\mathbf{\Gamma}$. The descending ordering of the coefficients a_k was chosen to get this familiar form of the Hankel matrix. Taking $L = N$ equations in Equation (5.23) yields the N unknowns a_k. Once the coefficients a_k are found, they can be substituted into the first equation. The approximant matches the first $2N + 1$ values of the impulse response f_k exactly. This is called partial realization or Padé approximation. The transfer function of the approximant is

$$H(e^{i\omega\Delta t}) = f_0 + \frac{\sum_{k=1}^{N} b_k\, e^{-i\omega\Delta t}}{\sum_{k=0}^{N} a_k\, e^{-i\omega\Delta t}} \tag{5.24}$$

Because in the discrete-time case the Markov parameters equal the impulse response they can effectively be calculated by the FFT (Equation 4.105). The underlying transfer function is periodic as shown in Figure 5.3 and the same problems as described earlier (page 79) arise. Specifically, cutting the tail of the original transfer function leads to a non-causal impulse response.

The Padé approximation is shown for the same example as before (Equation 2.66). Figure 5.6 shows the impulse response and its approximation with $N = 4$. The time scale is shifted to show the anti-causal behavior. The original impulse response is calculated using the FFT with $N_F = 512$, $c\lambda\Delta t = 0.5$ and a corresponding $\Delta\omega/(c\lambda) = 2\pi/(0.5N_F)$. The approximated impulse response is calculated by the recursive scheme. Figure 5.7 shows the original (continuous-time) transfer function and the (rational) approximation. The approximation is not quite satisfactory, specially for low frequencies. However, it turns out to be stable.

Instead of solving Equation (5.23) for $L = N$ equations, it can be written for a larger set of sampling points using $L > N$ and solving it by a least-squares method. This leads to

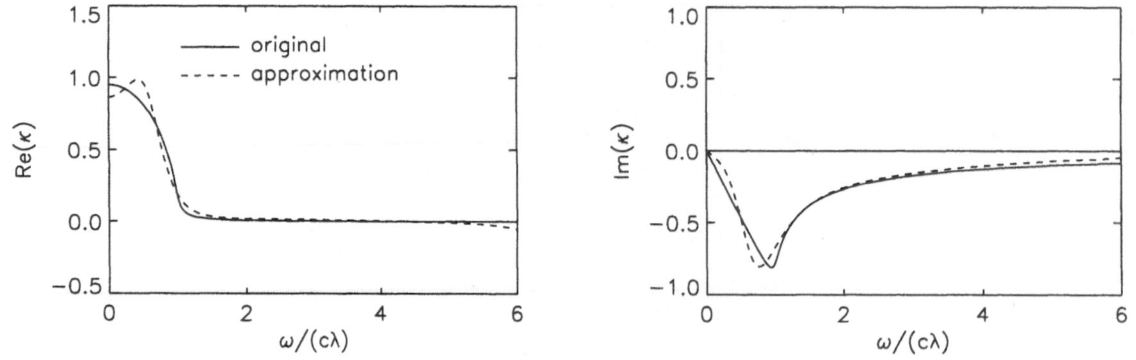

Figure 5.7: Padé approximation ($N = 4$): transfer function

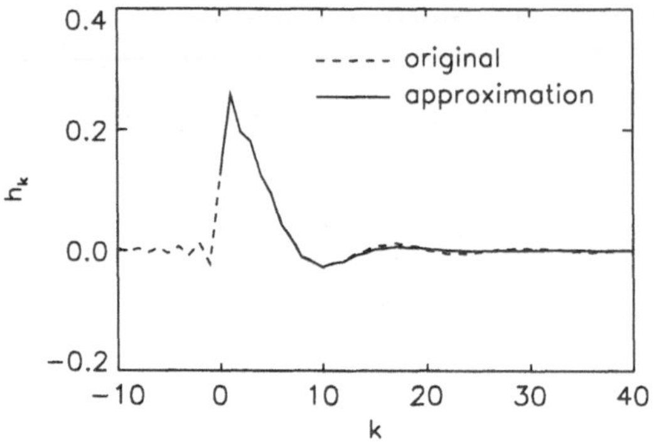

Figure 5.8: Least-squares Padé approximation ($N = 4$, $L = 20$): Markov parameters

an approximation which not only matches the first few values of f_k but approximates more values in an average sense. The method is applied to the above example taking $L = 20$. Figure 5.8 shows the approximation of the impulse response which is now better than before, resulting in a better transfer function shown in Figure 5.9. This approximation is relatively good and it is stable. Because the method is similar to the Padé approximation we call it the least-squares Padé approximation. For a general discussion on stability of the least-squares approximation of Markov parameters see also [MR76].

As in the frequency sampling, solving the least-squares problem for Equation (5.23) does not minimize the error

$$e_k = f_k - h_k = \frac{1}{2\pi i} \oint_C \Big(F(z) - H(z)\Big)\, z^{k-1}\, dz \tag{5.25}$$

but a weighted error corresponding to Equation (5.12)

$$e'_k = \frac{1}{2\pi i} \oint_C \Big(P(z) - Q(z)\, F(i\omega)\Big)\, z^{k-1}\, dz \tag{5.26}$$

Similar to Equations (5.19 and 5.20), the weighted error is

$$e'_k = \begin{cases} b_k - \sum_{j=0}^{k} f_{k-j}\, a_j = 0 & \text{when } k = 1, 2, \ldots, N \\ f_k + \sum_{j=1}^{N} f_{k-j}\, a_j & \text{when } k = N+1, N+2, \ldots \end{cases} \tag{5.27}$$

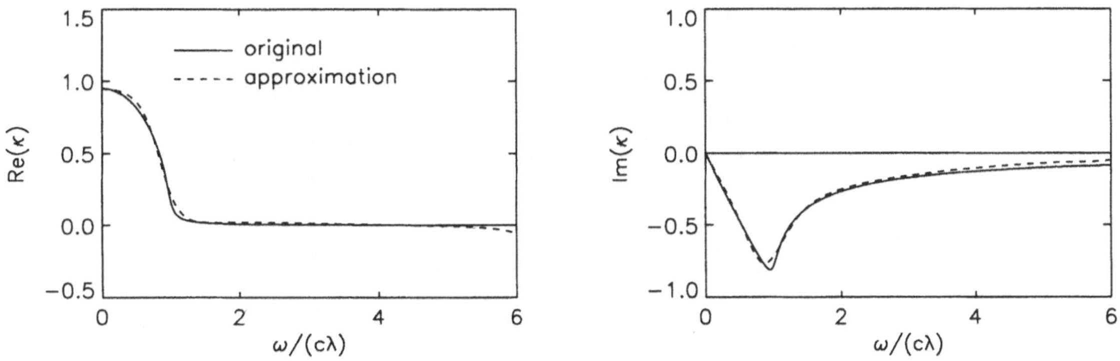

Figure 5.9: Least-squares Padé approximation ($N = 4$, $L = 20$): transfer function

In the first equation the error vanishes because the coefficients b_k are selected to satisfy Equation (5.21) identically. Noting that the error norms

$$\|e\|_2^2 = \sum_{k=0}^{\infty} e_k^2 \tag{5.28}$$

and

$$\|E(z)\|_2^2 = \frac{1}{2\pi i} \oint_C |E(z)|^2 \, dz \tag{5.29}$$

are identical by Parseval's equality (Equation 4.62), approximating the transfer function or approximating the Markov parameters is seen to be equivalent for a large number of sampling points. One advantage of approximating the Markov parameters is that the anti-causal part that leads to unstable results can be removed explicitly.

5.2.2 Markov parameters of mapped discrete-time system

For the continuous-time transfer function $F(s)$, the Markov parameters are not readily available. Therefore, we use the bilinear transform to get a mapped discrete-time problem with a transfer function $\tilde{F}(z)$. The transfer function of the mapped system is defined by

$$\tilde{F}(z) = \tilde{F}\left(\frac{\alpha + s}{\alpha - s}\right) = F(s) \tag{5.30}$$

The variable $z = e^{i\Omega}$ on the unit circle corresponds to the variable $s = i\omega$ on the imaginary axis and the two are related by (Equation 4.110) $\omega = \alpha \tan(\Omega/2)$. Thus we obtain the transfer function of the mapped discrete-time system by

$$\tilde{F}(e^{i\Omega}) = F(i\omega) = F\left(i\alpha \tan \frac{\Omega}{2}\right) \tag{5.31}$$

that is by a simple distortion of the frequency axis as already explained in Section 4.4.3. The transfer function of the mapped system is periodic and covers the whole frequency range because varying Ω on the the whole unit circle corresponds to varying ω on the whole imaginary s-axis from $-\infty$ to $+\infty$. The Markov paramters (impulse response) are calulated using the discrete Fourier transform Equation (4.55). With the techniques used in the previous section an approximation for the transfer function of the mapped discrete-time is obtained. This approximation is then transformed back to a continuous-time system by Equations (4.125–4.127).

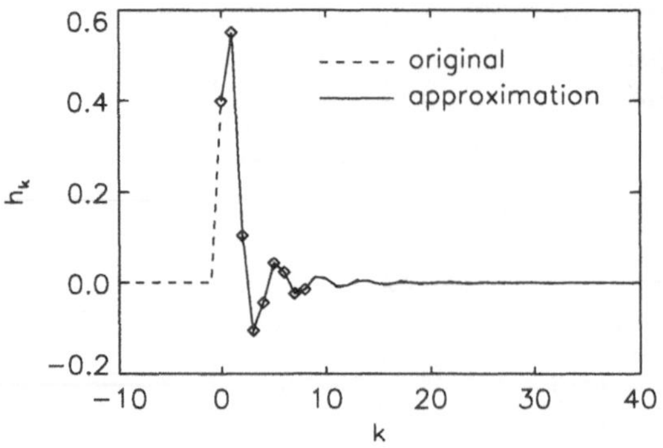

Figure 5.10: Padé approximation of mapped discrete-time system ($N = 4$): Markov parameters

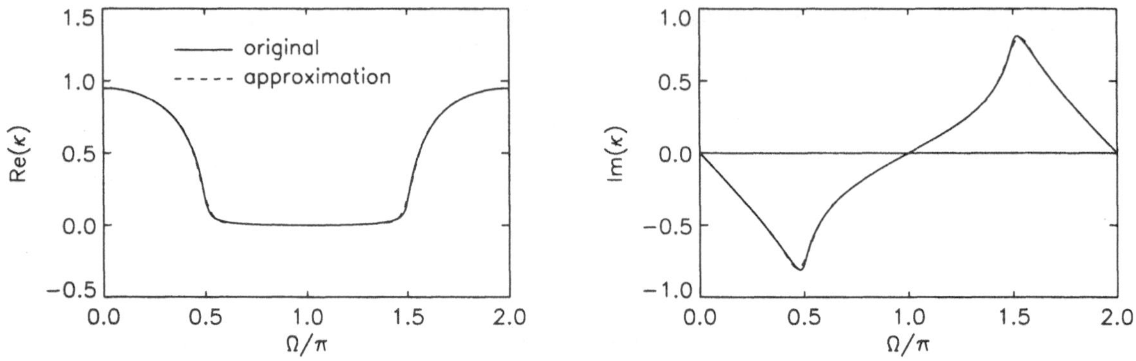

Figure 5.11: Padé approximation of mapped discrete-time system ($N = 4$): transfer function of discrete-time system

The previous example of the rod on a visco-elastic foundation is used for demonstration. The Markov parameters (impulse response coefficients) are obtained by first applying the bilinear transform with $\alpha = 1$ and the FFT with $N_F = 512$ to the transfer function. Then the Padé approximation is applied to these Markov parameters. For $L = N = 4$ the first 9 Markov parameters are matched by applying Equations (5.21 and 5.23). The approximation of the Markov parameters is shown in Figure 5.10. This figure should be compared to the discrete-time approximation Figure 5.6. Because the original transfer function is a periodic function obtained from the continuous-time transfer function, the impulse response is causal. As expected for a causal impulse resonse, the approximation is stable. Figure 5.11 shows the approximation in the mapped frequency domain. The agreement is very good. For later reference, the error curve is shown in Figure 5.12. The error is plotted in the complex plane with Ω as parameter. Because the transformation between $\tilde{F}(z)$ and $F(s)$ is only a distortion of the parameters Ω and ω, the same error curve is obtained for the continuous-time system. It is one of the advantages of the bilinear transform that an approximation in the mapped discrete-time domain is just as good for the continuous-time case when considering the maximum error. As can be seen, the maximum error is about 0.03. The error curve also shows that the error magnitude varies strongly with frequency. Later examples will show an error that is closer to a constant value (circle) but lower in magnitude. While the transformation from the mapped discrete-time system to the original continuous-time system could be obtained just by stretching the frequency axis according to Equation (4.110), the system needed for the time-domain

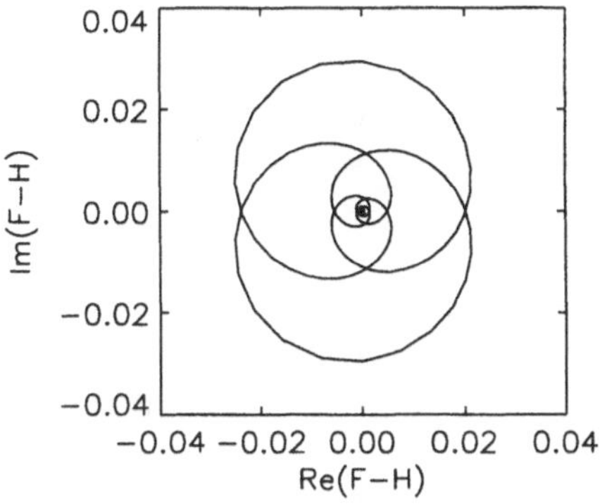

Figure 5.12: Padé approximation of mapped discrete-time system ($N = 4$): error curve

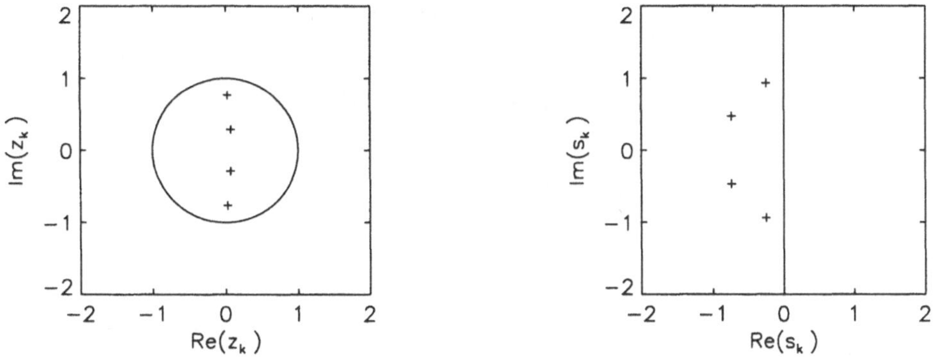

Figure 5.13: Padé approximation of mapped discrete-time system ($N = 4$). Left: poles of mapped discrete-time system. Right: poles of original continuous-time system.

analysis must be obtained by Equations (4.125–4.127). This inlcudes the transformation of the poles as shown in Figure 5.13 . Figure 5.14 finally shows the continuous-time transfer functions. This approximation is excellent. The improvement, compared to the earlier version, Figure 5.7, is remarkable. The figure also verifies the back transformation by the bilinear transform from the mapped discrete-time system to the continuous-time system.

A second example is calculated using a least-squares method for the approximation of the Markov parameters. $L = 20$ equations have been used. The approximation is slightly better than for the Padé approximation as can be seen from the error curve in Figure 5.15. It also turns out to be stable. The transfer functions are not shown because the figures are practically identical to the Padé approximation.

As can be concluded from the examples considered so far, the approximation of a continuous-time transfer function by a continuous-time system, leads to much better results than using a discrete-time system, both for accuracy and for stability. However, it should not be forgotten that the continuous-time system has to be discretized afterwards for use in an actual time-domain analysis (Hughes α-method). Time discretization introduces an additional error that might be even larger than the one obtained by directly approximating a discrete-time system. The advantage of first finding a continuous-time system and then introducing the time discretization is that the result is stable (if a stable discretization scheme is used) and, furthermore, that the time step does not have to be defined in advance but can be adjusted during the time-domain analysis.

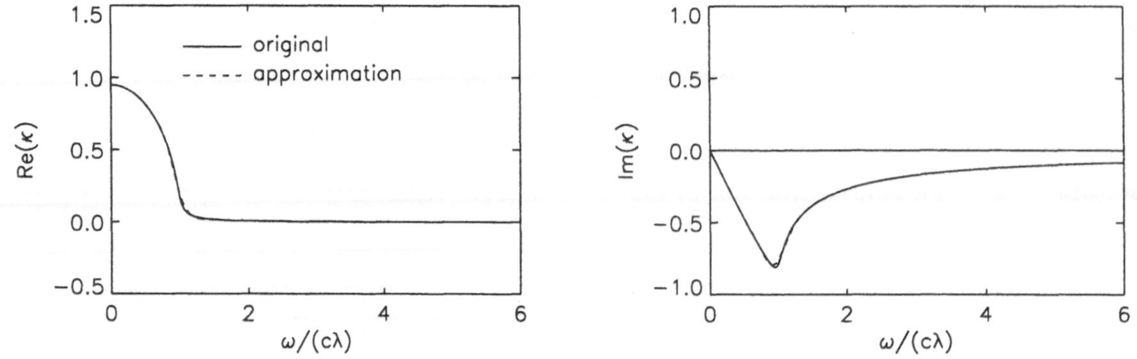

Figure 5.14: Padé approximation of mapped discrete-time system ($N = 4$): transfer function

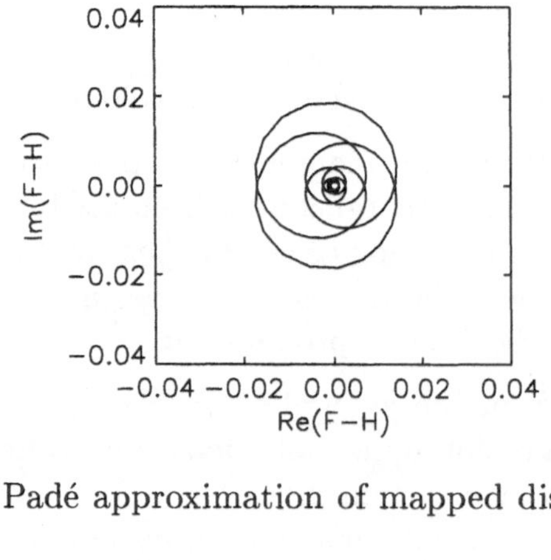

Figure 5.15: Least-squares Padé approximation of mapped discrete-time system ($N = 4$, $L = 20$): error curve

5.2.3 Potential problems

If the degree N of the approximation is chosen too high, the results become unstable. For example, taking an approximant of degree $N = 10$ in the previous example leads to one unstable pole. The corresponding constant in the partial fraction expansion is, however, quite small and can be disregarded without loss of accuracy. A similar conclusion is stated in [ST74].

The problem is investigated here for a smaller example. Consider the first-order system

$$F(z) = \frac{z^{-1}}{1 - 1/2\, z^{-1}} = z^{-1} + 1/2\, z^{-2} + 1/4\, z^{-3} + \cdots \tag{5.32}$$

The Hankel matrix has rank one, but if we introduce a small perturbation, it has full rank, although it is almost singular. The system of linear equations $\mathbf{\Gamma a} = \mathbf{f}$ is

$$\begin{bmatrix} 1.00 & 0.51 \\ 0.51 & 0.26 \end{bmatrix} \begin{bmatrix} a_2 \\ a_1 \end{bmatrix} = \begin{bmatrix} 0.26 \\ 0.13 \end{bmatrix} \tag{5.33}$$

Solving this equation, the coefficients are $a_0 = 1$, $a_1 = 26.00$, $a_2 = -13.00$. This corresponds to a system of order $N = 2$. The coefficients b_k are found by Equation (5.21) as

$$\begin{aligned} \begin{bmatrix} b_2 \\ b_1 \end{bmatrix} &= \begin{bmatrix} f_1 & f_2 \\ & f_1 \end{bmatrix} \begin{bmatrix} a_1 \\ a_0 \end{bmatrix} \\ &= \begin{bmatrix} 1.00 & 0.51 \\ & 1.00 \end{bmatrix} \begin{bmatrix} 26.00 \\ 1.00 \end{bmatrix} = \begin{bmatrix} 26.51 \\ 1.00 \end{bmatrix} \end{aligned} \tag{5.34}$$

The transfer function is then

$$\begin{aligned} H(z) &= \frac{b_1\, z^{-1} + b_2\, z^{-2}}{a_0 + a_1\, z^{-1} + a_2\, z^{-2}} \\ &= \frac{z^{-1} + 26.51\, z^{-2}}{1.00 + 26.00\, z^{-1} - 13.00\, z^{-2}} = \frac{z^{-1}(1 + 26.51\, z^{-1})}{(1 - 0.4907\, z^{-1})(1 + 26.49\, z^{-1})} \end{aligned} \tag{5.35}$$

The terms $(1 + 26.51\, z^{-1})$ in the numerator and $(1 + 26.49\, z^{-1})$ in the denominator almost cancel. If they were neglected, a good approximation would results. If they are kept, however, they lead to an unstable transfer function. The unstable part can be removed after expanding the transfer function into partial fractions.

$$H(z) = \frac{1.0007}{1 - 0.4907\, z^{-1}} - \frac{-0.0007}{1 + 26.49\, z^{-1}} \tag{5.36}$$

Clearly, the unstable part makes only a small contribution.

We can repeat the above calculations using a different perturbation. The linear equations for the coefficients a_1 and a_2 are

$$\begin{bmatrix} 1.00 & 0.51 \\ 0.51 & 0.25 \end{bmatrix} \begin{bmatrix} a_2 \\ a_1 \end{bmatrix} = \begin{bmatrix} 0.25 \\ 0.13 \end{bmatrix} \tag{5.37}$$

with the solution $a_1 = -0.2475$ and $a_2 = 0.3762$. The equations for the coefficients b_1 and b_2 are

$$\begin{bmatrix} b_2 \\ b_1 \end{bmatrix} = \begin{bmatrix} 1.00 & 0.51 \\ & 1.00 \end{bmatrix} \begin{bmatrix} -0.2475 \\ 1.00 \end{bmatrix} = \begin{bmatrix} 0.2625 \\ 1.00 \end{bmatrix} \tag{5.38}$$

The transfer function is

$$H(z) = \frac{z^{-1} + 0.2625\, z^{-2}}{1.00 - 0.2475\, z^{-1} + 0.3762\, z^{-2}} \tag{5.39}$$

The roots of the denominator are $z_{1,2} = 0.1238 \pm 0.6008i$. This time we obtain two complex conjugate poles that are stable and hence a system of degree two. That the degree of the system depends so much on small perturbations of the Markov parameters is not desirable. Later in this chapter (Section 5.4.1) we will present a method, where the degree of the approximation can be chosen depending on accuracy requirements.

5.3 Singular value decomposition (SVD)

The singular value decomposition (SVD) is an essential ingredient of the Hankel methods introduced in the next section. It has already been introduced in Section 4.2.4 in the context of infinite-dimensional matrices. Here we consider the SVD of finite-dimensional matrices. The SVD is a very powerful tool, both theoretically and numerically. It is the most reliable numerical method to determine the rank of a matrix, it deals with systems of linear equations and least-squares approximations even in the singular case and, last but not least, there exist robust numerical algorithms to perform the computations.

Although the singular value decomposition works equally well for complex and for non-square matrices, we only consider real square matrices. Any $L \times L$ matrix $\mathbf{A}$ can be decomposed as follows [GvL83].

$$\mathbf{A} = \mathbf{U}\mathbf{\Sigma}\mathbf{V}^{\mathrm{T}} = \begin{bmatrix} \mathbf{u}_1 & \cdots & \mathbf{u}_L \end{bmatrix} \begin{bmatrix} \sigma_1 & & \\ & \ddots & \\ & & \sigma_L \end{bmatrix} \begin{bmatrix} \mathbf{v}_1^{\mathrm{T}} \\ \vdots \\ \mathbf{v}_L^{\mathrm{T}} \end{bmatrix} = \sum_{k=1}^{L} \sigma_k \mathbf{u}_k \mathbf{v}_k^{\mathrm{T}} \tag{5.40}$$

The matrices $\mathbf{U}$ and $\mathbf{V}$ consist of the singular vectors $\mathbf{u}_k$ and $\mathbf{v}_k, k = 1 \ldots L$. The singular vectors are orthonormal, that is

$$\mathbf{U}^{\mathrm{T}}\mathbf{U} = \mathbf{I} \quad \text{and} \quad \mathbf{V}^{\mathrm{T}}\mathbf{V} = \mathbf{I} \tag{5.41}$$

The matrix $\mathbf{\Sigma}$ is diagonal and contains the singular values $\sigma_1 \geq \sigma_2 \geq \cdots \geq \sigma_L \geq 0$. The rank of the matrix is determined by the number N of non-zero singular values. If $\sigma_{N+1} = \cdots = \sigma_L = 0$, the matrix $\mathbf{A}$ can be written as

$$\mathbf{A} = \sum_{k=1}^{N} \sigma_k \mathbf{u}_k \mathbf{v}_k^{\mathrm{T}} \tag{5.42}$$

or in matrix form with submatrices $\mathbf{U}_1 = [\mathbf{u}_1, \ldots, \mathbf{u}_N]$, $\mathbf{V}_1 = [\mathbf{v}_1, \ldots, \mathbf{v}_N]$ and $\mathbf{\Sigma}_1 = \mathbf{diag}(\sigma_1, \ldots, \sigma_N)$ as

$$\mathbf{A} = \mathbf{U}_1 \mathbf{\Sigma}_1 \mathbf{V}_1^{\mathrm{T}} \tag{5.43}$$

In a numerical application, the higher singular values are usually not zero but small. A numerical rank is then determined by considering singular values up to a certain tolerance. Note the orthogonality properties of the submatrices $\mathbf{U}_1, \mathbf{U}_2, \mathbf{V}_1, \mathbf{V}_2$:

$$\mathbf{U}\mathbf{U}^{\mathrm{T}} = \begin{bmatrix} \mathbf{U}_1 & \mathbf{U}_2 \end{bmatrix} \begin{bmatrix} \mathbf{U}_1^{\mathrm{T}} \\ \mathbf{U}_2^{\mathrm{T}} \end{bmatrix} = \mathbf{U}_1\mathbf{U}_1^{\mathrm{T}} + \mathbf{U}_2\mathbf{U}_2^{\mathrm{T}} = \mathbf{I} \tag{5.44}$$

and

$$\mathbf{U}^{\mathrm{T}}\mathbf{U} = \begin{bmatrix} \mathbf{U}_1^{\mathrm{T}} \\ \mathbf{U}_2^{\mathrm{T}} \end{bmatrix} \begin{bmatrix} \mathbf{U}_1 & \mathbf{U}_2 \end{bmatrix} = \begin{bmatrix} \mathbf{I} & \\ & \mathbf{I} \end{bmatrix} \tag{5.45}$$

Therefore

$$\mathbf{U}_1^{\mathrm{T}}\mathbf{U}_1 = \mathbf{U}_2^{\mathrm{T}}\mathbf{U}_2 = \mathbf{I} \quad \text{and} \quad \mathbf{U}_1^{\mathrm{T}}\mathbf{U}_2 = \mathbf{U}_2^{\mathrm{T}}\mathbf{U}_1 = \mathbf{0} \tag{5.46}$$

but

$$\mathbf{U}_1\mathbf{U}_1^{\mathrm{T}} \neq \mathbf{I} \quad \mathbf{U}_1\mathbf{U}_1^{\mathrm{T}} \neq \mathbf{I} \quad \text{and} \quad \mathbf{U}_1\mathbf{U}_2^{\mathrm{T}} \neq \mathbf{0} \quad \mathbf{U}_2\mathbf{U}_1^{\mathrm{T}} \neq \mathbf{0} \tag{5.47}$$

Analogous relations hold for $\mathbf{V}_1$ and $\mathbf{V}_2$.

Formally, the decomposition corresponds to the two eigenvalue problems

$$(\mathbf{A}^{\mathrm{T}}\mathbf{A})\,\mathbf{V} = \mathbf{V}\mathbf{\Sigma}^2 \tag{5.48}$$

$$\mathbf{U}^{\mathrm{T}}\,(\mathbf{A}\mathbf{A}^{\mathrm{T}}) = \mathbf{\Sigma}^2\,\mathbf{U}^{\mathrm{T}} \tag{5.49}$$

Because the matrices $\mathbf{A}^{\mathrm{T}}\mathbf{A}$ and $\mathbf{A}\mathbf{A}^{\mathrm{T}}$ are real symmetric and positive semi-definite, the eigenvalues are non-negative and the eigenvectors are orthogonal. The eigenvectors have to be scaled to unit length to make them orthonormal (unitary). For real symmetric matrices the eigenvalue problem can be written as

$$\mathbf{A} = \mathbf{V}\mathbf{\Lambda}\mathbf{V}^{\mathrm{T}} \tag{5.50}$$

with real eigenvalues and (if properly scaled) orthonormal eigenvectors. This equation already looks like an SVD except that the eigenvalues are not necessarily positive. To make the singular values positive we introduce the sign matrix $\mathbf{\Theta} = \mathbf{diag}(\pm 1, \pm 1, \ldots)$ and write

$$\mathbf{A} = \mathbf{U}\mathbf{\Sigma}\mathbf{V}^{\mathrm{T}} = \mathbf{U}\mathbf{\Theta}^{-1}\mathbf{\Theta}\mathbf{\Lambda}\mathbf{V}^{\mathrm{T}} \tag{5.51}$$

By appropriately choosing the signs in $\mathbf{\Theta}$ the singular values

$$\mathbf{\Sigma} = \mathbf{\Theta}\mathbf{\Lambda} \tag{5.52}$$

are positive. The singular vectors

$$\mathbf{U} = \mathbf{V}\mathbf{\Theta}^{-1} \tag{5.53}$$

are, up to the sign, equal to $\mathbf{V}$. Note that the explanations given here are only for theoretical derivations. For numerical computations special accurate and robust algorithms should be used [GvL83].

The SVD is not only valuable for determining the rank of a matrix but it also gives a means of solving the least-squares problem. The classical least-squares method breaks down when the equations are singular or close to singular. This can happen if the estimated degree N of the approximant is higher than the actual degree N. If the matrix does not have full rank then an infinite number of solutions exist. The solution with minimum $\|\mathbf{x}\|_2$ is given by [GvL83]

$$\mathbf{x} = \mathbf{A}^{+}\mathbf{b} = \mathbf{V}_1\mathbf{\Sigma}_1^{-1}\mathbf{U}_1^{\mathrm{T}}\mathbf{b} = \sum_{i=1}^{N}(\mathbf{u}_i^{\mathrm{T}}\mathbf{b}/\sigma_i)\mathbf{v}_i \tag{5.54}$$

where $\mathbf{A}^{+} = \mathbf{V}_1\mathbf{\Sigma}_1^{-1}\mathbf{U}_1^{\mathrm{T}}$ is called the pseudo-inverse of $\mathbf{A}$.

A useful property of the singular value decomposition is that it determines the range and the null space of a matrix. The range of a matrix is given by $\mathbf{U}_1$ because all consistent right-hand sides are given by

$$\mathbf{A}\mathbf{x} = \sum_{i=1}^{N}\sigma_i\mathbf{u}_i\mathbf{v}_i^{\mathrm{T}}\mathbf{x} = \sum_{i=1}^{N}\mathbf{u}_i(\sigma_i\mathbf{v}_i^{\mathrm{T}}\mathbf{x}) = \sum_{i=1}^{N}\alpha_i\mathbf{u}_i \tag{5.55}$$

with any numbers α_i. The null space is defined by $\mathbf{V}_2$ because $\mathbf{Ax} = \mathbf{0}$ for any

$$\mathbf{x} = \sum_{i=N+1}^{L} \beta_i \mathbf{v}_i \tag{5.56}$$

with arbitrary numbers β_i. Indeed, the following relation holds

$$\mathbf{Ax} = \sum_{i=1}^{N} \left(\sigma_i \mathbf{u}_i \mathbf{v}_i^{\mathrm{T}} \sum_{j=N+1}^{L} \beta_j \mathbf{v}_j \right) = \mathbf{0} \tag{5.57}$$

because $\mathbf{v}_i^{\mathrm{T}} \mathbf{v}_j = 0$ for $i \neq j$.

5.4 Methods based on SVD of Hankel matrix

In the preceding sections, the rational approximation problem has been solved using simple mathematics, systems of linear equations and least-squares methods. Although leading to acceptable results, these techniques are merely curve-fitting procedures and do not give insight into the mathematical challenge of rational approximation. Methods that are more closely related to systems are the ones based on the singular value decomposition of the Hankel matrix. They not only yield more accurate results but also determine the degree of the system as part of the approximation process. The specific method we present here is the Carathéodory-Fejér (CF) method suggested by Gutknecht and Trefethen [GST83, GT80]. The derivations are in the framework of this chapter, that is, for scalar systems using rational functions notation. The extension to multivariable systems is not easy and is not undertaken in this work. Another method, applicable to multivariable systems, is shown later in Chapter 7.

5.4.1 Carathéodory-Fejér (CF) method

In this section we present a first method based on the singular value decomposition of the Hankel matrix, postponing a second one to later chapters after the matrix form of the state equations has been introduced. The method has been proposed by Gutknecht and Trefethen [GST83, GT80] who applied the theory by Takagi [Tak24, Tak25] to the design of digital filters. They call it the Carathéodory-Fejér method because the very first ideas go back to these authors.

Instead of solving Equation (5.23), the right-hand vector can be included in the Hankel matrix as

$$\begin{bmatrix} f_1 & f_2 & \cdots & f_{L+1} \\ f_2 & f_3 & \cdots & f_{L+2} \\ \vdots & \vdots & & \vdots \\ f_{L+1} & f_{L+2} & \cdots & f_{2L+1} \end{bmatrix} \begin{bmatrix} a_L \\ a_{L-1} \\ \vdots \\ a_0 \end{bmatrix} = \mathbf{0} \tag{5.58}$$

where also the number of equations has been increased by one to retain a quadratic matrix. The singular value decomposition yields

$$\mathbf{\Gamma} = \mathbf{U} \mathbf{\Sigma} \mathbf{V}^{\mathrm{T}} \tag{5.59}$$

Because the singular vectors are orthogonal, we have the relationship

$$\mathbf{\Gamma} \mathbf{v}_{N+1} = \sigma_{N+1} \mathbf{u}_{N+1} \tag{5.60}$$

If the degree of the system is L, the $(L+1) \times (L+1)$ Hankel matrix has rank $N = L$. Thus the singular value σ_{L+1} will be zero and the corresponding vector $\mathbf{v}_{L+1}$ is the desired solution for $\mathbf{a}$. It only remains to scale it for $a_0 = 1$.

As an approximation we can neglect small singular values and choose N such that σ_{N+1} is sufficiently small. For a real symmetric Hankel matrix, the singular vectors $\mathbf{u}_{N+1}$ and $\mathbf{v}_{N+1}$ are real and, up to the sign, equal. Taking the solution vector $\mathbf{a} = \mathbf{v}_{N+1}$ yields with Equation (5.60)

$$\begin{bmatrix} f_1 & f_2 & \cdots & f_{L+1} \\ f_2 & f_3 & \cdots & f_{L+2} \\ \vdots & \vdots & & \vdots \\ f_{L+1} & f_{L+2} & \cdots & f_{2L+1} \end{bmatrix} \begin{bmatrix} a_L \\ a_{L-1} \\ \vdots \\ a_0 \end{bmatrix} = \pm\sigma_{N+1} \begin{bmatrix} a_L \\ a_{L-1} \\ \vdots \\ a_0 \end{bmatrix} \tag{5.61}$$

Rewriting Equation (5.18) and substituting Equation (5.21) and Equation (5.61) we get

$$\begin{aligned} &(f_1\, z^{-1} + f_2\, z^{-2} + \cdots)\,(a_L\, z^{-L} + \cdots + a_1\, z^{-1} + a_0) \\ &\quad = b_1\, z^{-1} + \cdots + b_L\, z^{-L} \\ &\qquad \pm\sigma_{N+1}\,(a_L\, z^{-(L+1)} + a_{L-1}\, z^{-(L+2)} + \cdots + a_0\, z^{-(2L+1)}) \end{aligned} \tag{5.62}$$

Dividing by $(a_L\, z^{-L} + \cdots + a_1\, z^{-1} + a_0)$, we get back the original approximation problem

$$\begin{aligned} &(f_1\, z^{-1} + f_2\, z^{-2} + \cdots) \\ &\quad = \frac{b_1\, z^{-1} + \cdots + b_L\, z^{-L}}{a_L\, z^{-L} + \cdots + a_1\, z^{-1} + a_0} \\ &\qquad \pm\sigma_{N+1}\, \frac{a_L\, z^{-(L+1)} + a_{L-1}\, z^{-(L+2)} + \cdots + a_0\, z^{-(2L+1)}}{a_L\, z^{-L} + \cdots + a_1\, z^{-1} + a_0} \end{aligned} \tag{5.63}$$

or

$$F(z) = H(z) + E(z) \tag{5.64}$$

with the error term

$$E(z) = \pm\sigma_{N+1}\, z^{-(L+1)}\, \frac{a_L + a_{L-1}\, z^{-1} + \cdots + a_0\, z^{-L}}{a_0 + a_1\, z^{-1} + \cdots + a_L\, z^{-L}} \tag{5.65}$$

On the unit circle, $|z| = 1$, the error has an absolute value of

$$|E(z)| = \sigma_{N+1} \tag{5.66}$$

This can be seen by noting that $z^{-1} = \bar{z}$ (complex conjugate) on the unit circle. For real coefficients a_k we write

$$\frac{a_L + a_{L-1}\, z^{-1} + \cdots + a_0\, z^{-L}}{a_0 + a_1\, z^{-1} + \cdots + a_L\, z^{-L}} = z^{-L}\, \frac{\overline{a_0 + a_1\, z^{-1} + \cdots + a_L\, z^{-L}}}{a_0 + a_1\, z^{-1} + \cdots + a_L\, z^{-L}} \tag{5.67}$$

which clearly has unit amplitude. Plotting the error in the complex plane yields a circular error curve with radius σ_{N+1}. The circular error curve is characteristic for an approximation which is optimal in the sense of the maximum norm. Here the error norm is $\|E(z)\|_\infty = \sigma_{N+1}$. Because we have selected a small singular value, the error will be small and a good approximation is obtained.

Of course, this approximation is not the final solution yet. For one, it has degree L, whereas the Hankel matrix has been assumed to have approximately rank N. For the other, the question of stability has not been looked at yet. One surprising fact is that the transfer function of the approximant has N stable poles. It can be shown [Tak24] that the singular vector $\mathbf{v}_N$ corresponding to the Hankel matrix of the truncated impulse response

$$\begin{bmatrix} f_1 & f_2 & \cdots & \cdots & f_{L+1} \\ f_2 & f_3 & \cdots & f_{L+1} & \\ \vdots & \vdots & & & \\ f_L & f_{L+1} & & & \\ f_{L+1} & & & & \end{bmatrix} \tag{5.68}$$

leads to a polynomial

$$a_0 + a_1\, z^{-1} + \cdots + a_L\, z^{-L} = 0 \tag{5.69}$$

that has exactly N stable poles. In practice, the Hankel matrix of the infinite sequence (Equation 5.61) leads to the same result. If the number of stable poles is slightly different from N, we can still use the actual number without any drawbacks.

The procedure is thus to find a partial fraction expansion of $H(z)$ (Equation 5.63) and only retain the stable part. This is the desired stable approximation of degree N. It turns out that neglecting the unstable part increases the maximum error only slightly and the approximation is still valid. The error curve becomes somewhat distorted but is still close to circular. To find the stable part of the approximation, the zeroes of the denominator polynomial have to be evaluated. For large-degree polynomials this might cause numerical problems. Gutknecht and coworkers [GT80] suggest using a method by Henrici (see reference). For the example problems below we simply used standard libraries. Only the stable poles were retained for the denominator polynomial and the numerator polynomial was determined by Equation (5.21).

At first sight, it is not obvious that the approximation should be better than the least-squares solution. In both cases we start with the Hankel matrix and use the SVD to determine the coefficients. However in the least-squares method, the solution vector is optimal only for one column of the Hankel matrix (range). In the CF method, the solution vector is optimal for all columns (null space).

Examples. As a first example, consider again the sequence $f(z) = z^{-1} + 1/2\, z^{-2} + \cdots$ with the Hankel matrix

$$\mathbf{\Gamma} = \begin{bmatrix} 1 & 1/2 & 1/4 \\ 1/2 & 1/4 & 1/8 \\ 1/4 & 1/8 & 1/16 \end{bmatrix} \tag{5.70}$$

and the singular value decomposition

$$\mathbf{U} = \mathbf{V} = \begin{bmatrix} 0.8729 & 0.4880 & 0.0000 \\ 0.4364 & -0.7807 & 0.4472 \\ 0.2182 & -0.3904 & -0.8944 \end{bmatrix} \qquad \mathbf{\Sigma} = \begin{bmatrix} 1.3125 & & \\ & 0 & \\ & & 0 \end{bmatrix} \tag{5.71}$$

The polynomials corresponding to the singular vectors $\mathbf{v}_k$ (scaled to $a_0 = 1$) have the roots

$$k = 1: \qquad z^2 + 2\,z + 4 = 0 \;\Rightarrow\; z_{1,2} = -1 \pm 1.732\, i \tag{5.72}$$

$$k = 2: \qquad z^2 + 2\,z - 1.25 = 0 \;\Rightarrow\; z_1 = 0.5, \quad z_2 = -2.5 \tag{5.73}$$

$$k = 3: \qquad z^2 - 0.5\,z = 0 \;\Rightarrow\; z_1 = 0, \quad z_2 = 0.5 \tag{5.74}$$

As expected, the polynomial corresponding to $\mathbf{v}_k$ has $k-1$ poles inside the unit circle. Choosing $N = 1$ ($k = N + 1 = 2$), solving for b_l by Equation (5.21) and writing the transfer function in partial fractions form leads to the exact solution

$$H(z) = \frac{z^{-1}}{1 - 0.5\, z^{-1}} \tag{5.75}$$

The second term with root $z_1 = -2.5$ disappears ($c_2 = 0$).

As a second example, consider the same sequence slightly perturbed. The Hankel matrix

$$\mathbf{\Gamma} = \begin{bmatrix} 1.00 & 0.51 & 0.25 \\ 0.51 & 0.25 & 0.13 \\ 0.25 & 0.13 & 0.06 \end{bmatrix} \tag{5.76}$$

has the SVD

$$\begin{aligned} \mathbf{U} &= \begin{bmatrix} 0.8706 & 0.3261 & -0.3684 \\ 0.4410 & -0.8492 & 0.2905 \\ 0.2181 & 0.4154 & 0.8831 \end{bmatrix} \qquad \mathbf{V} = \begin{bmatrix} 0.8706 & -0.3261 & 0.3684 \\ 0.4410 & 0.8492 & -0.2905 \\ 0.2181 & -0.4154 & -0.8831 \end{bmatrix} \\ \mathbf{\Sigma} &= \mathbf{diag}\,(1.3210,\ 0.0095,\ 0.0015) \end{aligned} \tag{5.77}$$

The Hankel matrix still has approximately rank one. Due to the perturbation the higher singular values are not zero but small compared to the first one. The transfer function can therefore still be approximated by a first-order system. The polynomials corresponding to the singular vectors $\mathbf{v}_k$ are

$$k = 1: \quad z^2 + 2.0225\,z + 3.9923 = 0 \quad \Rightarrow \quad z_{1,2} = -1.0112 \pm 1.7233\,i \tag{5.78}$$

$$k = 2: \quad z^2 - 2.0441\,z + 0.7850 = 0 \quad \Rightarrow \quad z_1 = 0.5126, \quad z_2 = 1.5315 \tag{5.79}$$

$$k = 3: \quad z^2 + 0.3290\,z - 0.4172 = 0 \quad \Rightarrow \quad z_1 = -0.8310, \quad z_2 = 0.5020 \tag{5.80}$$

Also in this example the number of stable poles coincides with the ordinal number of the corresponding singular vector. Taking $N = 1$ ($k = 2$), the transfer function is

$$H(z) = \frac{1.0026\,z^{-1}}{1 - 0.5126\,z^{-1}} + \frac{-0.0026\,z^{-1}}{1 - 1.5315\,z^{-1}} \tag{5.81}$$

Clearly the second term which is unstable has only a small contribution and can be discarded. It reaches its maximum value at $z = -1$ where it is 0.0049, which is of the order of the overall error based on the singular value $\sigma_2 = 0.0095$.

The third example is the rod on a visco-elastic foundation. The approximation is performed in the mapped discrete-time domain. The Markov parameters have been calculated using the FFT with $N_F = 512$. The 20×20 Hankel matrix is used for the CF approximation. The first 10 singular values of the Hankel matrix are plotted in Figure 5.16. As can be seen the values fall rapidly. Taking $N = 4$ seems to give a good approximation. As expected the system has four stable poles. The CF approximation for $N = 4$ is shown in Figure 5.17. The two curves are almost identical. A closer look gives the error curve in Figure 5.18. Most cycles are circular with a radius equal to the singular value $\sigma_5 = 5 \cdot 10^{-3}$. Compared to the earlier approximations this is the best one.

5.5 Additional topics

5.5.1 Transfer function in the complex plane

An interesting question is the following. If we find a good approximation of the transfer function on the axis $s = i\omega$, then how is the approximation on the whole s-plane? From the theory of analytical functions it is clear by Cauchy's formula [Pap62, p. 295] that if the transfer function is analytic in a region, then it can be determined from its values on the boundary. If the approximation is good on the whole axis $s = i\omega$, then we expect that the approximation is also good in the right half-plane $\Re(s) > 0$ where both the

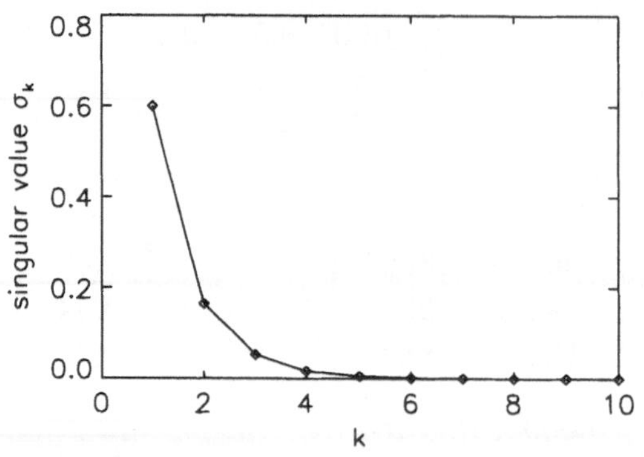

Figure 5.16: CF approximation: singular values

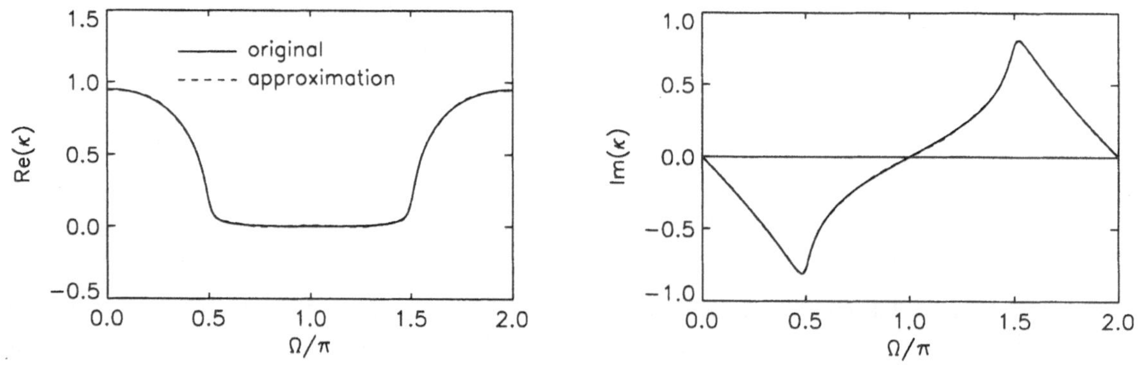

Figure 5.17: CF approximation ($N = 4$, $L = 20$): transfer function

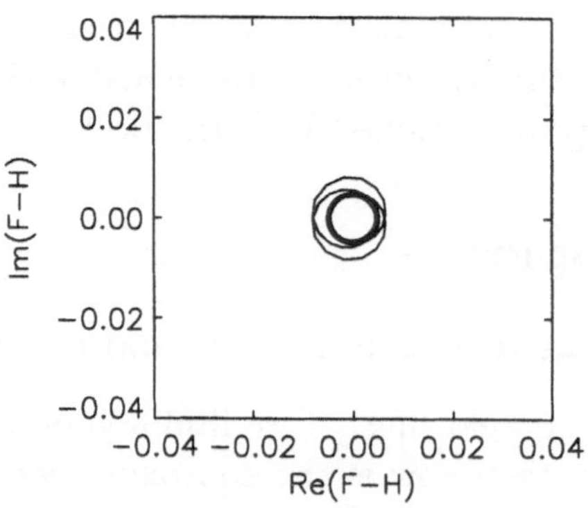

Figure 5.18: CF approximation ($N = 4$, $L = 20$): error curve

original transfer function and the approximation are analytic. The right half-plane with the imaginary axis as boundary can be viewed as a special case of a closed region. In fact, if we take the bilinear transform with $\alpha < 0$, the right half-plane is mapped to the interior of the unit circle. (The customary bilinear transform with $\alpha > 0$ maps the *left* half-plane into the interior of the unit circle.)

This theoretical consideration is illustrated in the following figures. In Figure 5.19 the real and imaginary parts of the original transfer function for the rod on a visco-elastic foundation are shown. The solid line indicates the curve on the imaginary s-axis. This plot has already been shown earlier (Figure 4.6) with a different view angle. In Figure 5.20 the same figure is shown for the approximation. Clearly, the approximation is not only valid on the imaginary axis but also on the whole right (real) half-plane. On the other hand, the approximation is completely different on the left half-plane. This is, however, of no concern since only the right half-plane is used in the Laplace transform. The rational approximation has four poles, two larger and two smaller ones, which are clearly visible. Each pole is simple and extends therefore from $-\infty$ to $+\infty$.

5.5.2 Causality and stability

In several examples we have encountered approximations that were unstable, but the unstable part made only a small contribution to the transfer function. Removing the unstable part solved the stability problem without affecting the approximation much. In the following we provide an intuitive reasoning for that.

A causal stable system $F(z) = \sum_{k=0}^{\infty} f_k \, z^{-k}$ is approximated by a rational transfer function. If it is a good approximation we expect of course that $h_k \approx f_k$ and hence the approximation to be causal. If it is causal then it is necessarily also stable. But it may well be that some of the h_k, $k < 0$ are not quite zero and the approximation is not causal and hence unstable. The error may be introduced by round-off or more typically when the original function $F(z)$ is not correctly captured in the high frequency range, leading to an original impulse response which already has some anti-causal terms. It is clear that setting $h_k = 0$ for $k < 0$ does not introduce any additional error. In the rational transfer function the anti-causal part can easily be removed by taking the partial fractions form and neglecting any terms with unstable poles.

5.5.3 General remarks

Discrete-time and continuous-time formulation. Even if we can find a very accurate approximation in the continuous-time domain, we afterwards have to discretize it using some time integration scheme. It seems, at first sight, advantageous to find an approximation directly for the discrete-time domain which minimizes simultaneously the system error and the discretization error and is hence more accurate. This has, however, two major drawbacks.

Firstly, the time step has to be determined in advance and is fixed for the subsequent finite element analysis. Typically, however, in a nonlinear analysis a flexible time step is necessary in the case of convergence problems.

Secondly, for a discrete-time approximation the time step has to be small enough such that the transfer function is able to approximate the original function. The highest frequency which can be represented by a discrete-time transfer function is the Nyquist frequency $\omega_\nu = \pi/\Delta t$, half the sampling frequency. If the transfer function has some contribution for higher frequencies this contribution will be cut off yielding possibly a non-causal and hence unstable approximation.

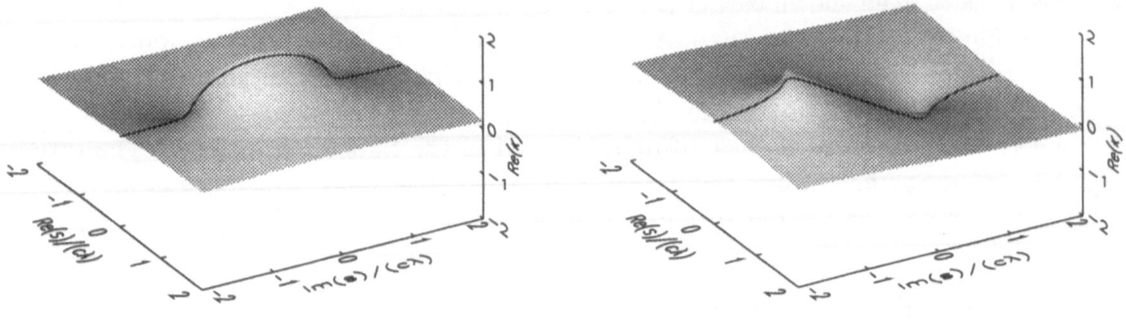

Figure 5.19: Transfer function in complex plane: analytical

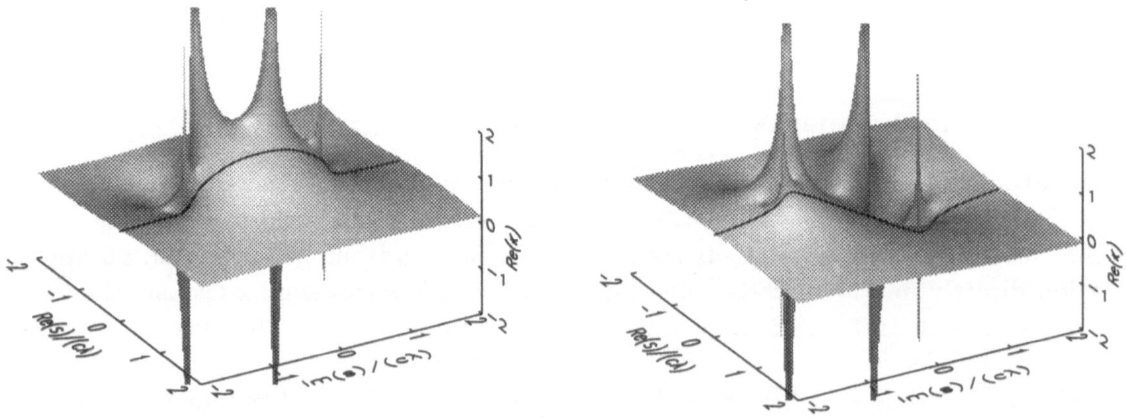

Figure 5.20: Transfer function in complex plane: approximation

Rational and polynomial formulation. The two formulations, the rational and the polynomial formulation, may be considered from various viewpoints.

- □ The problem may be viewed as an approximation problem. It is well known that rational approximations, although more complicated to determine, lead to much better results than polynomial approximations.
- □ The rational formulation can be implemented as a recursive scheme or by state variables (internal variables, see Equation (4.16) and next chapter).
- □ In a non-recursive realization, only past input is used to predict the present response, in a recursive formulation past input and past output is used. Therefore, in a non-recursive formulation a long input history is needed whereas in a recursive formulation only a short input and output history is necessary. The influence of the early input history is somehow included in the recent output history.
- □ In filter design, the non-recursive filters are also called finite impulse response (FIR) filters, the recursive filters are called infinite response (IIR) filters. From this viewpoint, a non-recursive formulation describes a finite impulse response given by its coefficients. A recursive formulation generates an infinite impulse response, that is, a few coefficients describe an infinitely long impulse response. This is analogous to the preceding viewpoint, where the influence of the past history on the present response was considered. Here we consider the influence of a present impulse on the future behaviour of the system.
- □ The discrete-time polynomial formulation is equivalent to a non-recursive difference scheme (Equation 4.6 with $N = 0$) of the form of a convolution, the rational formulation to a recursive difference equation (Equation 4.6), or to a differential equation (Equation 4.9).

Chapter 6

Multivariable Systems, State-Variable Description

For the further development we use the state-variable description of linear time-invariant systems. This is the standard form used in linear system theory. Introducing this notation brings some advantages. The state-variable description is the most general form and is equally applicable for scalar and for multivariable systems. Matrix calculus is employed instead of working with polynomials. The whole theory of linear systems is available.

In addition to the state-variable description of systems, the chapter introduces the concepts of observability and controllability and the balanced systems which are important for the method introduced in the next chapter. As for the scalar case, the duality between discrete-time and continuous-time systems, and the bilinear transform connecting the two, are also covered. Only the most important topics that are used directly in the further development, are introduced. For more details the reader is referred to text books on linear systems [Che84, Kai80]. Many topics related to balancing and approximation of systems are covered in Glover [Glo84] and Kung [KL81a, LK82]. Since most derivations are easier for discrete-time systems, we start with them, introducing the continuous-time counterparts later.

6.1 Discrete-time systems

6.1.1 Scalar systems

Any higher-order differential equation can be transformed into a set of first-order differential equations and analogously any difference equation into a set of first-order difference equations by introducing internal variables, the state variables. Many different forms exist, but we will only introduce the ones that are needed later.

The general form for scalar (single-input, single-output) discrete-time systems is

$$\mathbf{x}(n+1) = \mathbf{A}\,\mathbf{x}(n) + \mathbf{b}\,u(n) \tag{6.1}$$

$$y(n) = \mathbf{c}\,\mathbf{x}(n) \tag{6.2}$$

It is customary to use $\mathbf{x}$ for the state variables, u for the input and y for the output of a system. The matrix $\mathbf{A}$ is sometimes called the state-feedback matrix. Since this work employs theories from several different fields, each with its own notation, some overlapping of notation is unavoidable. Using customary but perhaps ambiguous notation seems to be less confusing than trying to find unique symbols by changing letters or using indices and accents. Thus we stick to customary notation as long as the meaning is clear from the context.

The transfer function of a system is easily obtained by replacing $\mathbf{x}(n)$ by $\mathbf{x}(z)$ and $\mathbf{x}(n+1)$ by $z\,\mathbf{x}(z)$ in the system equations:

$$\begin{aligned} \mathbf{x}(z) &= (z\mathbf{I}-\mathbf{A})^{-1}\mathbf{b}\,u(z) && (6.3)\\ y(z) &= \mathbf{c}\,\mathbf{x}(z) = \mathbf{c}\,(z\mathbf{I}-\mathbf{A})^{-1}\mathbf{b}\,u(z) && (6.4) \end{aligned}$$

Thus the transfer function is

$$H(z) = \mathbf{c}\,(z\mathbf{I}-\mathbf{A})^{-1}\mathbf{b} \tag{6.5}$$

The impulse response of a discrete-time system is the response due to the input sequence $u(0) = 1, u(1) = u(2) = \cdots = 0$. Applying this sequence to Equations (6.1 and 6.2) the following relation is obtained

$$h_n = \mathbf{c}\,\mathbf{x}(n) = \mathbf{c}\,\mathbf{A}^{n-1}\mathbf{b} \quad \text{for } n = 1, 2, \ldots \tag{6.6}$$

For $n = 0$, the impulse response is zero. The system is therefore strictly causal. A causal system with $h_0 \neq 0$ has a slightly different form. An additional term has to be added in Equation (6.2) leading to

$$y(n) = \mathbf{c}\,\mathbf{x}(n) + d\,u(n) \tag{6.7}$$

The term $d\,u(n)$ yields an immediate output and involves no dynamics. It is therefore sometimes called the direct transmission term. With this term included, the transfer function reads

$$H(z) = \mathbf{c}\,(z\mathbf{I}-\mathbf{A})^{-1}\mathbf{b} + d \tag{6.8}$$

and the impulse response is

$$\begin{aligned} h_0 &= d && (6.9)\\ h_n &= \mathbf{c}\,\mathbf{A}^{n-1}\mathbf{b} \quad \text{for } n = 1, 2, \ldots && (6.10) \end{aligned}$$

Because most of the derivations are somewhat simpler for a strictly causal systems, we use that form whenever possible, keeping in mind that the extension to causal system can easily be done by adding a direct transmission term.

We now look at a particular realization of $H(z)$. The difference equation described by the rational function, Equation (4.17), can be expressed by the system matrices

$$\begin{aligned} \mathbf{A}_c &= \begin{bmatrix} 0 & & & & -a_N \\ 1 & 0 & & & -a_{N-1} \\ & 1 & & & -a_{N-2} \\ & & \ddots & & \vdots \\ & & & 1 & -a_1 \end{bmatrix} \qquad \mathbf{b}_c = \begin{bmatrix} 1 \\ 0 \\ 0 \\ \vdots \\ 0 \end{bmatrix} \\ \mathbf{c}_c &= \begin{bmatrix} h_1 & h_2 & h_3 & \ldots & h_N \end{bmatrix} && (6.11) \end{aligned}$$

The index c stands for controllability form, which will be explained later (Subsection 6.1.5).

The equivalence of the state-variable description to the rational form is first shown for the small system

$$H(z) = \frac{b_1\,z^{-1} + b_2\,z^{-2}}{a_0 + a_1\,z^{-1} + a_2\,z^{-2}} \tag{6.12}$$

The impulse response for this transfer function is

$$\begin{aligned} h_1 &= b_1 && (6.13)\\ h_2 &= b_2 - a_1\,h_1 = b_2 - a_1\,b_1 && (6.14)\\ h_3 &= -a_1\,h_2 - a_2\,h_1 && (6.15)\\ &\vdots \end{aligned}$$

The system matrices are expressed in terms of the coefficients a_k and b_k as

$$\begin{aligned}\mathbf{A} &= \begin{bmatrix} 0 & -a_2 \\ 1 & -a_1 \end{bmatrix} \qquad \mathbf{b} = \begin{bmatrix} 1 \\ 0 \end{bmatrix} \\ \mathbf{c} &= \begin{bmatrix} h_1 & h_2 \end{bmatrix} = \begin{bmatrix} b_1 & b_2 - a_1 b_1 \end{bmatrix}\end{aligned} \tag{6.16}$$

Clearly

$$\mathbf{c}\,\mathbf{b} = h_1 \tag{6.17}$$

and with

$$\mathbf{c}\,\mathbf{A} = \begin{bmatrix} h_2 & -a_2 h_1 - a_1 h_2 \end{bmatrix} = \begin{bmatrix} h_2 & h_3 \end{bmatrix} \tag{6.18}$$

follows

$$\mathbf{c}\,\mathbf{A}\,\mathbf{b} = h_2 \tag{6.19}$$

and so on.

For the general case we observe that (Equation 5.22)

$$h_n = -\sum_{k=1}^{N} a_k\, h_{n-k} \qquad n > N \tag{6.20}$$

As an intermediate result we calculate

$$\mathbf{c}_c = \begin{bmatrix} h_1 & h_2 & \dots & h_N \end{bmatrix} \tag{6.21}$$

$$\mathbf{c}_c\mathbf{A}_c = \begin{bmatrix} h_2 & h_3 & \dots & h_{N+1} \end{bmatrix} \tag{6.22}$$

$$\vdots$$

$$\mathbf{c}_c\mathbf{A}_c^{k-1} = \begin{bmatrix} h_k & h_{k+1} & \dots & h_{N+k-1} \end{bmatrix} \tag{6.23}$$

The impulse response is therefore

$$\mathbf{c}_c\mathbf{A}_c^{k-1}\mathbf{b}_c = h_k \quad \text{for } n = 1, 2, \dots \tag{6.24}$$

This verifies that $(\mathbf{A}_c, \mathbf{b}_c, \mathbf{c}_c)$ is indeed a realization of $H(z)$. It is clear that another realization is given by

$$\mathbf{A}_o = \mathbf{A}_c^{\mathrm{T}} \qquad \mathbf{b}_o = \mathbf{c}_c^{\mathrm{T}} \qquad \mathbf{c}_o = \mathbf{b}_c^{\mathrm{T}} \tag{6.25}$$

because $H(z) = \mathbf{c}_c(z\mathbf{I} - \mathbf{A}_c)^{-1}\mathbf{b}_c = \mathbf{b}_c^{\mathrm{T}}(z\mathbf{I} - \mathbf{A}_c^{\mathrm{T}})^{-1}\mathbf{c}_c^{\mathrm{T}} = \mathbf{c}_o(z\mathbf{I} - \mathbf{A}_o)^{-1}\mathbf{b}_o$. The subscript o refers to the observability form (see Subsection 6.1.5).

6.1.2 Multivariable systems

Multivariable systems, that is, systems with multiple inputs and multiple outputs, can be written in the same form as scalar systems if the state-variable description is used. Although many problems, especially the question of minimal realization, are much more involved for multivariable systems than for the scalar case, most of the concepts and definitions are the same. Whenever this is the case we will only describe the general, multivariable form, considering scalar systems as a special case of multivariable systems.

The general form of a multivariable system with q inputs and p outputs is

$$\mathbf{x}(n+1) = \mathbf{A}\,\mathbf{x}(n) + \mathbf{B}\,\mathbf{u}(n) \tag{6.26}$$

$$\mathbf{y}(n) = \mathbf{C}\,\mathbf{x}(n) \tag{6.27}$$

where the size of the matrices $\mathbf{A}$, $\mathbf{B}$, $\mathbf{C}$ is $N \times N$, $N \times q$, and $p \times N$, respectively. The input and output vectors are given by

$$\mathbf{u}(n) = \begin{bmatrix} u_1(n) \\ u_2(n) \\ \vdots \\ u_q(n) \end{bmatrix} \qquad \mathbf{y}(n) = \begin{bmatrix} y_1(n) \\ y_2(n) \\ \vdots \\ y_p(n) \end{bmatrix} \tag{6.28}$$

The transfer function is

$$\mathbf{H}(z) = \mathbf{C}\,(z\mathbf{I} - \mathbf{A})^{-1}\mathbf{B} \tag{6.29}$$

and the impulse response

$$\mathbf{h}_n = \mathbf{C}\,\mathbf{A}^{n-1}\mathbf{B} \quad \text{for } n = 1, 2, \ldots \tag{6.30}$$

Adding a direct transmission term $\mathbf{D}\,\mathbf{u}(n)$ to Equation (6.27), the transfer function changes to

$$\mathbf{H}(z) = \mathbf{C}\,(z\mathbf{I} - \mathbf{A})^{-1}\mathbf{B} + \mathbf{D} \tag{6.31}$$

and the impuls respose to

$$\mathbf{h}_0 = \mathbf{D} \tag{6.32}$$

$$\mathbf{h}_n = \mathbf{C}\,\mathbf{A}^{n-1}\mathbf{B} \quad \text{for } n = 1, 2, \ldots \tag{6.33}$$

As seen already in Chapter 4 the impulse response of a discrete-time system equals the Markov parameters. This also applies to the multivariable case and the transfer function of a strictly causal system can therefore be written as

$$\mathbf{H}(z) = \sum_{k=1}^{\infty} \mathbf{h}_k\, z^{-k} = \begin{bmatrix} H_{11}(z) & H_{12}(z) & \ldots & H_{1q}(z) \\ H_{21}(z) & H_{22}(z) & \ldots & H_{2q}(z) \\ \vdots & \vdots & & \vdots \\ H_{p1}(z) & H_{p2}(z) & \ldots & H_{pq}(z) \end{bmatrix} \tag{6.34}$$

where $\mathbf{h}_k$ are the $p \times q$ impulse response matrices and each of the $H_{ij}(z)$ is a scalar rational transfer function. Let

$$a(z) = z^M + a_1 z^{M-1} + \cdots + a_M = \sum_{k=0}^{M} a_k\, z^{M-k} \tag{6.35}$$

with $a_0 = 1$ be the least common denominator of all $H_{ij}(z)$ Then we can write

$$\mathbf{H}(z) = \frac{\mathbf{N}(z)}{a(z)} \tag{6.36}$$

where $\mathbf{N}(z) = \sum_{n=1}^{M} \mathbf{N}_n\, z^{M-n}$ is a matrix polynomial of degree $M - 1$. Expanding the polynomials and multiplying by $a(z)$ yields

$$\sum_{j=1}^{\infty} \mathbf{h}_j\, z^{-j} \sum_{k=0}^{M} a_k\, z^{M-k} = \sum_{n=1}^{M} \mathbf{N}_n\, z^{M-n} \tag{6.37}$$

Setting $j = n - k$ and comparing equal powers of z yields

$$\sum_{k=0}^{M} a_k\, \mathbf{h}_{n-k} = \begin{cases} \mathbf{N}_n & n = 1, \ldots, M \\ \mathbf{0} & n > M \end{cases} \tag{6.38}$$

or

$$\mathbf{h}_n = -\sum_{k=1}^{M} a_k \, \mathbf{h}_{n-k} \qquad n > M \tag{6.39}$$

This equation is analogous to Equation (6.20) and the multivariable system matrices can therefore be constructed analogous to the scalar system matrices as

$$\begin{aligned} \mathbf{A}_c &= \begin{bmatrix} \mathbf{0} & & & & -a_M \mathbf{I}_q \\ \mathbf{I}_q & \mathbf{0} & & & -a_{M-1}\mathbf{I}_q \\ & \mathbf{I}_q & & & -a_{M-2}\mathbf{I}_q \\ & & \ddots & & \vdots \\ & & & \mathbf{I}_q & -a_1 \mathbf{I}_q \end{bmatrix} \qquad \mathbf{B}_c = \begin{bmatrix} \mathbf{I}_q \\ \mathbf{0} \\ \mathbf{0} \\ \vdots \\ \mathbf{0} \end{bmatrix} \\ \mathbf{C}_c &= \begin{bmatrix} \mathbf{h}_1 & \mathbf{h}_2 & \mathbf{h}_3 & \dots & \mathbf{h}_M \end{bmatrix} \end{aligned} \tag{6.40}$$

where $\mathbf{I}_q$ is the $q \times q$ unit matrix. Another realization is given by

$$\begin{aligned} \mathbf{A}_o &= \begin{bmatrix} \mathbf{0} & \mathbf{I}_p & & & \\ & \mathbf{0} & \mathbf{I}_p & & \\ & & & \ddots & \\ & & & & \mathbf{I}_p \\ -a_M \mathbf{I}_p & -a_{M-1}\mathbf{I}_p & -a_{M-2}\mathbf{I}_p & \dots & -a_1 \mathbf{I}_p \end{bmatrix} \qquad \mathbf{B}_o = \begin{bmatrix} \mathbf{h}_1 \\ \mathbf{h}_2 \\ \mathbf{h}_3 \\ \vdots \\ \mathbf{h}_M \end{bmatrix} \\ \mathbf{C}_o &= \begin{bmatrix} \mathbf{I}_p & \mathbf{0} & \mathbf{0} & \dots & \mathbf{0} \end{bmatrix} \end{aligned} \tag{6.41}$$

Analogously to the scalar case it is easy to show that both systems above have the impulse response $\mathbf{h}_1, \mathbf{h}_2, \dots$ and are therefore realizations of Equation (6.34). The systems constructed in this section are generally not minimal realizations, that is, there are other realizations using a smaller number of states. This topic is taken up again in Subsection 6.1.5.

6.1.3 Similarity transformation

One of the great advantages of the state-variable description is the possibility to transform a system to a different form by simple matrix multiplications. Consider the system

$$\mathbf{x}(n+1) = \mathbf{A}\,\mathbf{x}(n) + \mathbf{B}\,\mathbf{u}(n) \tag{6.42}$$

$$\mathbf{y}(n) = \mathbf{C}\,\mathbf{x}(n) \tag{6.43}$$

Substituting the transformed state variables $\mathbf{x} = \mathbf{T}\,\hat{\mathbf{x}}$ and premultiplying the first equation by $\mathbf{T}^{-1}$ leads to

$$\hat{\mathbf{x}}(n+1) = \mathbf{T}^{-1}\mathbf{A}\,\mathbf{T}\,\hat{\mathbf{x}}(n) + \mathbf{T}^{-1}\mathbf{B}\,\mathbf{u}(n) \tag{6.44}$$

$$\mathbf{y}(n) = \mathbf{C}\,\mathbf{T}\,\hat{\mathbf{x}}(n) \tag{6.45}$$

Therefore the transformed system is

$$\hat{\mathbf{A}} = \mathbf{T}^{-1}\mathbf{A}\,\mathbf{T} \tag{6.46}$$

$$\hat{\mathbf{B}} = \mathbf{T}^{-1}\mathbf{B} \tag{6.47}$$

$$\hat{\mathbf{C}} = \mathbf{C}\,\mathbf{T} \tag{6.48}$$

The transfer function is

$$\begin{aligned} \hat{\mathbf{H}}(z) &= \hat{\mathbf{C}}\,(z\mathbf{I} - \hat{\mathbf{A}})^{-1}\hat{\mathbf{B}} = \mathbf{C}\,\mathbf{T}\,(z\mathbf{I} - \mathbf{T}^{-1}\mathbf{A}\,\mathbf{T})^{-1}\mathbf{T}^{-1}\mathbf{B} \\ &= \mathbf{C}\,(z\mathbf{I} - \mathbf{A})^{-1}\mathbf{B} = \mathbf{H}(z) \end{aligned} \tag{6.49}$$

and the impulse response

$$\begin{aligned}\hat{\mathbf{h}}_n &= \hat{\mathbf{C}}\,\hat{\mathbf{A}}^{n-1}\hat{\mathbf{B}} = \mathbf{C}\,\mathbf{T}\,(\mathbf{T}^{-1}\mathbf{A}\mathbf{T})^{n-1}\mathbf{T}^{-1}\mathbf{B} \\ &= \mathbf{C}\,\mathbf{A}^{n-1}\mathbf{B} = \mathbf{h}_n \end{aligned} \tag{6.50}$$

Thus the similarity transformation only changes the internal description of the system, the external descriptions, transfer function and impulse response, remain unchanged. Generally, a system can be represented by many different system matrices; the observability and the controllability form are only two specific forms.

6.1.4 Stability

The matrix $\mathbf{A}$ in Equation (6.11) has a special form, a companion matrix, that is, the characteristic polynomial is (with $a_0 = 1$)

$$\det(z\mathbf{I} - \mathbf{A}) = \sum_{k=0}^{N} a_k\, z^{N-k} = 0 \tag{6.51}$$

Thus the eigenvalues of $\mathbf{A}$ are the zeroes of Equation (4.19) and hence the poles of the system. Because the similarity transform does not change the eigenvalues, this is true for any system matrix, even if does not have the special companion form. The eigenvalue problem

$$\mathbf{A}\mathbf{T} = \mathbf{T}\boldsymbol{\Lambda} \tag{6.52}$$

yields the poles and the transformation matrix $\mathbf{T}$. If the eigenvalues are distinct then $\mathbf{T}$ transforms the system to diagonal form.

$$\mathbf{x}(n+1) = \boldsymbol{\Lambda}\,\mathbf{x}(n) + \mathbf{T}^{-1}\mathbf{B}\,\mathbf{u}(n) \tag{6.53}$$

$$\mathbf{y} = \mathbf{C}\,\mathbf{T}\,\mathbf{x}(n) \tag{6.54}$$

For a scalar system, the diagonal form is the same as the partial fractions form Equation (4.18) with c_0=0.

$$\mathbf{c}\,\mathbf{T}\,(z\mathbf{I} - \boldsymbol{\Lambda})^{-1}\mathbf{T}^{-1}\mathbf{b} = \sum_{k=1}^{N} \frac{\mathbf{c}\,\mathbf{b}}{z - z_k} \tag{6.55}$$

The product $\mathbf{c}\,\mathbf{b}$ is an outer vector product because $\mathbf{c}$ is a row vector and $\mathbf{b}$ is a column vector.

The stability criterion that all poles have to lie within the unit disk can also be formulated that the spectral radius $\rho(\mathbf{A}) = |\lambda_{max}| < 1$. This also ensures that $h_k = \mathbf{C}\mathbf{A}^{k-1}\mathbf{B} \to 0$ for $k \to \infty$. If the eigenvalues are not distinct, the matrix $\boldsymbol{\Lambda}$ is not necessarily diagonal but the stability criterion remains the same.

6.1.5 Observability, controllability

If we are only interested in the input-output relationship (given by the impulse response or the transfer function), the state variables are of minor interest and in fact we have seen that there exist infinitely many different realizations of the same system, each with different state variables. On the other hand, a system has also properties that depend on its internal realization. Two important properties are the observability and the controllability of a system. In loose terms, a system is observable if all state variables have an influence on the output, and a system is controllable if all state variables are influenced by the input.

Observability. To investigate the influence of the state variables on the output, consider the output due to an initial state $\mathbf{x}(0)$ and no other input:

$$\begin{aligned} \mathbf{y}(0) &= \mathbf{C}\,\mathbf{x}(0) && (6.56)\\ \mathbf{y}(1) &= \mathbf{C}\,\mathbf{x}(1) = \mathbf{C}\,\mathbf{A}\,\mathbf{x}(0) && (6.57)\\ \mathbf{y}(2) &= \mathbf{C}\,\mathbf{x}(2) = \mathbf{C}\,\mathbf{A}^2\mathbf{x}(0) && (6.58)\\ &\vdots \end{aligned}$$

In matrix form, for L outputs, this is

$$\begin{bmatrix} \mathbf{y}(0) \\ \mathbf{y}(1) \\ \vdots \\ \mathbf{y}(L-1) \end{bmatrix} = \begin{bmatrix} \mathbf{C} \\ \mathbf{CA} \\ \vdots \\ \mathbf{CA}^{L-1} \end{bmatrix} \mathbf{x}(0) = \mathbf{W}_o\,\mathbf{x}(0) \tag{6.59}$$

For $L = N$, the matrix $\mathbf{W}_o$ is called the observability matrix, for $L > N$, the extended observability matrix. The system is observable if and only if the $Np \times N$ observability matrix $\mathbf{W}_o$ has full rank N. Full rank of a non-square matrix is equivalent to the Gramian being non-singular. The observability Gramian is the $N \times N$ square matrix

$$\mathbf{Q} = \mathbf{W}_o^{\mathrm{T}}\,\mathbf{W}_o \tag{6.60}$$

The observability Gramian can be used to calculate the initial state variables from the response because

$$\mathbf{W}_o^{\mathrm{T}} \begin{bmatrix} \mathbf{y}(0) \\ \mathbf{y}(1) \\ \vdots \\ \mathbf{y}(L-1) \end{bmatrix} = \mathbf{W}_o^{\mathrm{T}}\mathbf{W}_o\,\mathbf{x}(0) = \mathbf{Q}\,\mathbf{x}(0) \tag{6.61}$$

If the observability Gramian is singular, we cannot solve for $\mathbf{x}(0)$, meaning that there are state variables, which have no influence on the response and thus are not observable. For the system $(\mathbf{A}_o, \mathbf{C}_o)$ given in Equation (6.41), the observability Gramian is $\mathbf{Q} = \mathbf{I}$ and the system is therefore observable.

The observability Gramian is also related to the energy released from a given state $\mathbf{x}(0)$. The energy is

$$\sum_{n=0}^{L-1} |\mathbf{y}(n)|^2 = \mathbf{x}^{\mathrm{T}}(0)\,\mathbf{W}_o^{\mathrm{T}}\,\mathbf{W}_o\,\mathbf{x}(0) = \mathbf{x}^{\mathrm{T}}(0)\,\mathbf{Q}\,\mathbf{x}(0) \tag{6.62}$$

A small $\mathbf{Q}$ indicates that the state variables transfer little energy into the output.

If the system $(\mathbf{A}, \mathbf{B}, \mathbf{C})$ is transformed to $(\hat{\mathbf{A}}, \hat{\mathbf{B}}, \hat{\mathbf{C}})$ by the similarity transform (Equations 6.46–6.48), the observability matrix is transformed as follows

$$\hat{\mathbf{W}}_o = \begin{bmatrix} \mathbf{CT} \\ \mathbf{CTT}^{-1}\mathbf{AT} \\ \vdots \end{bmatrix} = \mathbf{W}_o\mathbf{T} \tag{6.63}$$

and the observability Gramian as

$$\hat{\mathbf{Q}} = \mathbf{T}^{\mathrm{T}}\mathbf{W}_o^{\mathrm{T}}\mathbf{W}_o\mathbf{T} = \mathbf{T}^{\mathrm{T}}\mathbf{Q}\mathbf{T} \tag{6.64}$$

Controllability. For the controllability we investigate the influence of the input to the state variables. As reference we choose the time zero, so that we have to find the state variables $\mathbf{x}(0)$ due to past input $\mathbf{u}(-1), \mathbf{u}(-2), \ldots$. For example, starting with $\mathbf{u}(-3)$ we obtain

$$\begin{aligned}
\mathbf{x}(-3) &= \mathbf{0} & (6.65)\\
\mathbf{x}(-2) &= \mathbf{B}\,\mathbf{u}(-3) & (6.66)\\
\mathbf{x}(-1) &= \mathbf{A}\,\mathbf{x}(-2) + \mathbf{B}\,\mathbf{u}(-2) = \mathbf{AB}\,\mathbf{u}(-3) + \mathbf{B}\,\mathbf{u}(-2) & (6.67)\\
\mathbf{x}(0) &= \mathbf{A}\,\mathbf{x}(-1) + \mathbf{B}\,\mathbf{u}(-1) = \mathbf{A}^2\mathbf{B}\,\mathbf{u}(-3) + \mathbf{AB}\,\mathbf{u}(-2) + \mathbf{B}\,\mathbf{u}(-1) & (6.68)
\end{aligned}$$

In matrix form, for L inputs, we have

$$\mathbf{x}(0) = \begin{bmatrix} \mathbf{B} & \mathbf{AB} & \cdots & \mathbf{A}^{L-1}\mathbf{B} \end{bmatrix} \begin{bmatrix} \mathbf{u}(-1) \\ \mathbf{u}(-2) \\ \vdots \\ \mathbf{u}(-L) \end{bmatrix} = \mathbf{W}_c \begin{bmatrix} \mathbf{u}(-1) \\ \mathbf{u}(-2) \\ \vdots \\ \mathbf{u}(-L) \end{bmatrix} \qquad (6.69)$$

For $L = N$, the matrix $\mathbf{W}_c$ is called the controllability matrix, for $L > N$, the extended controllability matrix. A system is controllable if and only if the $N \times Nq$ controllability matrix has full rank N or, equivalently, if the $N \times N$ controllability Gramian

$$\mathbf{P} = \mathbf{W}_c \mathbf{W}_c^{\mathrm{T}} \qquad (6.70)$$

is non-singular. With the controllability Gramian we can calculate the necessary input that leads to a given state $\mathbf{x}(0)$ as follows. Assume the input vector to be a linear combination of the rows of $\mathbf{W}_c$:

$$\begin{bmatrix} \mathbf{u}(-1) \\ \mathbf{u}(-2) \\ \vdots \\ \mathbf{u}(-L) \end{bmatrix} = \mathbf{W}_c^{\mathrm{T}} \boldsymbol{\eta} \qquad (6.71)$$

Then the initial states are

$$\mathbf{x}(0) = \mathbf{W}_c\, \mathbf{W}_c^{\mathrm{T}} \boldsymbol{\eta} = \mathbf{P}\boldsymbol{\eta} \qquad (6.72)$$

and the input which leads to that states is

$$\begin{bmatrix} \mathbf{u}(-1) \\ \mathbf{u}(-2) \\ \vdots \\ \mathbf{u}(-L) \end{bmatrix} = \mathbf{W}_c^{\mathrm{T}} \boldsymbol{\eta} = \mathbf{W}_c^{\mathrm{T}}\, \mathbf{P}^{-1}\, \mathbf{x}(0) \qquad (6.73)$$

The solution is, of course, not unique, because there are generally more input values $(qL, q \geq 1, L \geq N)$ than there are initial states (N). If the controllability Gramian is singular, then it is not possible to reach a given state by the input, because some of the state variables are not influenced by the input, they are not controllable. It can easily be checked that the system $(\mathbf{A}_c, \mathbf{B}_c)$ given by Equation (6.40) has the controllability Gramian $\mathbf{W}_c = \mathbf{I}$ and is thus controllable.

In terms of energy, the controllability Gramian describes the energy that has to be put into a system to obtain a given state $\mathbf{x}(0)$. The energy is

$$\sum_{n=0}^{L-1} |\mathbf{u}(-n)|^2 = \mathbf{x}^{\mathrm{T}}(0)\, \mathbf{P}^{-1}\, \mathbf{W}_c \mathbf{W}_c^{\mathrm{T}}\, \mathbf{P}^{-1}\, \mathbf{x}(0) = \mathbf{x}^{\mathrm{T}}(0)\, \mathbf{P}^{-1}\, \mathbf{x}(0) \qquad (6.74)$$

A small $\mathbf{P}$ means that the state variables are little affected by the input because a large energy is needed to obtain a certain state. It can be shown [Glo84, p. 1118] that the input values defined by Equation (6.73) yield the minimum energy.

A similarity transformation has the following effect on the controllability matrix

$$\hat{\mathbf{W}}_c = \begin{bmatrix} \mathbf{T}^{-1}\mathbf{B} & \mathbf{T}^{-1}\mathbf{A}\mathbf{T}\mathbf{T}^{-1}\mathbf{B} & \dots \end{bmatrix} = \mathbf{T}^{-1}\mathbf{W}_c \tag{6.75}$$

The controllability Gramian changes as

$$\hat{\mathbf{P}} = \mathbf{T}^{-1}\mathbf{W}_c\mathbf{W}_c^{\mathrm{T}}\mathbf{T}^{-\mathrm{T}} = \mathbf{T}^{-1}\mathbf{P}\mathbf{T}^{-\mathrm{T}} \tag{6.76}$$

Minimal realization. One of the important theorems of linear system theory [Kai80, p. 127] states that a realization is minimal if and only if the system is observable and controllable. This fact is derived in the following. An observable and controllable system has full rank observability and controllability matrices. On the other hand, a system is minimal if its Hankel matrix is non-singular. The Hankel matrix can be written as (Equation 6.97, Subsection 6.1.7) the product of the observability and the controllability matrices:

$$\mathbf{\Gamma} = \mathbf{W}_o\mathbf{W}_c \tag{6.77}$$

Thus if both the observability and the controllability matrices have full rank then the Hankel matrix also has full rank and the system is minimal.

Example. The following small example shows two realizations, one that is not observable and one that is not controllable. Hence both realizations are not minimal. Consider the transfer function

$$\mathbf{H}(z) = \begin{bmatrix} \dfrac{1}{z-1/2} & \dfrac{2}{z-1/2} \\ \dfrac{2}{z-1/2} & \dfrac{4}{z-1/2} \end{bmatrix} \tag{6.78}$$

The realization Equation (6.40) yields

$$\mathbf{A}_c = \begin{bmatrix} 1/2 & \\ & 1/2 \end{bmatrix} \quad \mathbf{B}_c = \begin{bmatrix} 1 & \\ & 1 \end{bmatrix} \quad \mathbf{C}_c = \begin{bmatrix} 1 & 2 \\ 2 & 4 \end{bmatrix} \tag{6.79}$$

The controllability and observability matrices are

$$\mathbf{W}_c = \begin{bmatrix} 1 & \\ & 1 \end{bmatrix} \quad \mathbf{W}_o = \begin{bmatrix} 1 & 2 \\ 2 & 4 \end{bmatrix} \tag{6.80}$$

Clearly $\mathbf{W}_c$ has rank 2 and $\mathbf{W}_o$ rank 1. The system is thus controllable but not observable. This can be made more clear by transforming the system with the transformation matrix

$$\mathbf{T} = \begin{bmatrix} 1 & -2 \\ 0 & 1 \end{bmatrix} \quad \mathbf{T}^{-1} = \begin{bmatrix} 1 & 2 \\ 0 & 1 \end{bmatrix} \tag{6.81}$$

into the form

$$\hat{\mathbf{A}}_c = \begin{bmatrix} 1/2 & \\ & 1/2 \end{bmatrix} \quad \hat{\mathbf{B}}_c = \begin{bmatrix} 1 & 2 \\ 0 & 1 \end{bmatrix} \quad \hat{\mathbf{C}}_c = \begin{bmatrix} 1 & 0 \\ 2 & 0 \end{bmatrix} \tag{6.82}$$

The second row of $\hat{\mathbf{C}}_c$ is zero and thus the second state is not observed in the output.

On the other hand, the realization Equation (6.41) is given by $\mathbf{A}_o = \mathbf{A}_c$, $\mathbf{B}_o = \mathbf{C}_c$ and $\mathbf{C}_o = \mathbf{B}_c$. The observability and the controllability matrices change their role and the system is thus observable but not controllable.

Both systems are not minimal. A minimal realization is obviously

$$A = 1/2 \quad \mathbf{B} = \begin{bmatrix} 1 & 2 \end{bmatrix} \quad \mathbf{C} = \begin{bmatrix} 1 \\ 2 \end{bmatrix} \tag{6.83}$$

6.1.6 Lyapunov equations

The Gramians introduced in the last section, clearly depend on the lentgh of the interval $[0, L-1]$ over which the they are taken. In addition to the properties derived so far we state some relations which only hold for infinite intervals ($L \to \infty$). We indicate the infinite time interval for the Gramians by a subscript whenever it is essential. For relations that hold for finite as well as for infinite time intervals the subscript is neglected.

The observability Gramian Equation (6.60) for $L \to \infty$ is written as

$$\mathbf{Q}_\infty = \sum_{k=0}^{\infty} (\mathbf{A}^{\mathrm{T}})^k \mathbf{C}^{\mathrm{T}} \mathbf{C} \mathbf{A}^k \tag{6.84}$$

The series converges for stable systems and satisfies the discrete-time Lyapunov equation

$$\mathbf{A}^{\mathrm{T}} \mathbf{Q}_\infty \mathbf{A} - \mathbf{Q}_\infty + \mathbf{C}^{\mathrm{T}} \mathbf{C} = \mathbf{0} \tag{6.85}$$

This equation can be used to determine the observability Gramian from the system matrices. The easiest (although not the numerically most efficient) way is to bring the system into diagonal form by a similarity transform. The Lyapunov equation for the transformed system is

$$\mathbf{\Lambda} \hat{\mathbf{Q}}_\infty \mathbf{\Lambda} - \hat{\mathbf{Q}}_\infty + \hat{\mathbf{C}}^{\mathrm{T}} \hat{\mathbf{C}} = \mathbf{0} \tag{6.86}$$

Because $\mathbf{\Lambda}$ is diagonal, each entry of $\hat{\mathbf{Q}}_\infty$ can be determined individually by

$$(\hat{\mathbf{Q}}_\infty)_{ij} = \frac{(\hat{\mathbf{C}}^{\mathrm{T}} \hat{\mathbf{C}})_{ij}}{1 - \lambda_i \lambda_j} \tag{6.87}$$

Transforming the system back to the original state variables, the observability Gramian is

$$\mathbf{Q}_\infty = \mathbf{T}^{-\mathrm{T}} \hat{\mathbf{Q}}_\infty \mathbf{T}^{-1} \tag{6.88}$$

Similarly, for the controllability Gramian the series

$$\mathbf{P}_\infty = \sum_{k=0}^{\infty} \mathbf{A}^k \mathbf{B} \mathbf{B}^{\mathrm{T}} (\mathbf{A}^{\mathrm{T}})^k \tag{6.89}$$

converges for stable systems and satisfies the discrete-time Lyapunov equation

$$\mathbf{A} \mathbf{P}_\infty \mathbf{A}^{\mathrm{T}} - \mathbf{P}_\infty + \mathbf{B} \mathbf{B}^{\mathrm{T}} = \mathbf{0} \tag{6.90}$$

This equation can be solved in the diagonal form by

$$(\hat{\mathbf{P}}_\infty)_{ij} = \frac{(\hat{\mathbf{B}} \hat{\mathbf{B}}^{\mathrm{T}})_{ij}}{1 - \lambda_i \lambda_j} \tag{6.91}$$

After the back-transformation the controllability Gramian is

$$\mathbf{P}_\infty = \mathbf{T} \hat{\mathbf{P}}_\infty \mathbf{T}^{\mathrm{T}} \tag{6.92}$$

The Lyapunov equations are useful in the model reduction problem for determining the Gramians from the system matrices without calculating the series. They are also employed in various theoretical derivations.

Stability has been assumed for the series in Equations (6.84) and (6.89) to converge. On the other hand, the Lyapunov equations can be used as a stability criterion not involving the eigenvalues. The Lypunov theorem says that $\mathbf{A}$ is a stability matrix, that is, $\rho(\mathbf{A}) < 1$, if for any positive-definite (symmetric) matrix $\mathbf{W}$ the Lyapunov equation

$$\mathbf{A}^{\mathrm{T}}\mathbf{P}_{\infty}\mathbf{A} - \mathbf{P}_{\infty} + \mathbf{W} = \mathbf{0} \tag{6.93}$$

has a positive-definite solution $\mathbf{P}_{\infty}$. The proof can be found in the literature [Kai80, p. 179] and is omitted here. A slightly different version is the following (see also [MR76]). A system $(\mathbf{A}, \mathbf{B})$ is stable if it is controllable and if for a positive semi-definite matrix $\mathbf{W}$ the Lyapunov equation Equation (6.93) has a positive-definite solution $\mathbf{P}_{\infty}$. This version is more useful because the matrix $\mathbf{W} = \mathbf{B}\mathbf{B}^{\mathrm{T}}$ is generally not positive definite but only positive semi-definite (because $\mathbf{B}$ is not square). The Lyapunov equation is not used directly to verify stability of a specific system but can be used in theoretical stability proofs.

6.1.7 Hankel matrix

The Hankel matrix of a scalar system has been defined in Equation (5.23) as

$$\mathbf{\Gamma} = \begin{bmatrix} h_1 & h_2 & \dots & h_L \\ h_2 & h_3 & \dots & h_{L+1} \\ \vdots & \vdots & & \vdots \\ h_L & h_{L+1} & \dots & h_{2L-1} \end{bmatrix} \tag{6.94}$$

For the multivariable case, we define analogously the block Hankel matrix as

$$\mathbf{\Gamma} = \begin{bmatrix} \mathbf{h}_1 & \mathbf{h}_2 & \dots & \mathbf{h}_L \\ \mathbf{h}_2 & \mathbf{h}_3 & \dots & \mathbf{h}_{L+1} \\ \vdots & \vdots & & \vdots \\ \mathbf{h}_L & \mathbf{h}_{L+1} & \dots & \mathbf{h}_{2L-1} \end{bmatrix} \tag{6.95}$$

The impulse response matrices $\mathbf{h}_k$ can be expressed in terms of the system matrices. Then the Hankel matrix is

$$\mathbf{\Gamma} = \begin{bmatrix} \mathbf{CB} & \mathbf{CAB} & \dots & \mathbf{CA}^{L-1}\mathbf{B} \\ \mathbf{CAB} & \mathbf{CA}^2\mathbf{B} & \dots & \mathbf{CA}^{L}\mathbf{B} \\ \vdots & \vdots & & \vdots \\ \mathbf{CA}^{L-1}\mathbf{B} & \mathbf{CA}^{L}\mathbf{B} & \dots & \mathbf{CA}^{2L-2}\mathbf{B} \end{bmatrix} \tag{6.96}$$

which can also be written in factored form as

$$\mathbf{\Gamma} = \begin{bmatrix} \mathbf{C} \\ \mathbf{CA} \\ \vdots \\ \mathbf{CA}^{L-1} \end{bmatrix} \begin{bmatrix} \mathbf{B} & \mathbf{AB} & \dots & \mathbf{A}^{L-1}\mathbf{B} \end{bmatrix} = \mathbf{W}_o\mathbf{W}_c \tag{6.97}$$

In view of Equations (6.59) and (6.69), this shows the important fact that the Hankel matrix relates the future output to the past input

$$\begin{bmatrix} \mathbf{y}(0) \\ \mathbf{y}(1) \\ \vdots \\ \mathbf{y}(L-1) \end{bmatrix} = \mathbf{W}_o\,\mathbf{x}(0) = \mathbf{W}_o\mathbf{W}_c \begin{bmatrix} \mathbf{u}(-1) \\ \mathbf{u}(-2) \\ \vdots \\ \mathbf{u}(-L) \end{bmatrix} = \mathbf{\Gamma} \begin{bmatrix} \mathbf{u}(-1) \\ \mathbf{u}(-2) \\ \vdots \\ \mathbf{u}(-L) \end{bmatrix} \tag{6.98}$$

The Hankel matrix is invariant under a similarity transformation. This can be seen by considering

$$\hat{\mathbf{\Gamma}} = \mathbf{W}_o\mathbf{T}\mathbf{T}^{-1}\mathbf{W}_c = \mathbf{W}_o\mathbf{W}_c = \mathbf{\Gamma} \tag{6.99}$$

or more directly by recalling that the entries of the Hankel matrix are the Markov parameters which are independent of the actual realization.

Singular values. Because of the relation Equation (6.97), the singular values of the Hankel matrix can also be calculated from the Gramians. This is shown in this paragraph. The product of the observability and controllability Gramian depends on a similarity transformation as follows.

$$\hat{\mathbf{P}}\hat{\mathbf{Q}} = \mathbf{T}^{-1}\mathbf{P}\mathbf{T}^{-\mathrm{T}}\mathbf{T}^{\mathrm{T}}\mathbf{Q}\mathbf{T} = \mathbf{T}^{-1}\mathbf{P}\mathbf{Q}\mathbf{T} \tag{6.100}$$

Although the matrix $\mathbf{PQ}$ changes with a similarity transformation its eigenvalues are invariant because the transformation $\mathbf{T}^{-1}\mathbf{PQT}$ does not change the eigenvalues of $\mathbf{PQ}$. These eigenvalues equal the non-zero singular values of the Hankel matrix because

$$\lambda_k(\mathbf{PQ}) = \lambda_k(\mathbf{W}_c\mathbf{W}_c^{\mathrm{T}}\mathbf{W}_o^{\mathrm{T}}\mathbf{W}_o) = \lambda_k(\mathbf{W}_c^{\mathrm{T}}\mathbf{W}_o^{\mathrm{T}}\mathbf{W}_o\mathbf{W}_c) = \lambda_k(\mathbf{\Gamma}^{\mathrm{T}}\mathbf{\Gamma}) = \sigma_k^2 \tag{6.101}$$

In the above equations the fact was used that changing the order of the matrices does not change the eigenvalues except for additional zero eigenvalues if the matrices do not have equal dimensions. In fact, there are L singular values of which only $N \leq L$ are non-zero.

6.1.8 Balanced systems

The similarity transform can be used to bring a system into a special form, the balanced form suggested by Moore [Moo78] and developed further by Laub [Lau80], and Pernebo and Silverman [PS82]. A system is balanced if the Gramians $\bar{\mathbf{P}}$ and $\bar{\mathbf{Q}}$ are equal and diagonal where the bar indicates the balanced form. Balanced systems play an important role in model reduction problems. Although the model reduction problem is different from our realization problem, there are some common aspects. In fact, in the next chapter we will use the balancing transformation to derive a new realization method.

The balancing transformation is derived following [Lau80] and [Glo84, p. 1129]. In these references, the balancing is performed over the whole time axis, but the same methodology can be used for a finite time interval. The results concerning stability, however, are only valid for the infinite interval. Because it is somewhat easier, we first consider the balancing over a finite interval. Later we present the balancing over the whole time axis and a result concerning stability.

Partially balanced systems. Balancing over a finite time interval is called partial balancing. The starting point are the Gramians $\mathbf{P}$ and $\mathbf{Q}$. They are given in Equations (6.60) and (6.70) and are repeated here for convenience:

$$\mathbf{P} = \mathbf{W}_c\mathbf{W}_c^{\mathrm{T}} \qquad \mathbf{Q} = \mathbf{W}_o^{\mathrm{T}}\mathbf{W}_o \tag{6.102}$$

Further, the Hankel matrix can be expressed as $\mathbf{\Gamma} = \mathbf{W}_o\mathbf{W}_c$. Using the singular value decomposition $\mathbf{\Gamma} = \mathbf{U}\mathbf{\Sigma}\mathbf{V}^{\mathrm{T}}$ we can write

$$\mathbf{W}_c^{\mathrm{T}}\mathbf{W}_o^{\mathrm{T}}\mathbf{W}_o\mathbf{W}_c = \mathbf{\Gamma}^{\mathrm{T}}\mathbf{\Gamma} = \mathbf{V}\mathbf{\Sigma}\mathbf{U}^{\mathrm{T}}\mathbf{U}\mathbf{\Sigma}\mathbf{V}^{\mathrm{T}} = \mathbf{V}\mathbf{\Sigma}^2\mathbf{V}^{\mathrm{T}} \tag{6.103}$$

The balancing transformation is given by the transformation matrix

$$\mathbf{T} = \mathbf{W}_c\mathbf{V}\mathbf{\Sigma}^{-1/2} \tag{6.104}$$

and its inverse

$$\mathbf{T}^{-1} = \mathbf{\Sigma}^{1/2}\mathbf{V}^{\mathrm{T}}\mathbf{W}_c^{-1} \tag{6.105}$$

Here it has been assumed that the system is minimal, so that the inverses of the matrices $\mathbf{\Sigma}$ and $\mathbf{W}_c$ do exist. The transformed Gramians are

$$\bar{\mathbf{P}} = \mathbf{T}^{-1}\mathbf{P}\mathbf{T}^{-\mathrm{T}} = \mathbf{\Sigma}^{1/2}\underbrace{\mathbf{V}^{\mathrm{T}}\mathbf{W}_c^{-1}(\mathbf{W}_c\mathbf{W}_c^{\mathrm{T}})\mathbf{W}_c^{-\mathrm{T}}\mathbf{V}}_{\mathbf{I}}\mathbf{\Sigma}^{1/2} = \mathbf{\Sigma} \tag{6.106}$$

and, considering Equation (6.103),

$$\bar{\mathbf{Q}} = \mathbf{T}^{\mathrm{T}}\mathbf{Q}\mathbf{T} = \mathbf{\Sigma}^{-1/2}\underbrace{\mathbf{V}^{\mathrm{T}}\mathbf{W}_c^{\mathrm{T}}\mathbf{W}_o^{\mathrm{T}}\mathbf{W}_o\mathbf{W}_c\mathbf{V}}_{\mathbf{\Sigma}^2}\mathbf{\Sigma}^{-1/2} = \mathbf{\Sigma} \tag{6.107}$$

Clearly, they are equal and diagonal and the system is therefore balanced in the interval $[0, L-1]$.

Fully balanced systems. The balancing over the whole time axis is called full balancing. Again, we start with the Gramians $\mathbf{P}_\infty$ and $\mathbf{Q}_\infty$. For the infinite time interval they can be found by solving the Lyapunov equations Equation (6.85 and 6.90). While other matrices such as the Hankel matrix $\mathbf{\Gamma}$ and the observability and controllability matrices $\mathbf{W}_o$ and $\mathbf{W}_c$ have infinite dimensions, the Gramians have finite size for a finite-dimensional system (size of matrix $\mathbf{A}$).

To find the transformation matrix, first factor

$$\mathbf{P}_\infty = \mathbf{W}\mathbf{W}^{\mathrm{T}} \tag{6.108}$$

The factoring can be done by solving the eigenvalue problem or more efficiently by a Cholesky decomposition. Then solve the eigenvalue problem

$$(\mathbf{W}^{\mathrm{T}}\mathbf{Q}_\infty\mathbf{W})\,\mathbf{V} = \mathbf{V}\,\mathbf{\Sigma}^2 \tag{6.109}$$

with $\mathbf{V}^{\mathrm{T}}\,\mathbf{V} = \mathbf{I}$. The balancing transformation is now given by the transformation matrix

$$\mathbf{T} = \mathbf{W}\mathbf{V}\mathbf{\Sigma}^{-1/2} \tag{6.110}$$

The transformed Gramians are

$$\bar{\mathbf{P}}_\infty = \mathbf{T}^{-1}\mathbf{P}_\infty\mathbf{T}^{-\mathrm{T}} = \mathbf{\Sigma}^{1/2}\underbrace{\mathbf{V}^{\mathrm{T}}\mathbf{W}^{-1}(\mathbf{W}\mathbf{W}^{\mathrm{T}})\mathbf{W}^{-\mathrm{T}}\mathbf{V}}_{\mathbf{I}}\mathbf{\Sigma}^{1/2} = \mathbf{\Sigma} \tag{6.111}$$

and with Equation (6.109)

$$\bar{\mathbf{Q}}_\infty = \mathbf{T}^{\mathrm{T}}\mathbf{Q}_\infty\mathbf{T} = \mathbf{\Sigma}^{-1/2}\underbrace{\mathbf{V}^{\mathrm{T}}\mathbf{W}^{\mathrm{T}}\mathbf{Q}\mathbf{W}\mathbf{V}}_{\mathbf{\Sigma}^2}\mathbf{\Sigma}^{-1/2} = \mathbf{\Sigma} \tag{6.112}$$

Because they are equal and diagonal, the system is indeed balanced.

Although these derivations look very similar to the ones for the balancing over a finite time interval, there is an essential difference between the two. Because here the observability and controllability matrices have infinite dimensions, the finite matrix $\mathbf{W}$ is not the controllability matrix and and $\mathbf{V}$ is not the matrix of singular vectors. The matrix $\mathbf{\Sigma}$, however, contains the non-zero singular values of the infinite Hankel matrix because (Equation 6.101) $\lambda_k(\mathbf{P}_\infty\mathbf{Q}_\infty) = \sigma_k^2$.

Reduced models. Balanced systems are useful for model reduction. We have seen earlier that the Gramians $\mathbf{P}$ and $\mathbf{Q}$ are related to the input and output energy. $\mathbf{P}^{-1}$ is a measure for the input energy that is necessary to reach a state. Thus small values in $\mathbf{P}$ are related to state variables that are hard to influence by the input. $\mathbf{Q}$ is a measure for the output energy released from a state. Small values in $\mathbf{Q}$ are related to state variables that have little influence on the output. In a balanced realization, state variables corresponding to small singular values are both hard to influence by the input and have little effect on the output. It is intuitively clear that neglecting these state variables has little effect on the input-output relationship of the system. Therefore a subsystem corresponding to the large singular values is a good approximation of the full system.

An advantage of this model reduction method is that the reduced model is stable if the original system is stable and if the balancing is over the whole time axis. Unfortunately, for the realization problem we can only consider a finite time interval and the stability is not guaranteed. Nevertheless, for a balancing over a long enough interval, we can hope to get a stable system.

For the proof of stability of the reduced model we note that for a fully balanced system $\mathbf{\Sigma} = \bar{\mathbf{P}}_\infty$ satisfies the Lyapunov equation. Written in block form this is

$$\begin{bmatrix} \bar{\mathbf{A}}_{11} & \bar{\mathbf{A}}_{12} \\ \bar{\mathbf{A}}_{21} & \bar{\mathbf{A}}_{22} \end{bmatrix} \begin{bmatrix} \mathbf{\Sigma}_1 & \\ & \mathbf{\Sigma}_2 \end{bmatrix} \begin{bmatrix} \bar{\mathbf{A}}_{11}^T & \bar{\mathbf{A}}_{21}^T \\ \bar{\mathbf{A}}_{12}^T & \bar{\mathbf{A}}_{22}^T \end{bmatrix} - \begin{bmatrix} \mathbf{\Sigma}_1 & \\ & \mathbf{\Sigma}_2 \end{bmatrix} = - \begin{bmatrix} \bar{\mathbf{B}}_1 \\ \bar{\mathbf{B}}_2 \end{bmatrix} \begin{bmatrix} \bar{\mathbf{B}}_1^T & \bar{\mathbf{B}}_2^T \end{bmatrix} \tag{6.113}$$

The upper left block is

$$\bar{\mathbf{A}}_{11}\,\mathbf{\Sigma}_1\,\bar{\mathbf{A}}_{11}^T + \bar{\mathbf{A}}_{12}\,\mathbf{\Sigma}_2\,\bar{\mathbf{A}}_{12}^T - \mathbf{\Sigma}_1 = -\bar{\mathbf{B}}_1\,\bar{\mathbf{B}}_1^T \tag{6.114}$$

The matrix $\mathbf{\Sigma}_1$ does not satisfy the Lyapunov equation because of the additional term $\bar{\mathbf{A}}_{12}\,\mathbf{\Sigma}_2\,\bar{\mathbf{A}}_{12}^T$. The subsystem of a balanced discrete-time system is thus not balanced. However, stability is guaranteed because $\bar{\mathbf{A}}_{12}\,\mathbf{\Sigma}_1\,\bar{\mathbf{A}}_{12}^T$ is positive semi-definite (see Equation 6.93).

6.1.9 Norms

Although not used directly, the norms given in Chapter 4 are generalized here for matrices. The Euclidean norm, or mean-square norm is defined by [KL81a]

$$\|\mathbf{E}(z)\|_2 = \rho\left(\frac{1}{2\pi}\int_0^{2\pi} \mathbf{E}^*(e^{i\Omega})\,\mathbf{E}(e^{i\Omega})\,d\Omega\right)^{1/2} = \rho\left(\sum_{k=-\infty}^{\infty} \mathbf{e}_k^*\,\mathbf{e}_k\right)^{1/2} = \|\mathbf{e}(z)\|_2 \tag{6.115}$$

where $()^*$ denotes the conjugate transpose and ρ the spectral radius. The spectral radius of a matrix is defined as the modulus of the largest eigenvalue. Here the eigenvalues are real and positive, because the matrices are real and symmetric. The maximum norm or Chebyshev norm is

$$\|\mathbf{E}(z)\|_\infty = \max_\Omega \rho\Big(\mathbf{E}^*(e^{i\Omega})\,\mathbf{E}(e^{i\Omega})\Big)^{1/2} = \max_\Omega \bar{\sigma}\Big(\mathbf{E}(e^{i\Omega})\Big) \tag{6.116}$$

where $\bar{\sigma}$ denotes the largest singular value.

6.2 Continuous-time systems

As observed earlier, there exists a close analogy between discrete-time and continuous-time systems. One formulation is obtained from the other merely by interchanging z and s and by using the Markov parameters instead of the impulse response. However, there are some differences in the energy related definitions because a formal analogy would not imply a physical analogy. Most of the features described in this section are not used directly later because all numerical approximations are performed in the discrete-time domain. However, when using discrete-time properties of matrices it seems to be useful to have some understanding of the analogous continuous-time counterparts. Most of the results presented in this section are described in [Glo84, LK82].

6.2.1 Multivariable systems

The general state-variable description is

$$\dot{\mathbf{x}}(t) = \mathbf{A}\,\mathbf{x}(t) + \mathbf{B}\,\mathbf{u}(t) \tag{6.117}$$

$$\mathbf{y}(t) = \mathbf{C}\,\mathbf{x}(t) \tag{6.118}$$

where the matrices **A**, **B** and **C** are defined as in the discrete-time case, Equation (6.40) or Equation (6.41), except that the coefficients a_k are now the coefficients of the continuous-time system and $\mathbf{h}_k$ are Markov parameters.

The transfer function is

$$\mathbf{H}(s) = \mathbf{C}\,(s\mathbf{I} - \mathbf{A})^{-1}\mathbf{B} \tag{6.119}$$

and the Markov parameters are

$$\mathbf{h}_n = \mathbf{C}\,\mathbf{x}(n) = \mathbf{C}\mathbf{A}^{n-1}\mathbf{B} \tag{6.120}$$

For continuous-time systems, the impulse response is different from the Markov parameters. It can be written as

$$\mathbf{h}(t) = \mathbf{C}\,e^{\mathbf{A}t}\,\mathbf{B} \tag{6.121}$$

The matrix function $e^{\mathbf{A}t}$ is defined analogously to the scalar exponential function as

$$e^{\mathbf{A}t} = \mathbf{I} + \mathbf{A}t + \frac{\mathbf{A}^2t^2}{2!} + \cdots \tag{6.122}$$

The equations for the similarity transform are the same as in the discrete-time case (Equations 6.46–6.48). The eigenvalues of **A** are the poles of the system

$$\det(s\mathbf{I} - \mathbf{A}) = \sum_{k=0}^{N} a_k\, s^{N-k} = 0 \tag{6.123}$$

and the system is stable if all poles have a negative real part. The diagonal form corresponds to the partial fraction expansion.

6.2.2 Observability and controllability

Observability and controllability matrices. The (extended) observability matrix $\mathbf{W}_o$ and the (extended) controllability matrix $\mathbf{W}_c$ are defined as before, Equations (6.59) and (6.69). As in the discrete-time case they give the influence of the past input on the state variables and the influence of the state variables on the output, but in slightly different form. Replacing discrete values by differentials yields

$$\begin{bmatrix} \mathbf{y} \\ \dot{\mathbf{y}} \\ \vdots \\ \mathbf{y}^{(L-1)} \end{bmatrix} = \begin{bmatrix} \mathbf{C} \\ \mathbf{CA} \\ \vdots \\ \mathbf{CA}^{L-1} \end{bmatrix} \mathbf{x}(0) = \mathbf{W}_o\,\mathbf{x}(0) \tag{6.124}$$

and

$$\mathbf{x}(0) = \begin{bmatrix} \mathbf{B} & \mathbf{AB} & \cdots & \mathbf{A}^{L-1}\mathbf{B} \end{bmatrix} \begin{bmatrix} \mathbf{u} \\ \dot{\mathbf{u}} \\ \vdots \\ \mathbf{u}^{(L)} \end{bmatrix} = \mathbf{W}_c \begin{bmatrix} \mathbf{u} \\ \dot{\mathbf{u}} \\ \vdots \\ \mathbf{u}^{(L)} \end{bmatrix} \tag{6.125}$$

Gramians. The Gramians are defined differently than for the discrete-time case. The energy necessary to obtain a given state $\mathbf{x}(0)$ is

$$\int_0^T |\mathbf{u}(-t)|^2\, dt = \mathbf{x}^{\mathrm{T}}(0)\, \mathbf{P}^{-1}\, \mathbf{x}(0) \tag{6.126}$$

where

$$\mathbf{P} = \int_0^T e^{\mathbf{A}t}\mathbf{B}\mathbf{B}^{\mathrm{T}}e^{\mathbf{A}^{\mathrm{T}}t}\, dt \tag{6.127}$$

The energy released from a given state $\mathbf{x}(0)$ is

$$\int_0^T |\mathbf{y}(t)|^2\, dt = \mathbf{x}^{\mathrm{T}}(0)\, \mathbf{Q}\, \mathbf{x}(0) \tag{6.128}$$

where

$$\mathbf{Q} = \int_0^T e^{\mathbf{A}^{\mathrm{T}}t}\mathbf{C}^{\mathrm{T}}\mathbf{C}e^{\mathbf{A}t}\, dt \tag{6.129}$$

To see the effect of a similarity transformation on the Gramians we first consider

$$e^{\mathbf{T}^{-1}\mathbf{A}\mathbf{T}\,t} = \mathbf{T}^{-1}(\mathbf{I} + \mathbf{A}t + \frac{\mathbf{A}^2t^2}{2!} + \cdots)\mathbf{T} = \mathbf{T}^{-1}e^{\mathbf{A}t}\mathbf{T} \tag{6.130}$$

Then clearly

$$\hat{\mathbf{P}} = \int_0^{\mathrm{T}} \mathbf{T}^{-1}\, e^{\mathbf{A}t}\, \mathbf{T}\, \mathbf{T}^{-1}\mathbf{B}\mathbf{B}^{\mathrm{T}}\mathbf{T}^{-T}\, \mathbf{T}^{\mathrm{T}}\, e^{\mathbf{A}^{\mathrm{T}}t}\, \mathbf{T}^{-\mathrm{T}}\, dt = \mathbf{T}^{-1}\mathbf{P}\mathbf{T}^{-\mathrm{T}} \tag{6.131}$$

and similarly

$$\hat{\mathbf{Q}} = \int_0^T \mathbf{T}^{\mathrm{T}}\, e^{\mathbf{A}^{\mathrm{T}}t}\, \mathbf{T}^{-\mathrm{T}}\, \mathbf{T}^{\mathrm{T}}\mathbf{C}^{\mathrm{T}}\, \mathbf{C}\mathbf{T}\, \mathbf{T}^{-1}e^{\mathbf{A}t}\, \mathbf{T}\, dt = \mathbf{T}^{\mathrm{T}}\mathbf{Q}\mathbf{T} \tag{6.132}$$

These are the same equations as in the discrete-time case (Equations 6.76 and 6.64). As for the discrete-time case the eigenvalues of the product $\mathbf{PQ}$ are invariant under a similarity transformation.

6.2.3 Lyapunov equations

Because the Gramians are defined differently for the continuous-time case, also the Lyapunov equations take a differnent form. For $T \to \infty$, $\mathbf{P}_\infty$ satisfies the continuous-time Lyapunov equation

$$\mathbf{A}\mathbf{P}_\infty + \mathbf{P}_\infty\mathbf{A}^{\mathrm{T}} = -\mathbf{B}\mathbf{B}^{\mathrm{T}} \tag{6.133}$$

because

$$\begin{aligned} \mathbf{A}\mathbf{P}_\infty + \mathbf{P}_\infty\mathbf{A}^{\mathrm{T}} &= \int_0^\infty \mathbf{A}e^{\mathbf{A}t}\mathbf{B}\mathbf{B}^{\mathrm{T}}e^{\mathbf{A}^{\mathrm{T}}t} + e^{\mathbf{A}t}\mathbf{B}\mathbf{B}^{\mathrm{T}}e^{\mathbf{A}^{\mathrm{T}}t}\mathbf{A}^{\mathrm{T}}\, dt \\ &= e^{\mathbf{A}t}\mathbf{B}\mathbf{B}^{\mathrm{T}}e^{\mathbf{A}^{\mathrm{T}}t}\Big|_0^\infty = -\mathbf{B}\mathbf{B}^{\mathrm{T}} \end{aligned} \tag{6.134}$$

To solve the Lyapunov equation we proceed as in the discrete-time case by first writing the equation for the diagonal form. The individual entries of the controllability Gramian are

$$(\hat{\mathbf{P}}_\infty)_{ij} = -\frac{(\hat{\mathbf{B}}\hat{\mathbf{B}}^{\mathrm{T}})_{ij}}{\lambda_i + \lambda_j} \tag{6.135}$$

and the back-transformation leads to

$$\mathbf{P}_\infty = \mathbf{T}\hat{\mathbf{P}}_\infty\mathbf{T}^{\mathrm{T}} \tag{6.136}$$

Similarly, $\mathbf{Q}_\infty$ satisfies the continuous-time Lyapunov equation

$$\mathbf{A}^{\mathrm{T}}\mathbf{Q}_\infty + \mathbf{Q}_\infty\mathbf{A} = -\mathbf{C}^{\mathrm{T}}\mathbf{C} \tag{6.137}$$

The equation can be solved in the diagonal form by

$$(\hat{\mathbf{Q}}_\infty)_{ij} = -\frac{(\hat{\mathbf{C}}^{\mathrm{T}}\hat{\mathbf{C}})_{ij}}{\lambda_i + \lambda_j} \tag{6.138}$$

and transformed back to

$$\mathbf{Q}_\infty = \mathbf{T}^{-\mathrm{T}}\hat{\mathbf{Q}}_\infty\mathbf{T}^{-1} \tag{6.139}$$

6.2.4 Hankel integral operator

The continuous counterpart of the discrete-time Hankel matrix is the Hankel integral operator. Although, subsequently, it is not used directly, we give some explanations to complete the picture. More details can be found in [Glo84, LK82]. A strong mathematical derivation of rational approximation using directly the Hankel integral operator is given in[GCP88].

Analogously to the scalar case (Subsection 4.3.5), the Hankel integral operator is defined as

$$(\mathbf{\Gamma x})(t) = \int_0^\infty \mathbf{h}(t+\tau)\,\mathbf{x}(\tau)\,d\tau \tag{6.140}$$

The singular value decomposition for the continuous-time Hankel integral operator is

$$(\mathbf{\Gamma x})(t) = \sum_{i=1}^{N} \sigma_i \left(\int_0^\infty \mathbf{v}_i(\tau)\,\mathbf{x}(\tau)\,d\tau\right)\mathbf{u}_i(t) \tag{6.141}$$

where the number of singular values is N. The singular vectors $\mathbf{u}(t)$ and $\mathbf{v}(t)$ are orthonormal, that is

$$\int_0^\infty \mathbf{u}_i(t)\,\mathbf{u}_j(t)\,dt = \begin{cases} 1 & i = j \\ 0 & i \neq j \end{cases} \tag{6.142}$$

and likewise for $\mathbf{v}(t)$. A scalar example has been given in Subsection 4.3.5.

For multivariable systems, the derivation of the singular values and singular vectors is involved. However, for a fully balanced system $(\bar{\mathbf{A}}, \bar{\mathbf{B}}, \bar{\mathbf{C}})$ the following result applies: The continuous-time impulse response is given by $\mathbf{h}(t) = \bar{\mathbf{C}}e^{\bar{\mathbf{A}}t}\bar{\mathbf{B}}$ and the Hankel integral operator by

$$(\mathbf{\Gamma x})(t) = \int_0^\infty \bar{\mathbf{C}}e^{\bar{\mathbf{A}}(t+\tau)}\bar{\mathbf{B}}\mathbf{x}(\tau)\,d\tau = \bar{\mathbf{C}}e^{\bar{\mathbf{A}}(t)}\int_0^\infty e^{\bar{\mathbf{A}}(\tau)}\bar{\mathbf{B}}\mathbf{x}(\tau)\,d\tau \tag{6.143}$$

The singular values are given by

$$\mathbf{\Sigma} = \mathbf{diag}(\sigma_1, \sigma_2, \ldots, \sigma_N) = \bar{\mathbf{P}}_\infty = \bar{\mathbf{Q}}_\infty \tag{6.144}$$

and the singular vectors by

$$\mathbf{U} = \left[\begin{array}{ccc} \mathbf{u}_1(t) & \mathbf{u}_2(t) \ldots \mathbf{u}_N(t) \end{array}\right] = \bar{\mathbf{C}} e^{\bar{\mathbf{A}}t} \mathbf{\Sigma}^{-1/2} \tag{6.145}$$

$$\mathbf{V} = \left[\begin{array}{ccc} \mathbf{v}_1(t) & \mathbf{v}_2(t) \ldots \mathbf{v}_N(t) \end{array}\right] = \bar{\mathbf{B}}^{\mathrm{T}} e^{\bar{\mathbf{A}}^{\mathrm{T}}t} \mathbf{\Sigma}^{-1/2} \tag{6.146}$$

This can be verified by

$$\begin{aligned}
&\sum_{i=1}^{N} \sigma_i \left(\int_0^\infty \mathbf{v}_i^{\mathrm{T}}(\tau)\, \mathbf{x}(\tau)\, d\tau \right) \mathbf{u}_i(t) \\
&= \mathbf{U}(t)\, \mathbf{\Sigma} \left(\int_0^\infty \mathbf{V}^{\mathrm{T}}(\tau)\, \mathbf{x}(\tau)\, d\tau \right) \\
&= \left(\bar{\mathbf{C}}\, e^{\bar{\mathbf{A}}t} \mathbf{\Sigma}^{-1/2} \right) \mathbf{\Sigma} \left(\mathbf{\Sigma}^{-1/2} \int_0^\infty e^{\bar{\mathbf{A}}\tau} \bar{\mathbf{B}}\, \mathbf{x}(\tau)\, d\tau \right) \\
&= \bar{\mathbf{C}}\, e^{\bar{\mathbf{A}}t} \int_0^\infty e^{\bar{\mathbf{A}}\tau} \bar{\mathbf{B}}\, \mathbf{x}(\tau)\, d\tau = \int_0^\infty \bar{\mathbf{C}}\, e^{\bar{\mathbf{A}}(t+\tau)} \bar{\mathbf{B}}\, \mathbf{x}(\tau)\, d\tau \\
&= \int_0^\infty \mathbf{h}(t+\tau)\, \mathbf{x}(\tau)\, d\tau = (\mathbf{\Gamma x})(t)
\end{aligned} \tag{6.147}$$

The eigenvectors $\mathbf{u}(t)$ and $\mathbf{v}(t)$ are orthonormal because

$$\mathbf{U}^{\mathrm{T}}(t)\, \mathbf{U}(t) = \mathbf{\Sigma}^{-1/2} \left(\int_0^\infty e^{\bar{\mathbf{A}}^{\mathrm{T}}t} \bar{\mathbf{C}}^{\mathrm{T}} \bar{\mathbf{C}} e^{\bar{\mathbf{A}}t}\, dt \right) \mathbf{\Sigma}^{-1/2} = \mathbf{\Sigma}^{-1/2}\, \bar{\mathbf{Q}}_\infty\, \mathbf{\Sigma}^{-1/2} = \mathbf{I} \tag{6.148}$$

and

$$\mathbf{V}^{\mathrm{T}}(t)\, \mathbf{V}(t) = \mathbf{\Sigma}^{-1/2} \left(\int_0^\infty e^{\bar{\mathbf{A}}t} \bar{\mathbf{B}} \bar{\mathbf{B}}^{\mathrm{T}} e^{\bar{\mathbf{A}}^{\mathrm{T}}t}\, dt \right) \mathbf{\Sigma}^{-1/2} = \mathbf{\Sigma}^{-1/2}\, \bar{\mathbf{P}}_\infty\, \mathbf{\Sigma}^{-1/2} = \mathbf{I} \tag{6.149}$$

The assumption of a fully balanced system is only needed in the last two equations to get orthonormal singular vectors. Because the Hankel integral operator is invariant under a similarity transformation (it depends only on the impulse response) the singular values are also invariant. And because also the eigenvalues of $\mathbf{P}_\infty \mathbf{Q}_\infty$ are invariant, it can be concluded that the singular values can be calculated from

$$\sigma_k^2 = \lambda_k(\mathbf{P}_\infty \mathbf{Q}_\infty) \tag{6.150}$$

6.2.5 Balanced systems

Because the continuous-time Gramians are transformed the same way as the discrete-time Gramians, the method used to bring discrete-time systems into a balanced form equally applies to continuous-time systems.

Reduced models. As in the discrete-time case, a reduced system based on the truncation of a fully balanced system is stable. For a continuous-time balanced system, the matrix $\mathbf{\Sigma} = \bar{\mathbf{P}}_\infty$ satisfies the the Lyapunov equation, which is written in block form as

$$\begin{bmatrix} \bar{\mathbf{A}}_{11} & \bar{\mathbf{A}}_{12} \\ \bar{\mathbf{A}}_{21} & \bar{\mathbf{A}}_{22} \end{bmatrix} \begin{bmatrix} \mathbf{\Sigma}_1 & \\ & \mathbf{\Sigma}_2 \end{bmatrix} + \begin{bmatrix} \mathbf{\Sigma}_1 & \\ & \mathbf{\Sigma}_2 \end{bmatrix} \begin{bmatrix} \bar{\mathbf{A}}_{11}^{\mathrm{T}} & \bar{\mathbf{A}}_{21}^{\mathrm{T}} \\ \bar{\mathbf{A}}_{12}^{\mathrm{T}} & \bar{\mathbf{A}}_{22}^{\mathrm{T}} \end{bmatrix} = - \begin{bmatrix} \bar{\mathbf{B}}_1 \\ \bar{\mathbf{B}}_2 \end{bmatrix} \begin{bmatrix} \bar{\mathbf{B}}_1^{\mathrm{T}} & \bar{\mathbf{B}}_2^{\mathrm{T}} \end{bmatrix} \tag{6.151}$$

The upper left block is

$$\bar{\mathbf{A}}_{11} \mathbf{\Sigma}_1 + \mathbf{\Sigma}_1 \bar{\mathbf{A}}_{11}^{\mathrm{T}} = -\bar{\mathbf{B}}_1 \bar{\mathbf{B}}_1^{\mathrm{T}} \tag{6.152}$$

Because $\mathbf{\Sigma}_1$ satisfies the continuous-time Lyapunov equation, a subsystem of a fully balanced continuous-time system is balanced. If the subsystem of a balanced continuous-time system is minimal (observable and controllable) then it is also stable because it satisfies the Lyapunov equation. It was shown by Pernebo and Silverman [PS82] that this is the case if $\mathbf{\Sigma}_1$ and $\mathbf{\Sigma}_2$ have no common singular values. This can always be guaranteed by choosing the subsystem appropriately, that is, such that multiple singular values are not split up. Typically, no multiple singular values occur in numerical applications. Unfortunately, for the realization problem, we get only a partially balanced system and the above stability consideration is not applicable.

The error introduced by truncating a fully balanced system to a system of degree N has been shown by Glover [Glo84, p. 1170] to be $\|\mathbf{E}(s)\|_\infty \leq 2(\sigma_{N+1} + \cdots)$, where multiple singular values are just counted once. For a partially balanced system, this twice-the-sum-of-the-tail rule is not strictly applicable as an error bound but is still useful as a criterion to determine the degree of the approximation. Such a criterion will be used in the next chapter.

6.2.6 Norms

For continuous-time systems the Euclidean or mean-square norm is

$$\|\mathbf{E}(s)\|_2 = \rho\left(\frac{1}{2\pi}\int_{-\infty}^{\infty} \mathbf{E}^*(i\omega)\,\mathbf{E}(i\omega)\,d\omega\right)^{1/2} = \rho\left(\int_{-\infty}^{\infty} \mathbf{e}^*(t)\,\mathbf{e}(t)\,dt\right)^{1/2} = \|\mathbf{e}(t)\|_2 \quad (6.153)$$

and the maximum norm or Chebyshev norm is

$$\|\mathbf{E}(s)\|_\infty = \max_\omega \rho\Big(\mathbf{E}^*(i\omega)\,\mathbf{E}(i\omega)\Big)^{1/2} = \max_\omega \bar{\sigma}\Big(\mathbf{E}(i\omega)\Big) \quad (6.154)$$

where the notation is the same as for the discrete-time case in Subsection 6.1.9.

6.3 Bilinear transform

As seen before, a discrete-time system can be transformed to a continuous-time system and vice versa. Because this transformation changes the constant part of the system, we have to use the more general form including the matrix $\mathbf{D}$.

The following development is similar to the one in Subsection 4.4.4. The transfer functions

$$\mathbf{H}(s) = \tilde{\mathbf{H}}(z) = \tilde{\mathbf{H}}\left(\frac{\alpha + s}{\alpha - s}\right) \quad (6.155)$$

are written with their system matrices as

$$\mathbf{C}\,(s\mathbf{I} - \mathbf{A})^{-1}\mathbf{B} + \mathbf{D} = \tilde{\mathbf{C}}\left(\frac{\alpha + s}{\alpha - s}\mathbf{I} - \tilde{\mathbf{A}}\right)^{-1}\tilde{\mathbf{B}} + \tilde{\mathbf{D}} \quad (6.156)$$

The right-hand side is expanded to

$$R.H.S. = (\alpha - s)\,\tilde{\mathbf{C}}\left(s\,(\mathbf{I} + \tilde{\mathbf{A}}) - \alpha\,(\tilde{\mathbf{A}} - \mathbf{I})\right)^{-1}\tilde{\mathbf{B}} + \tilde{\mathbf{D}} \quad (6.157)$$

taking the term $(\mathbf{I} + \tilde{\mathbf{A}})$ out of the inversion leads to

$$R.H.S. = (\alpha - s)\,\tilde{\mathbf{C}}\,(\mathbf{I} + \tilde{\mathbf{A}})^{-1}\left(s\mathbf{I} - \alpha\,(\mathbf{I} + \tilde{\mathbf{A}})^{-1}(\tilde{\mathbf{A}} - \mathbf{I})\right)^{-1}\tilde{\mathbf{B}} + \tilde{\mathbf{D}} \quad (6.158)$$

The term in the large parentheses now has the desired form and we write (not confusing $\mathbf{A}$ and $\tilde{\mathbf{A}}$)

$$R.H.S. = (\alpha - s)\,\tilde{\mathbf{C}}\,(\mathbf{I} + \tilde{\mathbf{A}})^{-1}\left(s\mathbf{I} - \mathbf{A}\right)^{-1}\tilde{\mathbf{B}} + \tilde{\mathbf{D}} \quad (6.159)$$

with $\mathbf{A} = \alpha\,(\mathbf{I} + \tilde{\mathbf{A}})^{-1}(\tilde{\mathbf{A}} - \mathbf{I})$. Separating the constant part

$$\tilde{\mathbf{C}}\,(\mathbf{I} + \tilde{\mathbf{A}})^{-1}\tilde{\mathbf{B}} = \tilde{\mathbf{C}}\,(\mathbf{I} + \tilde{\mathbf{A}})^{-1}\Big(s\mathbf{I} - \mathbf{A}\Big)^{-1}\Big(s\mathbf{I} - \mathbf{A}\Big)\,\tilde{\mathbf{B}} \tag{6.160}$$

the right-hand side is

$$R.H.S. = \tilde{\mathbf{C}}\,(\mathbf{I} + \tilde{\mathbf{A}})^{-1}\Big(s\mathbf{I} - \mathbf{A}\Big)^{-1}(\alpha\mathbf{I} - \mathbf{A})\,\tilde{\mathbf{B}} + \tilde{\mathbf{D}} - \tilde{\mathbf{C}}\,(\mathbf{I} + \tilde{\mathbf{A}})^{-1}\tilde{\mathbf{B}} \tag{6.161}$$

After substituting $\alpha\mathbf{I} - \mathbf{A} = 2\,(\mathbf{I} + \tilde{\mathbf{A}})^{-1}$ we get the final expression

$$R.H.S. = 2\alpha\,\tilde{\mathbf{C}}\,(\mathbf{I} + \tilde{\mathbf{A}})^{-1}\Big(s\mathbf{I} - \mathbf{A}\Big)^{-1}(\mathbf{I} + \tilde{\mathbf{A}})^{-1}\tilde{\mathbf{B}} + \tilde{\mathbf{D}} - \tilde{\mathbf{C}}\,(\mathbf{I} + \tilde{\mathbf{A}})^{-1}\tilde{\mathbf{B}} \tag{6.162}$$

Comparing with the left-hand side, $\mathbf{C}\,(s\mathbf{I} - \mathbf{A})^{-1}\mathbf{B} + \mathbf{D}$, we get [Glo84]

$$\begin{aligned}
\mathbf{A} &= \alpha\,(\mathbf{I} + \tilde{\mathbf{A}})^{-1}(\tilde{\mathbf{A}} - \mathbf{I}) && (6.163)\\
\mathbf{B} &= \sqrt{2\alpha}\,(\mathbf{I} + \tilde{\mathbf{A}})^{-1}\tilde{\mathbf{B}} && (6.164)\\
\mathbf{C} &= \sqrt{2\alpha}\,\tilde{\mathbf{C}}\,(\mathbf{I} + \tilde{\mathbf{A}})^{-1} && (6.165)\\
\mathbf{D} &= \tilde{\mathbf{D}} - \tilde{\mathbf{C}}\,(\mathbf{I} + \tilde{\mathbf{A}})^{-1}\tilde{\mathbf{B}} && (6.166)
\end{aligned}$$

Lyapunov equations. As shown earlier (Section 4.4), the bilinear transformation maps the stability region of continuous-time systems onto the one of discrete-time systems and vice versa. It is therefore to be expected that also the discrete-time and the continuous-time Lyapunov equations are related by the bilinear transform. The derivation follows [Glo84]. The continuous-time Lyapunov equation is multiplied from the left by $1/\alpha\,(\mathbf{I}+\tilde{\mathbf{A}})$ and from the right by $1/\alpha\,(\mathbf{I} + \tilde{\mathbf{A}}^{\mathrm{T}})$ and then the system matrices $\mathbf{A}, \mathbf{B}, \mathbf{C}$ are substituted by their discrete-time counterparts $\tilde{\mathbf{A}}, \tilde{\mathbf{B}}, \tilde{\mathbf{C}}$.

$$\begin{aligned}
&1/\alpha\,(\mathbf{I} + \tilde{\mathbf{A}})(\mathbf{A}\mathbf{P}_\infty + \mathbf{P}_\infty\mathbf{A}^{\mathrm{T}} + \mathbf{B}\mathbf{B}^{\mathrm{T}})(\mathbf{I} + \tilde{\mathbf{A}}^{\mathrm{T}})1/\alpha\\
&\quad= (\tilde{\mathbf{A}} - \mathbf{I})\mathbf{P}_\infty(\mathbf{I} + \tilde{\mathbf{A}}^{\mathrm{T}}) + (\mathbf{I} + \tilde{\mathbf{A}})\mathbf{P}_\infty(\tilde{\mathbf{A}}^{\mathrm{T}} - \mathbf{I}) + 2\tilde{\mathbf{B}}\tilde{\mathbf{B}}^{\mathrm{T}}\\
&\quad= 2(\tilde{\mathbf{A}}\mathbf{P}_\infty\tilde{\mathbf{A}}^{\mathrm{T}} - \mathbf{P}_\infty + \tilde{\mathbf{B}}\tilde{\mathbf{B}}^{\mathrm{T}}) = 0 && (6.167)
\end{aligned}$$

But this is the discrete-time Lyapunov equation. The derivation also shows that $\mathbf{P}_\infty = \tilde{\mathbf{P}}_\infty$. In the same way it can be shown that $\mathbf{Q}_\infty = \tilde{\mathbf{Q}}_\infty$. Thus the bilinear transform does not change the Gramians corresponding to an infinite time interval. Therefore, a fully balanced system remains balanced when transformed by the bilinear transform. However, since a truncated system of a balanced continuous-time system is balanced while a truncated system of a balanced discrete-time system is not, the subsystems are not related by the bilinear transform.

Singular values. An important aspect is the relation between the singular values of a continuous-time system and the discrete-time system obtained by the bilinear transform. Because, for an infinite time interval, the discrete-time Gramians equal the continuous-time Gramians, the singular values also are the same.

$$\sigma_k = \lambda_k(\mathbf{P}_\infty\mathbf{Q}_\infty) = \lambda_k(\tilde{\mathbf{P}}_\infty\tilde{\mathbf{Q}}_\infty) = \lambda_k(\tilde{\mathbf{\Gamma}}^{\mathrm{T}}\tilde{\mathbf{\Gamma}}) \tag{6.168}$$

But since the singular values are a measure for the approximation error it is justified to perform the approximation for continuous-time systems in the discrete-time domain. For the relationship between the discrete-time and the continuous-time singular vectors see [LK82].

6.4 Remarks

The following remarks summarize some of the results obtained in this chapter.

- □ For the continuous-time case, neither the Gramians $\mathbf{P}$ and $\mathbf{Q}$ nor the Hankel integral operator can directly be calculated from the observability and controllability matrices $\mathbf{W}_o$ and $\mathbf{W}_c$. This is only possible for the discrete-time case. However, the Gramians corresponding to an infinite time interval are related to the singular values of the Hankel operator by $\sigma_k = \lambda_k(\mathbf{P}_\infty \mathbf{Q}_\infty)$.

- □ The Gramians corresponding to an infinite time interval are the same for the discrete-time and the continuous-time case if the systems are related to each other by the bilinear transform. Therefore, also the singular values of the infinite Hankel matrix and the Hankel integral operator are equal.

- □ The maximum error is the same for a discrete-time system and a continuous-time system if the two systems are related by the bilinear transform, because the transformation is only a distortion of the frequency axis. Thus, an approximation done in the discrete-time domain is equally valid for the continuous-time domain.

- □ If a fully balanced discrete-time system is transformed to a continuous-time system by the bilinear transform, the continuous-time is also fully balanced and the discrete-time Gramians are equal to the continuous-time Gramians. However, balancing a discrete-time system over a certain number of terms is not the same as balancing a continuous-time over a certain time. There is no general relationship between a partially balanced discrete-time system and a partially balanced continuous-time system.

Some of these results are also illustrated by the example in Section 7.2.3 of the next chapter.

Chapter 7

Balanced Realization

In this chapter we derive a rational approximation method for multivariable systems. The method is formulated using the state-variable description presented in the previous chapter. The reason for introducing yet another method is that the methods treated so far are not easily extended to the multivariable case. This is particularly true for the CF method which is the preferred method so far investigated. One essential property used for that method is that the polynomial related to the $(N+1)$th singular vector has exactly N zeroes inside the unit circle. This property does not hold for multivariable systems. While optimal Hankel-norm approximations have been derived by Kung [KL81b, KL81a] and by Glover [Glo84] in the context of model reduction, a method for the realization problem has apparently not been found yet. On the other hand, there are methods available based on balancing which are not quite as accurate but numerically more attractive. A critical survey of different methods can be found in [KA87].

For the balanced realization we start, as in the CF method, with the Hankel matrix of the mapped discrete-time system. The singular value decomposition is used to construct the system matrices of a balanced system. Depending on the required accuracy, only a subsystem is kept as an approximation. The original, not balanced, realization algorithm is due to Ho and Kalman [HK66]. Zeiger and McEwen [ZM74] suggested using the singular value decomposition in the algorithm. Chen [Che84, p. 270] showed that using the singular value decomposition results in a balanced realization. A variant method is described in Kung [Kun78]. The algorithm used here is in the form given by Chen. Several detailed examples are presented which also give some insight into the theoretical background presented in the last chapter.

7.1 Algorithm

7.1.1 General derivation

We first state the algorithm starting from the (block) Hankel matrix. The Hankel matrix can be found as in Subsection 4.4.1 by first transforming the continuous-time system to a discrete-time system using the bilinear transform. As explained there, this step is merely a distortion of the frequency axis according to Equation (4.110). The impulse response is then found by the DFT, Equation (4.117). From the impulse response, the block Hankel matrix is constructed. For a system with q inputs and p outputs, each block is of size $p \times q$. All matrices used in this section are related to the discrete-time system and are not specially marked as such.

The Hankel matrix of the original system is used to construct the system matrices of an approximate rational system. The Hankel matrix can be written as the product of the observability and the controllability matrices of the approximate system (Equation (6.97):

$$\mathbf{\Gamma} = \begin{bmatrix} \mathbf{f}_1 & \dots & \mathbf{f}_L \\ \vdots & & \vdots \\ \mathbf{f}_L & \dots & \mathbf{f}_{2L-1} \end{bmatrix} = \begin{bmatrix} \mathbf{C} \\ \vdots \\ \mathbf{CA}^{L-1} \end{bmatrix} \begin{bmatrix} \mathbf{B} & \dots & \mathbf{A}^{L-1}\mathbf{B} \end{bmatrix} = \mathbf{W}_o \mathbf{W}_c \tag{7.1}$$

Similarly we write for the shifted Hankel matrix

$$\check{\mathbf{\Gamma}} = \begin{bmatrix} \mathbf{f}_2 & \dots & \mathbf{f}_{L+1} \\ \vdots & & \vdots \\ \mathbf{f}_{L+1} & \dots & \mathbf{f}_{2L} \end{bmatrix} = \begin{bmatrix} \mathbf{C} \\ \vdots \\ \mathbf{CA}^{L-1} \end{bmatrix} \mathbf{A} \begin{bmatrix} \mathbf{B} & \dots & \mathbf{A}^{L-1}\mathbf{B} \end{bmatrix} = \mathbf{W}_o \mathbf{A} \mathbf{W}_c \tag{7.2}$$

The next step is to find the singular value decomposition

$$\mathbf{\Gamma} = \mathbf{U}\mathbf{\Sigma}\mathbf{V}^{\mathrm{T}} \tag{7.3}$$

The matrices are partitioned according to the singular values as

$$\mathbf{\Gamma} = \begin{bmatrix} \mathbf{U}_1 & \mathbf{U}_2 \end{bmatrix} \begin{bmatrix} \mathbf{\Sigma}_1 & \\ & \mathbf{\Sigma}_2 \end{bmatrix} \begin{bmatrix} \mathbf{V}_1^{\mathrm{T}} \\ \mathbf{V}_2^{\mathrm{T}} \end{bmatrix} \tag{7.4}$$

where $\mathbf{\Sigma}_1 = \mathbf{diag}(\sigma_1, \dots, \sigma_N)$ and $\mathbf{\Sigma}_2 = \mathbf{diag}(\sigma_{N+1}, \dots, \sigma_L) \approx \mathbf{0}$. With these submatrices the Hankel matrix can be written as

$$\mathbf{\Gamma} = \mathbf{U}_1 \mathbf{\Sigma}_1 \mathbf{V}_1^{\mathrm{T}} \tag{7.5}$$

Comparing this equation with Equation (7.1), we can associate

$$\mathbf{W}_o = \mathbf{U}_1 \, \mathbf{\Sigma}_1^{1/2} \tag{7.6}$$

$$\mathbf{W}_c = \mathbf{\Sigma}_1^{1/2} \, \mathbf{V}_1^{\mathrm{T}} \tag{7.7}$$

and take $\mathbf{B}$ and $\mathbf{C}$ directly from these matrices. The matrix $\mathbf{A}$ can be found from the last part of Equation (7.2). Since $\mathbf{W}_o$ and $\mathbf{W}_c$ do not necessarily have full rank, we have to use the (left and right) pseudo-inverses (see Equation 5.54)

$$\mathbf{W}_o^+ = \mathbf{\Sigma}_1^{-1/2} \, \mathbf{U}_1^{\mathrm{T}} \tag{7.8}$$

$$\mathbf{W}_c^+ = \mathbf{V}_1 \, \mathbf{\Sigma}_1^{-1/2} \tag{7.9}$$

These do indeed satisfy $\mathbf{W}_o^+\mathbf{W}_o = \mathbf{I}$ and $\mathbf{W}_c\mathbf{W}_c^+ = \mathbf{I}$ (but $\mathbf{W}_o\mathbf{W}_o^+ \neq \mathbf{I}$ and $\mathbf{W}_c^+\mathbf{W}_c \neq \mathbf{I}$, see Section 5.3) because $\mathbf{U}$ and $\mathbf{V}$ are orthonormal and therefore $\mathbf{U}_1^{\mathrm{T}}\mathbf{U}_1 = \mathbf{I}$ and $\mathbf{V}_1^{\mathrm{T}}\mathbf{V}_1 = \mathbf{I}$. With these preliminaries we calculate the following system matrices:

$$\bar{\mathbf{A}} = \mathbf{W}_o^+ \, \check{\mathbf{\Gamma}} \, \mathbf{W}_c^+ = \mathbf{\Sigma}_1^{-1/2} \, \mathbf{U}_1^{\mathrm{T}} \, \check{\mathbf{\Gamma}} \, \mathbf{V}_1 \, \mathbf{\Sigma}_1^{-1/2} \tag{7.10}$$

$$\bar{\mathbf{B}} = \text{first block column of } \mathbf{W}_c = \text{first block column of } \mathbf{\Sigma}_1^{1/2} \, \mathbf{V}_1^{\mathrm{T}} \tag{7.11}$$

$$\bar{\mathbf{C}} = \text{first block row of } \mathbf{W}_o = \text{first block row of } \mathbf{U}_1 \, \mathbf{\Sigma}_1^{1/2} \tag{7.12}$$

$$\bar{\mathbf{D}} = \mathbf{f}_0 \tag{7.13}$$

where the "first block column" means the submatrix containing the first q columns, and the "first block row" means the submatrix containing the first p rows.

7.1.2 Full rank Hankel matrix

To gain more insight into the method, we will prove that this realization does in fact reproduce the impulse response used in the Hankel matrix. First we assume that somehow we know the minimal degree N of the system and can take $L = N$ such that $\mathbf{\Gamma}$ has full rank. Then it can be seen that

$$\mathbf{\Gamma}\mathbf{A}_c = \check{\mathbf{\Gamma}} \tag{7.14}$$

where $\mathbf{A}_c$ has been defined in Equation (6.11). Because $\mathbf{\Gamma}$ has full rank we find

$$\mathbf{A}_c = \mathbf{\Gamma}^{-1}\check{\mathbf{\Gamma}} \tag{7.15}$$

Also

$$\mathbf{B}_c = \begin{bmatrix} \mathbf{I}_q \\ \mathbf{0} \end{bmatrix} \tag{7.16}$$

and

$$\mathbf{C}_c = \begin{bmatrix} \mathbf{f}_1 & \dots & \mathbf{f}_L \end{bmatrix} \tag{7.17}$$

are readily available. This system is now transformed to a system balanced in the interval $[0 \dots L-1]$. The controllability matrix is given by

$$\mathbf{W}_c = \begin{bmatrix} \mathbf{B}_c & \mathbf{A}_c\mathbf{B}_c & \dots & \mathbf{A}_c^{L-1}\mathbf{B}_c \end{bmatrix} = \mathbf{I} \tag{7.18}$$

and the observability matrix by

$$\mathbf{W}_o = \begin{bmatrix} \mathbf{C}_c \\ \mathbf{C}_c\mathbf{A}_c \\ \vdots \\ \mathbf{C}_c\mathbf{A}_c^{L-1} \end{bmatrix} = \mathbf{\Gamma} = \mathbf{U\Sigma V} \tag{7.19}$$

The Gramians are

$$\mathbf{P} = \mathbf{W}_c\mathbf{W}_c^{\mathrm{T}} = \mathbf{I} \tag{7.20}$$

and

$$\mathbf{Q} = \mathbf{W}_o^{\mathrm{T}}\mathbf{W}_o = \mathbf{\Gamma}^{\mathrm{T}}\mathbf{\Gamma} = \mathbf{V\Sigma}^2\mathbf{V}^{\mathrm{T}} \tag{7.21}$$

To transform these into $\bar{\mathbf{P}}$ and $\bar{\mathbf{Q}}$, both equal and diagonal, we proceed as in Subsection 6.1.8. With $\mathbf{W}_c = \mathbf{I}$, the balancing transform is given by the transformation matrix (Equation 6.104)

$$\mathbf{T} = \mathbf{V\Sigma}^{-1/2} \tag{7.22}$$

and the inverse

$$\mathbf{T}^{-1} = \mathbf{\Sigma}^{1/2}\mathbf{V}^{\mathrm{T}} \tag{7.23}$$

The inverse $\mathbf{\Gamma}^{-1}$ is expressed in terms of the singular value decomposition as

$$\mathbf{\Gamma}^{-1} = \mathbf{V\Sigma}^{-1}\mathbf{U}^{\mathrm{T}} \tag{7.24}$$

Then the balanced system becomes

$$\begin{aligned} \bar{\mathbf{A}} &= \mathbf{T}^{-1}\mathbf{A}_c\mathbf{T} = (\mathbf{\Sigma}^{1/2}\,\mathbf{V}^{\mathrm{T}})(\mathbf{V\Sigma}^{-1}\mathbf{U}^{\mathrm{T}}\,\check{\mathbf{\Gamma}})(\mathbf{V\Sigma}^{-1/2}) \\ &= \mathbf{\Sigma}^{-1/2}\,\mathbf{U}^{\mathrm{T}}\,\check{\mathbf{\Gamma}}\,\mathbf{V}\,\mathbf{\Sigma}^{-1/2} \end{aligned} \tag{7.25}$$

$$\begin{aligned} \bar{\mathbf{B}} &= \mathbf{T}^{-1}\mathbf{B}_c = \mathbf{\Sigma}^{1/2}\,\mathbf{V}^{\mathrm{T}} \begin{bmatrix} \mathbf{I}_q \\ \mathbf{0} \end{bmatrix} \\ &= \text{first block column of } \mathbf{\Sigma}^{1/2}\,\mathbf{V}^{\mathrm{T}} \end{aligned} \tag{7.26}$$

$$\begin{aligned}\bar{\mathbf{C}} &= \mathbf{C}_c\mathbf{T} = \left[\begin{array}{ccc} \mathrm{f}_1 & \dots & \mathrm{f}_L \end{array}\right] \mathbf{V}\boldsymbol{\Sigma}^{-1/2} \\ &= (\text{first block row of } \boldsymbol{\Gamma})\, \mathbf{V}\boldsymbol{\Sigma}^{-1/2} \\ &= \text{first block row of } \mathbf{U}\boldsymbol{\Sigma}\mathbf{V}^{\mathrm{T}}\mathbf{V}\boldsymbol{\Sigma}^{-1/2} \\ &= \text{first block row of } \mathbf{U}\boldsymbol{\Sigma}^{1/2} \end{aligned} \tag{7.27}$$

These equations are identical to the proposed algorithm. Because they are equivalent to the system $(\mathbf{A}_c, \mathbf{B}_c, \mathbf{C}_c)$ they are also equivalent to the Hankel matrix. As expected, the Gramians become

$$\bar{\mathbf{P}} = \mathbf{T}^{-1}\mathbf{P}\mathbf{T}^{-\mathrm{T}} = (\boldsymbol{\Sigma}^{1/2}\mathbf{V}^{\mathrm{T}})\,\mathbf{I}\,(\mathbf{V}\boldsymbol{\Sigma}^{1/2}) = \boldsymbol{\Sigma} \tag{7.28}$$

$$\begin{aligned}\bar{\mathbf{Q}} &= \mathbf{T}^{\mathrm{T}}\mathbf{Q}\mathbf{T} = (\boldsymbol{\Sigma}^{-1/2}\mathbf{V}^{\mathrm{T}})(\boldsymbol{\Gamma}^{\mathrm{T}}\boldsymbol{\Gamma})(\mathbf{V}\boldsymbol{\Sigma}^{-1/2}) \\ &= (\boldsymbol{\Sigma}^{-1/2}\mathbf{V}^{\mathrm{T}})(\mathbf{V}\boldsymbol{\Sigma}\mathbf{U}^{\mathrm{T}}\mathbf{U}\boldsymbol{\Sigma}\mathbf{V}^{\mathrm{T}})(\mathbf{V}\boldsymbol{\Sigma}^{-1/2}) = \boldsymbol{\Sigma}\end{aligned} \tag{7.29}$$

Thus the system is minimal (by assumption) and balanced.

7.1.3 Rank deficient Hankel matrix

Next consider the case where we do not know the minimal degree N in advance or where the first N columns of the Hankel matrix are not linearly independent, so that we have to take more than N columns to get a Hankel matrix of rank N. We take L large enough such that the Hankel matrix has rank N. Again, observe that

$$\boldsymbol{\Gamma}\mathbf{A}_c = \check{\boldsymbol{\Gamma}} \tag{7.30}$$

But now $\boldsymbol{\Gamma}$ does not have full rank and cannot be inverted. However, we can find a solution using the pseudo-inverse (Section 5.3) $\boldsymbol{\Gamma}^+ = \mathbf{V}_1\,\boldsymbol{\Sigma}_1^{-1}\,\mathbf{U}_1^{\mathrm{T}}$, where $\mathbf{U}_1, \mathbf{V}_1, \boldsymbol{\Sigma}_1$ are the submatrices of the singular value decomposition of $\boldsymbol{\Gamma}$. The solution is

$$\mathbf{A}_{\bar{c}} = \mathbf{V}_1\,\boldsymbol{\Sigma}_1^{-1}\,\mathbf{U}_1^{\mathrm{T}}\,\check{\boldsymbol{\Gamma}} \tag{7.31}$$

We have written $\mathbf{A}_{\bar{c}}$ instead of $\mathbf{A}_c$ because the solution of the overdetermined Equation (7.30) is not unique and not necessarily of the form $\mathbf{A}_c$. The matrices $\mathbf{B}_c$ and $\mathbf{C}_c$ are as before. $\mathbf{A}_{\bar{c}}$ has dimension $Lq \times Lq$ but rank $N < Lq$. The system $(\mathbf{A}_{\bar{c}}, \mathbf{B}_c, \mathbf{C}_c)$ is therefore not minimal. To find a system of minimal degree we use a similarity transform that separates any uncontrollable and unobservable states. That this is possible is guaranteed by the general decomposition theorem ([Kai80, p. 132]). Also we would like the transformed system to be balanced. However, a balancing transformation only exists for systems of minimal degree. The following similarity transform separates the unobservable states and the remaining system is of minimal degree and balanced. The transformation is similar to the balancing transformation Equation (7.22), except that $\boldsymbol{\Sigma}_2 = \mathbf{0}$ does not appear because it is not invertible.

$$\mathbf{T} = \left[\begin{array}{cc} \mathbf{V}_1\boldsymbol{\Sigma}_1^{-1/2} & \mathbf{V}_2 \end{array}\right] \tag{7.32}$$

and the inverse

$$\mathbf{T}^{-1} = \left[\begin{array}{c} \boldsymbol{\Sigma}_1^{1/2}\,\mathbf{V}_1^{\mathrm{T}} \\ \mathbf{V}_2^{\mathrm{T}} \end{array}\right] \tag{7.33}$$

Then

$$\begin{aligned}
\mathbf{T}^{-1}\mathbf{A}_{\bar{c}}\mathbf{T} &= \begin{bmatrix} \boldsymbol{\Sigma}_1^{1/2}\mathbf{V}_1^{\mathrm{T}} \\ \mathbf{V}_2^{\mathrm{T}} \end{bmatrix} \mathbf{V}_1\,\boldsymbol{\Sigma}_1^{-1}\mathbf{U}_1^{\mathrm{T}}\,\check{\boldsymbol{\Gamma}} \begin{bmatrix} \mathbf{V}_1\boldsymbol{\Sigma}_1^{-1/2} & \mathbf{V}_2 \end{bmatrix} \\
&= \begin{bmatrix} \boldsymbol{\Sigma}_1^{-1/2}\,\mathbf{U}_1^{\mathrm{T}}\,\check{\boldsymbol{\Gamma}}\mathbf{V}_1\,\boldsymbol{\Sigma}_1^{-1/2} & \boldsymbol{\Sigma}_1^{-1/2}\,\mathbf{U}_1^{\mathrm{T}}\,\check{\boldsymbol{\Gamma}}\mathbf{V}_2 \\ \mathbf{0} & \mathbf{0} \end{bmatrix} = \begin{bmatrix} \bar{\mathbf{A}} & \mathbf{0} \\ \mathbf{0} & \mathbf{0} \end{bmatrix} \qquad (7.34)
\end{aligned}$$

To see that $\boldsymbol{\Sigma}_1^{-1/2}\,\mathbf{U}_1^{\mathrm{T}}\,\check{\boldsymbol{\Gamma}}\mathbf{V}_2$ is zero, we use the relationship

$$\check{\boldsymbol{\Gamma}} = \mathbf{A}_o\boldsymbol{\Gamma} \qquad (7.35)$$

with $\mathbf{A}_o$ of the form defined in Equation (6.41). Now, $\check{\boldsymbol{\Gamma}}\mathbf{V}_2 = \mathbf{A}_o\boldsymbol{\Gamma}\mathbf{V}_2 = \mathbf{A}_o\mathbf{U}_1\boldsymbol{\Sigma}_1\mathbf{V}_1^{\mathrm{T}}\mathbf{V}_2$. But $\mathbf{V}_1^{\mathrm{T}}\mathbf{V}_2 = \mathbf{0}$ because $\mathbf{V}$ is orthonormal. The other system matrices are

$$\begin{aligned}
\mathbf{T}^{-1}\mathbf{B}_c &= \begin{bmatrix} \boldsymbol{\Sigma}_1^{1/2}\mathbf{V}_1^{\mathrm{T}} \\ \mathbf{V}_2^{\mathrm{T}} \end{bmatrix} \begin{bmatrix} \mathbf{I}_q \\ \mathbf{0} \end{bmatrix} \\
&= \text{first block column of} \begin{bmatrix} \boldsymbol{\Sigma}_1^{1/2}\mathbf{V}_1^{\mathrm{T}} \\ \mathbf{V}_2^{\mathrm{T}} \end{bmatrix} = \begin{bmatrix} \bar{\mathbf{B}} \\ \mathsf{x} \end{bmatrix} \qquad (7.36)
\end{aligned}$$

where x denotes a matrix whose value is not important here, and

$$\begin{aligned}
\mathbf{C}_c\mathbf{T} &= \begin{bmatrix} \mathbf{f}_1 & \dots & \mathbf{f}_L \end{bmatrix} \begin{bmatrix} \mathbf{V}_1\boldsymbol{\Sigma}_1^{-1/2} & \mathbf{V}_2 \end{bmatrix} \\
&= \text{first block row of } \boldsymbol{\Gamma} \begin{bmatrix} \mathbf{V}_1\boldsymbol{\Sigma}_1^{-1/2} & \mathbf{V}_2 \end{bmatrix} \\
&= \text{first block row of} \left(\mathbf{U}_1\boldsymbol{\Sigma}_1\mathbf{V}_1^{\mathrm{T}} \begin{bmatrix} \mathbf{V}_1\boldsymbol{\Sigma}_1^{-1/2} & \mathbf{V}_2 \end{bmatrix}\right) \\
&= \text{first block row of} \begin{bmatrix} \mathbf{U}_1\boldsymbol{\Sigma}_1^{1/2} & \mathbf{0} \end{bmatrix} = \begin{bmatrix} \bar{\mathbf{C}} & \mathbf{0} \end{bmatrix} \qquad (7.37)
\end{aligned}$$

Obviously, the impulse response is

$$\mathbf{h}_k = \begin{bmatrix} \bar{\mathbf{C}} & \mathbf{0} \end{bmatrix} \begin{bmatrix} \bar{\mathbf{A}} & \mathbf{0} \\ \mathbf{0} & \mathbf{0} \end{bmatrix}^{k-1} \begin{bmatrix} \bar{\mathbf{B}} \\ \mathsf{x} \end{bmatrix} = \bar{\mathbf{C}}\bar{\mathbf{A}}^{k-1}\bar{\mathbf{B}} \qquad (7.38)$$

and thus $(\bar{\mathbf{A}}, \bar{\mathbf{B}}, \bar{\mathbf{C}})$ is a minimal (degree N) realization of the approximation $(\mathbf{A}_{\bar{c}}, \mathbf{B}_c, \mathbf{C}_c)$ and has the same Hankel matrix as the original system. This proof is different from and simpler than the one given in [Che84, p. 270].

To show that the system is balanced we have to calculate the observability and controllability matrices and their Gramians. To calculate the observability matrix of the system $(\bar{\mathbf{A}}, \bar{\mathbf{B}}, \bar{\mathbf{C}})$ note that (recalling $\mathbf{U}_1\mathbf{U}_1^{\mathrm{T}} \neq \mathbf{I}$)

$$\mathbf{U}_1\mathbf{U}_1^{\mathrm{T}}\,\check{\boldsymbol{\Gamma}} = \mathbf{U}_1\mathbf{U}_1^{\mathrm{T}}(\mathbf{U}_1\boldsymbol{\Sigma}_1\mathbf{V}_1^{\mathrm{T}}\mathbf{A}_c) = \mathbf{U}_1\boldsymbol{\Sigma}_1\mathbf{V}_1^{\mathrm{T}}\mathbf{A}_c = \check{\boldsymbol{\Gamma}} \qquad (7.39)$$

and

$$\boldsymbol{\Gamma}\mathbf{V}_1\boldsymbol{\Sigma}_1^{-1/2} = (\mathbf{U}_1\boldsymbol{\Sigma}_1\mathbf{V}_1^{\mathrm{T}})\mathbf{V}_1\boldsymbol{\Sigma}_1^{-1/2} = \mathbf{U}_1\boldsymbol{\Sigma}_1^{1/2} \qquad (7.40)$$

Then

$$\begin{aligned}
\bar{\mathbf{C}} &= \text{first block row of } \mathbf{U}_1\boldsymbol{\Sigma}_1^{1/2} \qquad (7.41) \\
\bar{\mathbf{C}}\bar{\mathbf{A}} &= \text{first block row of } \mathbf{U}_1\boldsymbol{\Sigma}_1^{1/2}(\boldsymbol{\Sigma}_1^{-1/2}\mathbf{U}_1^{\mathrm{T}}\check{\boldsymbol{\Gamma}}\mathbf{V}_1\boldsymbol{\Sigma}_1^{-1/2}) \\
&= \text{first block row of } \check{\boldsymbol{\Gamma}}\mathbf{V}_1\boldsymbol{\Sigma}_1^{-1/2} \\
&= \text{second block row of } \boldsymbol{\Gamma}\mathbf{V}_1\boldsymbol{\Sigma}_1^{-1/2} \\
&= \text{second block row of } \mathbf{U}_1\boldsymbol{\Sigma}_1^{1/2} \qquad (7.42) \\
&\vdots
\end{aligned}$$

Hence, the extended observability matrix is

$$\bar{\mathbf{W}}_o = \begin{bmatrix} \bar{\mathbf{C}} \\ \bar{\mathbf{C}}\bar{\mathbf{A}} \\ \vdots \\ \bar{\mathbf{C}}\bar{\mathbf{A}}^{L-1} \end{bmatrix} = \mathbf{U}_1\mathbf{\Sigma}_1^{1/2} \tag{7.43}$$

Clearly, it has rank N, where N is the dimension of the matrix $\bar{\mathbf{A}}$. Thus the system is observable. For the controllability matrix note that (recalling $\mathbf{V}_1\mathbf{V}_1^{\mathrm{T}} \neq \mathbf{I}$)

$$\check{\mathbf{\Gamma}}\mathbf{V}_1\mathbf{V}_1^{\mathrm{T}} = (\mathbf{A}_o\mathbf{U}_1\mathbf{\Sigma}_1\mathbf{V}_1^{\mathrm{T}})\mathbf{V}_1\mathbf{V}_1^{\mathrm{T}} = \mathbf{A}_o\mathbf{U}_1\mathbf{\Sigma}_1\mathbf{V}_1^{\mathrm{T}} = \check{\mathbf{\Gamma}} \tag{7.44}$$

and

$$\mathbf{\Sigma}_1^{-1/2}\mathbf{U}_1^{\mathrm{T}}\mathbf{\Gamma} = \mathbf{\Sigma}_1^{-1/2}\mathbf{U}_1^{\mathrm{T}}(\mathbf{U}_1\mathbf{\Sigma}_1\mathbf{V}_1^{\mathrm{T}}) = \mathbf{\Sigma}_1^{1/2}\mathbf{V}_1^{\mathrm{T}} \tag{7.45}$$

Then

$$\bar{\mathbf{B}} = \text{first block column of } \mathbf{\Sigma}_1^{1/2}\mathbf{V}_1^{\mathrm{T}} \tag{7.46}$$

$$\begin{aligned}
\bar{\mathbf{A}}\bar{\mathbf{B}} &= \text{first block column of } (\mathbf{\Sigma}_1^{-1/2}\mathbf{U}_1^{\mathrm{T}}\check{\mathbf{\Gamma}}\mathbf{V}_1\mathbf{\Sigma}_1^{-1/2})\mathbf{\Sigma}_1\mathbf{V}_1^{\mathrm{T}} \\
&= \text{first block column of } \mathbf{\Sigma}_1^{-1/2}\mathbf{U}_1^{\mathrm{T}}\check{\mathbf{\Gamma}} \\
&= \text{second block column of } \mathbf{\Sigma}_1^{-1/2}\mathbf{U}_1^{\mathrm{T}}\mathbf{\Gamma} \\
&= \text{second block column of } \mathbf{\Sigma}_1^{1/2}\mathbf{V}_1^{\mathrm{T}}
\end{aligned} \tag{7.47}$$

$$\vdots$$

Hence, the extended controllability matrix is

$$\bar{\mathbf{W}}_c = \begin{bmatrix} \bar{\mathbf{B}} & \bar{\mathbf{A}}\bar{\mathbf{B}} & \ldots & \bar{\mathbf{A}}^{L-1}\bar{\mathbf{B}} \end{bmatrix} = \mathbf{\Sigma}^{1/2}\mathbf{V}_1^{\mathrm{T}} \tag{7.48}$$

Obviously, it has rank N and the system is thus controllable. Because the system is observable and controllable it is also minimal (Subsection 6.1.5).

The Gramians are as expected

$$\mathbf{P} = \mathbf{W}_c\mathbf{W}_c^{\mathrm{T}} = \mathbf{\Sigma}_1^{1/2}\,\mathbf{V}_1^{\mathrm{T}}\,\mathbf{V}_1\,\mathbf{\Sigma}_1^{1/2} = \mathbf{\Sigma}_1 \tag{7.49}$$

$$\mathbf{Q} = \mathbf{W}_o^{\mathrm{T}}\mathbf{W}_o = \mathbf{\Sigma}_1^{1/2}\,\mathbf{U}_1^{\mathrm{T}}\,\mathbf{U}_1\,\mathbf{\Sigma}_1^{1/2} = \mathbf{\Sigma}_1 \tag{7.50}$$

This confirms that the system $(\bar{\mathbf{A}}, \bar{\mathbf{B}}, \bar{\mathbf{C}})$ is balanced in the interval $[0, \ldots, L-1]$ corresponding to the size of $\mathbf{\Gamma}$.

7.1.4 Reduced systems

As was explained in the last chapter (Subsection 6.1.8), a balanced system can be reduced by taking a subsystem and thereby neglecting the almost unobservable and uncontrollable states corresponding to small singular values. In the context of the balanced realization algorithm given in this section, taking a subsystem is the same as including only a smaller number of singular values and vectors in the realization as shown in the following. We partition the singular values and the corresponding singular vectors as

$$\mathbf{\Sigma} = \begin{bmatrix} \mathbf{\Sigma}_1 & & \\ & \mathbf{\Sigma}_{2A} & \\ & & \mathbf{\Sigma}_{2B} \end{bmatrix} \qquad \mathbf{U} = \begin{bmatrix} \mathbf{U}_1 & \mathbf{U}_{2A} & \mathbf{U}_{2B} \end{bmatrix} \qquad \mathbf{V} = \begin{bmatrix} \mathbf{V}_1 & \mathbf{V}_{1A} & \mathbf{V}_{1B} \end{bmatrix} \tag{7.51}$$

where index 1 pertains to the reduced system, index $2A$ to the neglected, small but non-zero singular values and index $2B$ to the zero singular values. Taking all non-zero singular values into account, the system matrices are

$$\begin{aligned}
\bar{\mathbf{A}} &= \begin{bmatrix} \boldsymbol{\Sigma}_1 & \\ & \boldsymbol{\Sigma}_{2A} \end{bmatrix}^{-1/2} \begin{bmatrix} \mathbf{U}_1^{\mathrm{T}} \\ \mathbf{U}_{2A}^{\mathrm{T}} \end{bmatrix} \check{\boldsymbol{\Gamma}} \begin{bmatrix} \mathbf{V}_1 & \mathbf{V}_{2A} \end{bmatrix} \begin{bmatrix} \boldsymbol{\Sigma}_1 & \\ & \boldsymbol{\Sigma}_{2A} \end{bmatrix}^{-1/2} \\
&= \begin{bmatrix} \boldsymbol{\Sigma}_1^{-1/2} \mathbf{U}_1^{\mathrm{T}} \\ \boldsymbol{\Sigma}_{2A}^{-1/2} \mathbf{U}_{2A}^{\mathrm{T}} \end{bmatrix} \check{\boldsymbol{\Gamma}} \begin{bmatrix} \mathbf{V}_1 \boldsymbol{\Sigma}_1^{-1/2} & \mathbf{V}_{2A} \boldsymbol{\Sigma}_{2A}^{-1/2} \end{bmatrix} \\
&= \begin{bmatrix} \boldsymbol{\Sigma}_1^{-1/2} \mathbf{U}_1^{\mathrm{T}} \check{\boldsymbol{\Gamma}} \mathbf{V}_1 \boldsymbol{\Sigma}_1^{-1/2} & \boldsymbol{\Sigma}_1^{-1/2} \mathbf{U}_1^{\mathrm{T}} \check{\boldsymbol{\Gamma}} \mathbf{V}_{2A} \boldsymbol{\Sigma}_{2A}^{-1/2} \\ \boldsymbol{\Sigma}_{2A}^{-1/2} \mathbf{U}_{2A}^{\mathrm{T}} \check{\boldsymbol{\Gamma}} \mathbf{V}_1 \boldsymbol{\Sigma}_1^{-1/2} & \boldsymbol{\Sigma}_{2A}^{-1/2} \mathbf{U}_{2A}^{\mathrm{T}} \check{\boldsymbol{\Gamma}} \mathbf{V}_{2A} \boldsymbol{\Sigma}_{2A}^{-1/2} \end{bmatrix} \qquad (7.52) \\
\bar{\mathbf{B}} &= \text{first block column of} \begin{bmatrix} \boldsymbol{\Sigma}_1^{1/2} \mathbf{V}_1^{\mathrm{T}} \\ \boldsymbol{\Sigma}_{2A}^{1/2} \mathbf{V}_{2A}^{\mathrm{T}} \end{bmatrix} \qquad (7.53) \\
\bar{\mathbf{C}} &= \text{first block row of} \begin{bmatrix} \mathbf{U}_1 \boldsymbol{\Sigma}_1^{1/2} & \mathbf{U}_{2A} \boldsymbol{\Sigma}_{2A}^{1/2} \end{bmatrix} \qquad (7.54)
\end{aligned}$$

Taking the subsystem 1 leads to the same reduced system as the one constructed directly from $\boldsymbol{\Sigma}_1, \mathbf{U}_1$ and $\mathbf{V}_1$ without first calculating the full matrices. The minimal balanced realization and the approximation are performed in one step and no distinction has to made between small singular values $\boldsymbol{\Sigma}_{2A}$ and zero singular values $\boldsymbol{\Sigma}_{2B}$ in the algorithm. (The distinction is, however, necessary in the proofs.) For dam-reservoir interaction, the finite-dimensional Hankel matrix is obtained from an infinite-dimensional transfer function by truncation and has therefore always full rank, although some singular values are quite small. In this case there are theoretically no zero singular values.

Numerical-rank criterion. In the numerical procedure, a criterion is needed for determining the number of singular values to be included in the realization. The criterion used here is a modification of the theoretical error bound given by Glover [Glo84, p. 1170] stating that the maximum error obtained by taking a size-N subsystem of a balanced system is $\|\mathbf{E}(s)\|_\infty \leq 2(\sigma_{N+1} + \cdots)$. Because the singular values of the finite Hankel matrix depend on the size of this matrix, we use an error relative to the 2-norm of the Hankel matrix. Because this norm equals the largest singular value, we can state the following criterion for selecting the size N of the subsystem

$$\sum_{i=N+1}^{N_\sigma} \sigma_i \leq tol\, \sigma_1 \qquad (7.55)$$

where tol is a prescribed tolerance and N_σ the total number of singular values. Thus we include as many terms necessary such that the sum of the tail is less than a given percentage of the first singular value.

Stability. In Section 6.1.8 it was shown that a subsystem of a balanced system is stable. Unfortunately, these results do not pertain to partially balanced systems. Nevertheless, if L is sufficiently large we expect a nearly balanced and stable system. If there are some unstable poles, they can be removed using the arguments given in Subsection 5.5.2.

7.2 Examples

7.2.1 Spherical cavity

The balanced realization described in the last section gives a fully balanced system only if we use infinite dimensional Hankel matrices. This is generally only possible in simple examples but not for practical applications. Such a simple problem is the one already used earlier

$$H(s) = \frac{1}{1+s} \tag{7.56}$$

This transfer function can be considered as the transfer function of the spherical cavity problem introduced in Section 2.2. The dynamic stress-displacement relation is given in Equation (2.31). The first term can directly be implemented by a dashpot, so only the second term will be considered. To simplify the algebra further, we scale the problem to its simplest form given above.

Using the bilinear transform with $\alpha = 1$ we obtain

$$\tilde{H}(z) = \frac{1}{1+\frac{z-1}{z+1}} = \frac{1}{2} + \frac{1}{2z} \tag{7.57}$$

The transfer function is already in the form of Equation (4.14) and therefore the impulse response is given by $h_0 = h_1 = 1/2$ and $h_k = 0$ for $k > 1$. Also note that $h_0 = H(\alpha) = 1/2$. If, as in more complicated problems, $H(s)$ is not given as a rational function but can only be evaluated at certain frequencies as $H(i\omega_k)$, the impulse response has to be evaluated numerically by the DFT as described in Subsection 4.4.3. Because the impulse response has only two terms, the following calculations are somewhat trivial but nevertheless instructive.

The infinite Hankel matrix and the shifted Hankel matrix are

$$\mathbf{\Gamma} = \begin{bmatrix} 1/2 & 0 & \dots \\ 0 & 0 & \\ \vdots & & \end{bmatrix} \qquad \check{\mathbf{\Gamma}} = \begin{bmatrix} 0 & 0 & \dots \\ 0 & 0 & \\ \vdots & & \end{bmatrix} \tag{7.58}$$

The Hankel matrix has rank 1 and the singular value decomposition yields

$$\mathbf{\Gamma} = \mathbf{U}_1 \mathbf{\Sigma}_1 \mathbf{V}_1^{\mathrm{T}} = \begin{bmatrix} 1 \\ 0 \\ \vdots \end{bmatrix} \begin{bmatrix} 1/2 \end{bmatrix} \begin{bmatrix} 1 & 0 & \dots \end{bmatrix} \tag{7.59}$$

Because the impulse response has degenerated to a finite sequence, there are no infinite sums involved in Equations (7.10–7.13). The balanced realization is

$$\tilde{A} = 0 \qquad \tilde{B} = \tilde{C} = 1/\sqrt{2} \qquad \tilde{D} = 1/2 \tag{7.60}$$

Transforming this back to the continuous-time case by Equations (6.163–6.166) yields

$$A = -1 \qquad B = C = 1 \qquad D = 0 \tag{7.61}$$

This corresponds to the original transfer function because $C(s-A)^{-1}B = 1/(1+s)$.

The Gramians of the discrete-time system can be found by the Lyapunov equations (6.133 and 6.137). Immediately we see that $\tilde{P} = \tilde{Q} = \Sigma = 1/2$. The continuous-time Gramians are given by the Lyapunov equations (6.85 and 6.90). They lead to $P = Q = \Sigma = 1/2$ which is, as expected, the same as for the discrete-time case.

To have a somewhat less degenerate case we take $\alpha = 3$ for the bilinear transformation as in example Equation (4.114). Then we obtain

$$\tilde{H}(z) = \frac{1}{4} + \frac{3}{8}\left(\frac{z^{-1}}{1 + 1/2\, z^{-1}}\right) = \frac{1}{4} + \frac{3}{8}\left(z^{-1} + \frac{1}{2}z^{-2} + \frac{1}{4}z^{-3} + \cdots\right) \tag{7.62}$$

Because of the simplicity of the problem we are able to work with infinite matrices. In a general numerical application this will not be possible. The infinite Hankel matrices are

$$\mathbf{\Gamma} = \frac{3}{8}\begin{bmatrix} 1 & 1/2 & \dots \\ 1/2 & 1/4 & \dots \\ \vdots & & \end{bmatrix} \qquad \check{\mathbf{\Gamma}} = \frac{3}{8}\begin{bmatrix} 1/2 & 1/4 & \dots \\ 1/4 & 1/8 & \dots \\ \vdots & & \end{bmatrix} = \frac{1}{2}\mathbf{\Gamma} \tag{7.63}$$

$\mathbf{\Gamma}$ has rank 1 and the singular value decomposition

$$\mathbf{\Gamma} = \mathbf{U}_1\mathbf{\Sigma}_1\mathbf{V}_1^{\mathrm{T}} = \left(\frac{\sqrt{3}}{2}\begin{bmatrix} 1 \\ 1/2 \\ 1/4 \\ \vdots \end{bmatrix}\right)\begin{bmatrix} 1/2 \end{bmatrix}\left(\frac{\sqrt{3}}{2}\begin{bmatrix} 1 & 1/2 & 1/4 & \dots \end{bmatrix}\right) \tag{7.64}$$

For the scaling of the singular vectors the infinite sum $1 + 1/4 + 1/16 + \cdots = 4/3$ was used. The minimal, fully balanced system is now (Equations 7.10–7.13)

$$\begin{aligned} \tilde{A} &= \sqrt{2}\left(\frac{\sqrt{3}}{2}\begin{bmatrix} 1 & 1/2 & \dots \end{bmatrix}\right)\left(\frac{3}{8}\begin{bmatrix} 1/2 & 1/4 & \dots \\ 1/4 & & \\ \vdots & & \end{bmatrix}\right)\left(\frac{\sqrt{3}}{2}\begin{bmatrix} 1 \\ 1/2 \\ \vdots \end{bmatrix}\right)\sqrt{2} \\ \tilde{B} &= \tilde{C} = \text{first element of } \left(\sqrt{\frac{1}{2}}\frac{\sqrt{3}}{2}\begin{bmatrix} 1 \\ 1/2 \\ \vdots \end{bmatrix}\right) \qquad \tilde{D} = h_0 \end{aligned} \tag{7.65}$$

Simplifying above expressions we obtain the discrete-time system matrices

$$\tilde{A} = \frac{1}{2} \qquad \tilde{B} = \tilde{C} = \frac{1}{2}\sqrt{\frac{3}{2}} \qquad \tilde{D} = \frac{1}{4} \tag{7.66}$$

The discrete-time Lyapunov equations (6.133 and 6.137) yield $\tilde{P} = \tilde{Q} = \tilde{\Sigma} = 1/2$. Transforming the discrete-time system back to the continuous-time case by Equation (6.163–6.166) yields

$$A = -1 \qquad B = C = 1 \qquad D = 0 \tag{7.67}$$

as before. Although the discrete-time system depends on the value α of the bilinear transform, the singular values, the Gramians and the continuous-time system are invariant.

In the next case we consider the same example but take only a (finite) 2×2 Hankel matrix.

$$\mathbf{\Gamma} = \frac{3}{8}\begin{bmatrix} 1 & 1/2 \\ 1/2 & 1/4 \end{bmatrix} \qquad \check{\mathbf{\Gamma}} = \frac{3}{8}\begin{bmatrix} 1/2 & 1/4 \\ 1/4 & 1/8 \end{bmatrix} = \frac{1}{2}\mathbf{\Gamma} \tag{7.68}$$

The singular value decomposition is

$$\mathbf{\Gamma} = \mathbf{U}_1\mathbf{\Sigma}_1\mathbf{V}_1^{\mathrm{T}} = \left(\frac{2}{\sqrt{5}}\begin{bmatrix} 1 \\ 1/2 \end{bmatrix}\right)\begin{bmatrix} 15/32 \end{bmatrix}\left(\frac{2}{\sqrt{5}}\begin{bmatrix} 1 & 1/2 \end{bmatrix}\right) \tag{7.69}$$

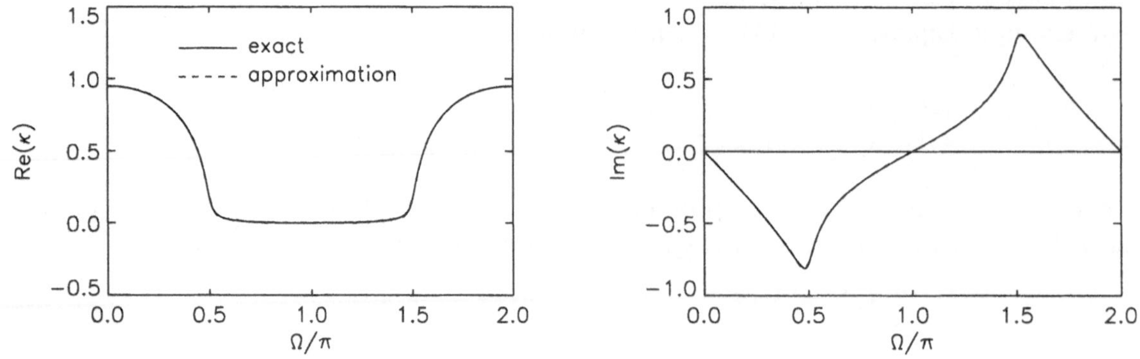

Figure 7.1: Balanced realization approximation ($N = 4$, $L = 20$): transfer function

and the system matrices are

$$\tilde{A} = \frac{1}{2} \qquad \tilde{B} = \tilde{C} = \frac{1}{2}\sqrt{\frac{3}{2}} \qquad \tilde{D} = \frac{1}{4} \tag{7.70}$$

Because the system matrices are the same as for the infinite case, also the Gramians are the same, $\tilde{P} = \tilde{Q} = 1/2$. Only the singular value is different.

The algorithm can also be used for a slightly perturbed sequence. We consider the Hankel matrices

$$\mathbf{\Gamma} = \frac{3}{8}\begin{bmatrix} 1 & 0.51 \\ 0.51 & 0.25 \end{bmatrix} \qquad \check{\mathbf{\Gamma}} = \frac{3}{8}\begin{bmatrix} 0.51 & 0.25 \\ 0.25 & 0.13 \end{bmatrix} \tag{7.71}$$

The singular value decomposition is

$$\mathbf{\Gamma} = \mathbf{U}_1\mathbf{\Sigma}_1\mathbf{V}_1^{\mathrm{T}} = \begin{bmatrix} 0.892 & -0.451 \\ 0.451 & 0.892 \end{bmatrix}\begin{bmatrix} 0.4718 & \\ & 0.0030 \end{bmatrix}\begin{bmatrix} 0.892 & 0.451 \\ 0.451 & -0.892 \end{bmatrix} \tag{7.72}$$

the discrete-time system matrices are

$$\tilde{A} = 0.504 \qquad \tilde{B} = \tilde{C} = 0.613 \qquad \tilde{D} = 0.25 \tag{7.73}$$

and the continuous-time system matrices are

$$A = -0.990 \qquad B = C = 0.998 \qquad D = 0 \tag{7.74}$$

The Gramians are

$$P = Q = 0.503 \tag{7.75}$$

In the case of a slightly perturbed sequence, the algorithm yields a good approximation. The procedure is robust, which is an advantage over the simple methods investigated in Subsection 5.2.3

7.2.2 Rod on visco-elastic foundation

As a second example we repeat the problem of approximating the dynamic stiffness of a rod on a visco-elastic foundation. We show the curves corresponding to the equivalent discrete-time system as in Chapter 5. Figure 7.1 shows the approximation for the real and the imaginary part. The agreement is excellent. In order to compare to other approximation methods, we show in Figure 7.2 the error curve. The balanced realization approximation is not quite as good as the CF approximation (Figure 5.18) but much better than the least-squares Padé approximation (Figure 5.15). Its main advantages are that it is numerically simple to construct and that it is also applicable to multivariable systems.

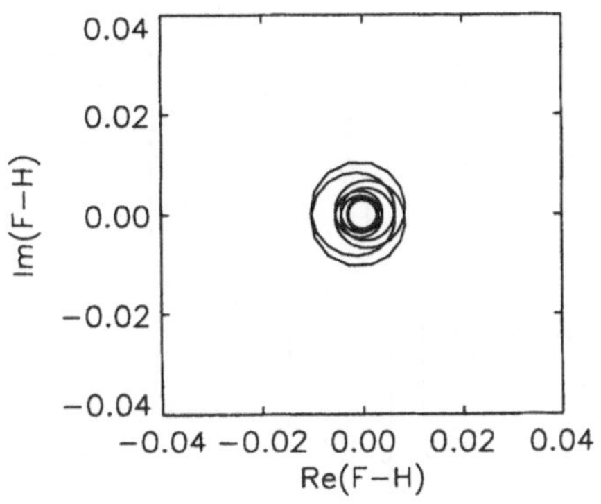

Figure 7.2: Balanced realization approximation ($N = 4$, $L = 20$): error curve

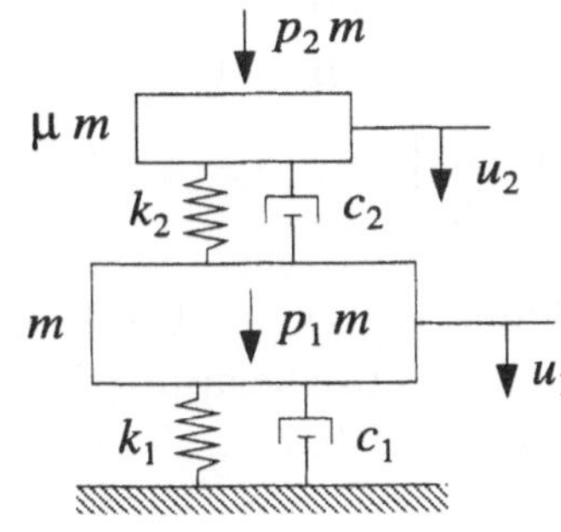

Figure 7.3: Two-degree-of-freedom system

7.2.3 Two-degree-of-freedom system

In this last, extensive example we want to show how the proposed method can be applied to a multivariable system. Here we also give the numerical values of the matrices to illustrate the theory.

Consider the two-degree-of-freedom system shown in Figure 7.3. The natural frequencies are given by $\omega_1^2 = k_1/m$, $\omega_2^2 = k_2/(\mu m)$ and the damping by $2\xi_1\omega_1 = c_1/m$, $2\xi_2\omega_2 = c_2/(\mu m)$ where μ is the mass ratio. The dynamic equations are written in the usual second-order form as

$$\begin{bmatrix} 1 & 0 \\ 0 & \mu \end{bmatrix}\begin{bmatrix} \ddot{u}_1 \\ \ddot{u}_2 \end{bmatrix} + \begin{bmatrix} 2(\xi_1\omega_1 + \mu\xi_2\omega_2) & -2\mu\xi_2\omega_2 \\ -2\mu\xi_2\omega_2 & 2\mu\xi_2\omega_2 \end{bmatrix}\begin{bmatrix} \dot{u}_1 \\ \dot{u}_2 \end{bmatrix} + \begin{bmatrix} \omega_1^2 + \mu\omega_2^2 & -\mu\omega_2^2 \\ -\mu\omega_2^2 & \mu\omega_2^2 \end{bmatrix}\begin{bmatrix} u_1 \\ u_2 \end{bmatrix} = \begin{bmatrix} p_1 \\ p_2 \end{bmatrix} \tag{7.76}$$

Taking the numerical values $\omega_1 = 2$, $\omega_2 = 4$, $\xi_1 = \xi_2 = 0.05$ and $\mu = 0.5$, the equation reads

$$\begin{bmatrix} 1 & 0 \\ 0 & 0.5 \end{bmatrix}\begin{bmatrix} \ddot{u}_1 \\ \ddot{u}_2 \end{bmatrix} + \begin{bmatrix} 0.4 & -0.2 \\ -0.2 & 0.2 \end{bmatrix}\begin{bmatrix} \dot{u}_1 \\ \dot{u}_2 \end{bmatrix} + \begin{bmatrix} 12 & -8 \\ -8 & 8 \end{bmatrix}\begin{bmatrix} u_1 \\ u_2 \end{bmatrix} = \begin{bmatrix} p_1 \\ p_2 \end{bmatrix} \tag{7.77}$$

These two second-order differential equations can be converted into four first-order equations. The four state variables are the displacements u_1 and u_2 and the velocities $\dot{u}_1$ and $\dot{u}_2$. In order to have a multivariable system we consider the two input variables p_1 and p_2.

The output variables are the two displacements u_1 and u_2. The state-variable form is

$$\begin{bmatrix} \dot{u}_1 \\ \ddot{u}_1 \\ \dot{u}_2 \\ \ddot{u}_2 \end{bmatrix} = \begin{bmatrix} 0 & 1 & 0 & 0 \\ -12 & -0.4 & 8 & 0.2 \\ 0 & 0 & 0 & 1 \\ 16 & 0.4 & -16 & -0.4 \end{bmatrix} \begin{bmatrix} u_1 \\ \dot{u}_1 \\ u_2 \\ \dot{u}_2 \end{bmatrix} + \begin{bmatrix} 0 & 0 \\ 1 & 0 \\ 0 & 0 \\ 0 & 2 \end{bmatrix} \begin{bmatrix} p_1 \\ p_2 \end{bmatrix}$$
$$\begin{bmatrix} u_1 \\ u_2 \end{bmatrix} = \begin{bmatrix} 1 & 0 & 0 & 0 \\ 0 & 0 & 1 & 0 \end{bmatrix} \begin{bmatrix} u_1 \\ \dot{u}_1 \\ u_2 \\ \dot{u}_2 \end{bmatrix} \tag{7.78}$$

We identify the system matrices

$$\mathbf{A}' = \begin{bmatrix} 0 & 1 & 0 & 0 \\ -12 & -0.4 & 8 & 0.2 \\ 0 & 0 & 0 & 1 \\ 16 & 0.4 & -16 & -0.4 \end{bmatrix} \qquad \mathbf{B}' = \begin{bmatrix} 0 & 0 \\ 1 & 0 \\ 0 & 0 \\ 0 & 2 \end{bmatrix}$$
$$\mathbf{C}' = \begin{bmatrix} 1 & 0 & 0 & 0 \\ 0 & 0 & 1 & 0 \end{bmatrix} \tag{7.79}$$

These matrices represent the original system and are marked by a prime to distinguish them from the matrices obtained later on by the approximation procedure. The eigenvalues of $\mathbf{A}'$ are $-0.0607 \pm 1.583i$ and $-0.339 \pm 5.037i$. They are used later on for comparison purposes.

Instead of calculating the discrete-time Hankel matrices by the Fourier transform, we apply the bilinear transform directly to the system matrices and calculate the Hankel matrices from the mapped discrete-time system. This procedure is not generally possible but is used here to simplify the example.

The discrete-time system matrices are given by inversion of Equations (6.163–6.166)

$$\tilde{\mathbf{A}}' = (\alpha\mathbf{I} + \mathbf{A}')(\alpha\mathbf{I} - \mathbf{A}')^{-1} \tag{7.80}$$
$$\tilde{\mathbf{B}}' = \sqrt{2\alpha}\,(\alpha\mathbf{I} - \mathbf{A}')^{-1}\mathbf{B}' \tag{7.81}$$
$$\tilde{\mathbf{C}}' = \sqrt{2\alpha}\,\mathbf{C}'\,(\alpha\mathbf{I} - \mathbf{A}')^{-1} \tag{7.82}$$
$$\tilde{\mathbf{D}}' = \mathbf{D}' + \mathbf{C}'\,(\alpha\mathbf{I} - \mathbf{A}')^{-1}\mathbf{B}' \tag{7.83}$$

Numerically with $\alpha = 1$ we obtain

$$\tilde{\mathbf{A}}' = \begin{bmatrix} -0.573 & 0.353 & 0.162 & 0.166 \\ -1.573 & -0.647 & 0.162 & 0.166 \\ 0.357 & 0.332 & -0.686 & 0.272 \\ 0.357 & 0.332 & -1.686 & -0.728 \end{bmatrix} \qquad \tilde{\mathbf{B}}' = \begin{bmatrix} 0.249 & 0.235 \\ 0.249 & 0.235 \\ 0.235 & 0.384 \\ 0.235 & 0.384 \end{bmatrix}$$
$$\tilde{\mathbf{C}}' = \begin{bmatrix} 0.302 & 0.249 & 0.115 & 0.118 \\ 0.252 & 0.235 & 0.222 & 0.192 \end{bmatrix} \qquad \tilde{\mathbf{D}}' = \begin{bmatrix} 0.176 & 0.166 \\ 0.166 & 0.272 \end{bmatrix} \tag{7.84}$$

The matrix $\tilde{\mathbf{D}}'$ arises from the bilinear transformation even if the original matrix $\mathbf{D}'$ was zero. Since there is no dynamics involved in this term, it is not considered in the approximation. It can be added again after the reduced system has been found.

The block Hankel matrices are given by

$$\mathbf{\Gamma} = \begin{bmatrix} \mathbf{f}_1 & \mathbf{f}_2 & \cdots \\ \mathbf{f}_2 & \mathbf{f}_3 & \cdots \\ \vdots & & \end{bmatrix} \qquad \check{\mathbf{\Gamma}} = \begin{bmatrix} \mathbf{f}_2 & \mathbf{f}_3 & \cdots \\ \mathbf{f}_3 & \mathbf{f}_4 & \cdots \\ \vdots & & \end{bmatrix} \tag{7.85}$$

and the impulse response matrices are

$$\mathbf{f}_k = \tilde{\mathbf{C}}' \, \tilde{\mathbf{A}}'^{k-1} \, \tilde{\mathbf{B}}' \tag{7.86}$$

The sequence for this example is

$$\begin{aligned}
\mathbf{f}_1 &= \begin{bmatrix} 0.192 & 0.219 \\ 0.219 & 0.273 \end{bmatrix} & \mathbf{f}_2 &= \begin{bmatrix} -0.150 & -0.166 \\ -0.166 & -0.221 \end{bmatrix} & \mathbf{f}_3 &= \begin{bmatrix} -0.050 & -0.074 \\ -0.074 & -0.059 \end{bmatrix} \\
\mathbf{f}_4 &= \begin{bmatrix} 0.176 & 0.225 \\ 0.225 & 0.240 \end{bmatrix} & \mathbf{f}_5 &= \begin{bmatrix} -0.095 & -0.127 \\ -0.127 & -0.130 \end{bmatrix} & \mathbf{f}_6 &= \begin{bmatrix} -0.090 & -0.096 \\ -0.096 & -0.129 \end{bmatrix} \\
\mathbf{f}_7 &= \begin{bmatrix} 0.167 & 0.194 \\ 0.194 & 0.237 \end{bmatrix} & \mathbf{f}_8 &= \begin{bmatrix} -0.056 & -0.069 \\ -0.069 & -0.080 \end{bmatrix}
\end{aligned} \tag{7.87}$$

The reader is again reminded that the above procedure of getting the Hankel matrix has been chosen for convenience in this example. Generally, only the transfer function matrix $\mathbf{F}(i\omega_k)$ is known at certain frequency points and the Hankel matrix has to be evaluated entry by entry by the DFT.

Since we know that $N = 4$ we have to take at least four columns ($L = 2$ block columns) of the block Hankel matrix. In a first step we take exactly the four columns. We perform the singular value decomposition of the Hankel matrix $\mathbf{\Gamma} = \mathbf{U}\,\mathbf{\Sigma}\,\mathbf{V}^{\mathrm{T}}$ and obtain

$$\begin{aligned}
\mathbf{U} &= \begin{bmatrix} 0.575 & 0.264 & -0.481 & -0.607 \\ 0.697 & 0.336 & 0.372 & 0.512 \\ -0.263 & 0.593 & 0.600 & -0.468 \\ -0.337 & 0.683 & -0.519 & 0.389 \end{bmatrix} \\
\mathbf{V} &= \begin{bmatrix} 0.575 & -0.264 & -0.481 & 0.607 \\ 0.697 & -0.336 & 0.372 & -0.512 \\ -0.263 & -0.593 & 0.600 & 0.468 \\ -0.337 & -0.683 & -0.519 & -0.389 \end{bmatrix} \\
\mathbf{\Sigma} &= \mathbf{diag}(0.6232,\ 0.2955,\ 0.0310,\ 0.0019)
\end{aligned} \tag{7.88}$$

As expected, all four singular values are non-zero because the system has degree four. As explained earlier (Equation 5.53), the columns of $\mathbf{U}$ and $\mathbf{V}$ are equal up to the sign.

With Equations (7.10–7.13) we find the discrete-time system matrices

$$\begin{aligned}
\tilde{\mathbf{A}} &= \begin{bmatrix} -0.180 & 0.900 & -0.035 & 0.007 \\ -0.900 & -0.651 & 0.079 & -0.015 \\ -0.035 & -0.079 & -1.162 & 0.454 \\ -0.007 & -0.015 & -0.454 & -0.641 \end{bmatrix} & \tilde{\mathbf{B}} &= \begin{bmatrix} 0.454 & 0.551 \\ -0.143 & -0.183 \\ -0.085 & 0.066 \\ 0.026 & -0.022 \end{bmatrix} \\
\tilde{\mathbf{C}} &= \begin{bmatrix} 0.454 & 0.143 & -0.085 & -0.026 \\ 0.551 & 0.183 & 0.066 & 0.022 \end{bmatrix} & \tilde{\mathbf{D}} &= \begin{bmatrix} 0.176 & 0.166 \\ 0.166 & 0.272 \end{bmatrix}
\end{aligned} \tag{7.89}$$

The matrix $\tilde{\mathbf{D}} = \tilde{\mathbf{D}}'$ has been added directly from Equation (7.84). Note the sign symmetries of $\tilde{\mathbf{A}}$ and of $\tilde{\mathbf{B}}$ and $\tilde{\mathbf{C}}$. This symmetry will always be present for input-output symmetric systems. There is actually a special sign pattern, because the different signs originate from the different signs in the singular vectors. The system can easily be changed to symmetric form by the transformation matrix $\mathbf{T} = \mathbf{\Theta}^{1/2}$, where $\mathbf{\Theta}$ is the sign matrix (Equation 5.53) such that $\mathbf{U} = \mathbf{V}\mathbf{\Theta}^{-1}$. The resulting matrices are symmetric consisting of real and pure imaginary numbers. This symmetry guarantees that the system can always be diagonalized which will be used to find the symmetric second-order form in Chapter 8.

As a verification of the theory we calculate the controllability Gramian as

$$\mathbf{P} = \mathbf{B}\mathbf{B}^{\mathrm{T}} + \mathbf{A}\mathbf{B}\mathbf{B}^{\mathrm{T}}\mathbf{A}^{\mathrm{T}} = \begin{bmatrix} 0.6232 & 0.0 & 0.0 & 0.0 \\ 0.0 & 0.2955 & 0.0 & 0.0 \\ 0.0 & 0.0 & 0.0310 & 0.0 \\ 0.0 & 0.0 & 0.0 & 0.0019 \end{bmatrix} \tag{7.90}$$

Clearly this matrix is diagonal and equal to the singular values. The same result is obtained for the observability Gramian.

If we transform this system back to the continuous-time domain by Equations (6.163–6.166) we get the original system, although not in the original form.

$$\begin{aligned} \mathbf{A} &= \begin{bmatrix} 0.349 & 1.636 & -0.279 & 0.435 \\ -1.636 & -0.493 & 0.115 & -0.173 \\ -0.279 & -0.115 & -3.854 & 6.141 \\ -0.435 & -0.173 & -6.141 & 3.198 \end{bmatrix} \quad \mathbf{B} = \begin{bmatrix} 0.350 & 0.485 \\ 0.384 & 0.436 \\ -0.327 & 0.416 \\ -0.287 & 0.466 \end{bmatrix} \\ \mathbf{C} &= \begin{bmatrix} 0.350 & -0.384 & -0.327 & 0.287 \\ 0.485 & -0.436 & 0.416 & -0.466 \end{bmatrix} \quad \mathbf{D} = \mathbf{0} \end{aligned} \tag{7.91}$$

To show that this system is equivalent to the original system one has to plot the transfer function. Here we just give the eigenvalues of $\bar{\mathbf{A}}$ which are $-0.0607 \pm 1.583i$ and $-0.339 \pm 5.037i$. They are identical to the ones of original system.

To see how well the system is balanced, the Gramians related to an infinite time interval are calculated by the Lyapunov equations. The Gramians are identical for the discrete-time system as for the continuous-time system namely

$$\mathbf{P}_\infty = \begin{bmatrix} 3.932 & -0.997 & -0.220 & 0.044 \\ -0.997 & 3.666 & -0.079 & -0.101 \\ -0.220 & -0.079 & 0.318 & 0.166 \\ 0.044 & -0.101 & 0.166 & 0.272 \end{bmatrix} \tag{7.92}$$

and

$$\mathbf{Q}_\infty = \begin{bmatrix} 3.932 & 0.997 & -0.220 & -0.044 \\ 0.997 & 3.666 & 0.079 & -0.101 \\ -0.220 & 0.079 & 0.318 & -0.166 \\ -0.044 & -0.101 & -0.166 & 0.272 \end{bmatrix} \tag{7.93}$$

As can be seen the sign symmetry of the system matrices propagates to the Gramians. Because we took only a 4×4 Hankel matrix, the system is only partially balanced. One should not be confused by the observation that the diagonal terms are quite different from the singular values. The singular values are those of the (small) finite Hankel matrix system while the Gramian obtained by the Lyapunov equation pertains to the infinite Hankel matrix.

Even though the system is not well balanced for an infinite time interval we can still try to find a lower order approximation by truncating the system to degree two. Taking submatrices of Equation (7.89) yields

$$\begin{aligned} \tilde{\mathbf{A}} &= \begin{bmatrix} -0.180 & 0.900 \\ -0.900 & -0.651 \end{bmatrix} \quad \tilde{\mathbf{B}} = \begin{bmatrix} 0.454 & 0.551 \\ -0.143 & -0.183 \end{bmatrix} \\ \tilde{\mathbf{C}} &= \begin{bmatrix} 0.454 & 0.143 \\ 0.551 & 0.183 \end{bmatrix} \quad \tilde{\mathbf{D}} = \begin{bmatrix} 0.176 & 0.166 \\ 0.166 & 0.272 \end{bmatrix} \end{aligned} \tag{7.94}$$

Again the matrix $\tilde{\mathbf{D}}$ has been copied unchanged.

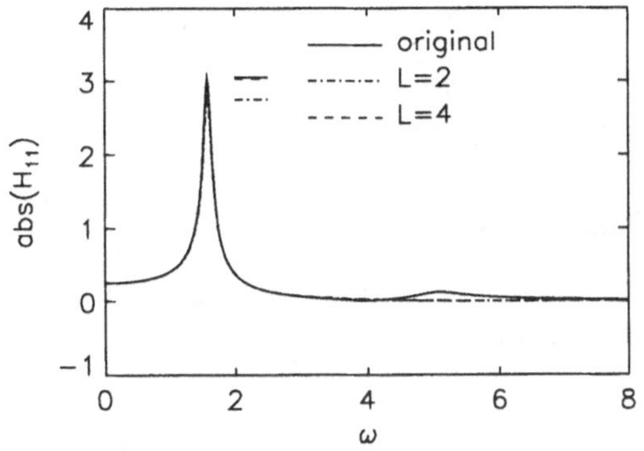

Figure 7.4: Balanced realization approximation of two-mass model: transfer function

The reduced system can be transformed back to the continuous-time domain by Equations (6.163–6.166). This leads to the system matrices

$$\begin{aligned} \mathbf{A} &= \begin{bmatrix} 0.364 & 1.642 \\ -1.642 & -0.496 \end{bmatrix} & \mathbf{B} = \begin{bmatrix} 0.371 & 0.460 \\ 0.376 & 0.446 \end{bmatrix} \\ \mathbf{C} &= \begin{bmatrix} 0.371 & -0.376 \\ 0.460 & -0.446 \end{bmatrix} & \mathbf{D} = \begin{bmatrix} 0.019 & -0.027 \\ -0.027 & 0.035 \end{bmatrix} \end{aligned} \tag{7.95}$$

Note, that the result is not quite the same as if we take the subsystem from the full continuous-time system Equation (7.91). Especially the matrix $\mathbf{D}$ does not change completely back to zero.

The comparison of the reduced system (Equation 7.95) with the original system is shown in Figure 7.4. The approximation, labeled as '$L = 2$' is relatively good for the first resonance frequency of the transfer function although the peak value of the original system is not quite obtained (the short horizontal lines near the peaks show the extreme values). The second peak cannot be captured because the two conjugate complex poles correspond to a single-degree-of-freedom system.

In the next step we take the 8×8 ($L = 4$ block columns) Hankel matrix to get a more balanced system and, hopefully, a better approximation. The singular value decomposition yields for $\mathbf{\Gamma} = \mathbf{U}_1\,\mathbf{\Sigma}_1\,\mathbf{V}_1^{\mathrm{T}}$

$$\begin{aligned} \mathbf{U}_1 &= \begin{bmatrix} 0.417 & 0.091 & 0.266 & -0.505 \\ 0.493 & 0.129 & -0.437 & 0.414 \\ -0.251 & 0.450 & -0.391 & 0.166 \\ -0.302 & 0.514 & 0.367 & -0.101 \\ -0.181 & -0.465 & 0.354 & 0.230 \\ -0.208 & -0.538 & -0.363 & -0.153 \\ 0.386 & -0.028 & -0.190 & -0.484 \\ 0.452 & -0.041 & 0.397 & 0.475 \end{bmatrix} \\ \mathbf{\Sigma} &= \mathbf{diag}(1.1261,\ 0.6452,\ 0.0681,\ 0.0132) \end{aligned} \tag{7.96}$$

The matrix $\mathbf{V}_1$ equals the matrix $\mathbf{U}_1$ with sign changes in columns 2 and 4. As expected only four singular values are non-zero. Only the blocks corresponding to the non-zero singular values are shown. With Equations (7.10–7.13) we find the balanced system matrices of degree four as

$$\tilde{\mathbf{A}} = \begin{bmatrix} -0.449 & 0.873 & -0.036 & -0.020 \\ -0.873 & -0.379 & -0.042 & 0.002 \\ -0.036 & 0.042 & -0.887 & -0.371 \\ 0.020 & 0.002 & 0.371 & -0.919 \end{bmatrix} \quad \tilde{\mathbf{B}} = \begin{bmatrix} 0.443 & 0.523 \\ -0.073 & -0.104 \\ 0.069 & -0.114 \\ 0.058 & -0.048 \end{bmatrix}$$

$$\tilde{\mathbf{C}} = \begin{bmatrix} 0.443 & 0.073 & 0.069 & -0.058 \\ 0.523 & 0.104 & -0.114 & 0.048 \end{bmatrix} \quad \tilde{\mathbf{D}} = \begin{bmatrix} 0.176 & 0.166 \\ 0.166 & 0.272 \end{bmatrix} \tag{7.97}$$

The matrix $\tilde{\mathbf{D}}$ has again be copied from Equation (7.84).

To see how well this system is balanced we solve the Lypunov equations to get

$$\mathbf{P}_\infty = \begin{bmatrix} 3.820 & 0.150 & 0.044 & 0.119 \\ 0.150 & 3.547 & 0.219 & -0.036 \\ 0.044 & 0.219 & 0.261 & 0.008 \\ 0.119 & -0.036 & 0.008 & 0.221 \end{bmatrix} \tag{7.98}$$

The matrix $\mathbf{Q}_\infty$ is again the same up to sign changes and is not shown. The system is now much better balanced. The subsystem of order two is again used as an approximation and transformed to the continuous-time case by Equations (6.163–6.166). This leads to the system matrices

$$\mathbf{A} = \begin{bmatrix} -0.125 & 1.581 \\ -1.581 & 0.003 \end{bmatrix} \quad \mathbf{B} = \begin{bmatrix} 0.433 & 0.532 \\ 0.444 & 0.512 \end{bmatrix}$$

$$\mathbf{C} = \begin{bmatrix} 0.433 & -0.444 \\ 0.532 & -0.512 \end{bmatrix} \quad \mathbf{D} = \begin{bmatrix} 0.018 & -0.027 \\ -0.027 & 0.038 \end{bmatrix} \tag{7.99}$$

The eigenvalues of $\mathbf{A}$ are $-0.0608 \pm 1.580i$. They are quite similar to the first two eigenvalues of the original system. Thus the approximation is, for this example, close to the first mode of the original system. The approximation is shown in Figure 7.4 as the curve labeled '$L = 4$'. The first peak is now very well captured, while the second peak cannot be approximated by a system of degree two.

To get an almost fully balanced system we have to take a 160×160 Hankel matrix. This leads to the singular values

$$\mathbf{\Sigma}_1 = \mathbf{diag}(3.800,\ 3.517,\ 0.243,\ 0.213) \tag{7.100}$$

and to the infinite-time controllability Gramian

$$\mathbf{P}_\infty = \begin{bmatrix} 3.814 & -0.001 & 0.000 & 0.000 \\ -0.001 & 3.532 & 0.000 & 0.000 \\ 0.000 & 0.000 & 0.247 & 0.000 \\ 0.000 & 0.000 & 0.000 & 0.216 \end{bmatrix} \tag{7.101}$$

Although a large Hankel matrix is necessary to get a fully balanced system, only a relatively small matrix is necessary to get a good approximation. The large size is only necessary to get the right singular values which is not our primary concern.

7.3 Conclusions

After having investigated a number of different approximation methods, it is now the point to give some interim conclusions. We have started with various methods approximating sampling points of the transfer function or the first few Markov parameters. These methods only involved simple mathematics, such as solving linear equations or least-squares problems. We have seen, that best results were obtained by approximating the Markov parameters of a mapped discrete-time system obtained by the bilinear transformation. The approximated system can then be transformed back to the continuous-time domain. The disadvantage of these methods is that we have to choose the degree of the approximant in advance. If the degree is to high, the problem becomes very sensitive to round-off errors and instabilities may arise.

A method that overcomes these drawbacks is the CF method. With this method we can choose the degree of the approximant using the singular values of the Hankel matrix as an indication. The approximation gives very accurate results. Although there exist extensions to the multivariable case, they are quite involved and have not been investigated here.

The balanced realization method considered in this chapter is slightly less accurate. On the other hand, it is numerically somewhat simpler, because we do not have to find the roots of a high-degree polynomial. Diagonalization of the system is always possible and therefore also the symmetric second-order form. The main advantage, however, is that this method is equally applicable to multivariable systems. The limiting factor is only the size of the block Hankel matrix that can be handled numerically.

Chapter 8

Symmetric Second-Order Systems

For the approximation problem it is practical to use the general first-order formulation. However, for the coupled system it is desirable to have a symmetric second-order formulation because this is the form of the finite element model. This allows us to use the standard finite element matrix formulation for the whole system. The symmetrization of the matrices is not absolutely necessary. The non-symmetric matrices could be treated as an explicit element in a implicit-explicit formulation [MFH89]. However, this solution may lead to a very small time step based on the stability requirement for the explicit scheme. The small time step dominates the whole analysis unless a more complicated strategy using different time steps for different parts of the model is employed. Thus the symmetrization of the matrices is a very desirable feature.

8.1 Scalar systems

8.1.1 Symmetric systems

As an introduction we start with scalar systems. The general second-order form is

$$H(s) = \frac{c_1 s + c_2}{a_0\, s^2 + a_1 s + a_2} \tag{8.1}$$

This form, however, is not the desired symmetric form. This can be seen when trying to couple this system to the system

$$M\ddot{x} + C\dot{x} + Kx = F \tag{8.2}$$

The form of Equation (8.1) corresponds to the following set of differential equations:

$$a_0\,\ddot{y} + a_1\,\dot{y} + a_2\,y = -x \tag{8.3}$$

$$F = -(c_1\,\dot{y} + c_2\,y) \tag{8.4}$$

The coupled system is therefore

$$\begin{bmatrix} M & \\ & a_0 \end{bmatrix}\begin{bmatrix} \ddot{x} \\ \ddot{y} \end{bmatrix} + \begin{bmatrix} C & c_1 \\ & a_1 \end{bmatrix}\begin{bmatrix} \dot{x} \\ \dot{y} \end{bmatrix} + \begin{bmatrix} K & c_2 \\ 1 & a_2 \end{bmatrix}\begin{bmatrix} x \\ y \end{bmatrix} = \begin{bmatrix} 0 \\ 0 \end{bmatrix} \tag{8.5}$$

Clearly, the matrices are not symmetric. The following form, however, will lead to symmetric matrices for the coupled problem:

$$H(s) = \frac{(b_1 s + b_2)^2}{a_0\, s^2 + a_1 s + a_2} - \frac{b_1^2}{a_0} \tag{8.6}$$

The second part, $-b_1^2/a_0$, must be introduced to account for the high-frequency behavior. The problem is that the first part of Equation (8.6) introduces a term with s^2 in the numerator which has to be removed by the second part. Expanding Equation (8.6) yields

$$\frac{(b_1 s + b_2)^2}{a_0\, s^2 + a_1 s + a_2} - \frac{b_1^2}{a_0} = \frac{(2b_1 b_2 - a_1 b_1^2/a_0)s + (b_2^2 - a_2 b_1^2/a_0)}{a_0\, s^2 + a_1 s + a_2} \tag{8.7}$$

which has the general form of Equation (8.1). The symmetric system, Equation (8.6), corresponds to the set of differential equations

$$a_0\, \ddot{y} + a_1\, \dot{y} + a_2\, y = -(b_1\, \dot{x} + b_2\, x) \tag{8.8}$$

$$F = -(b_1\, \dot{y} + b_2\, y) - (b_1^2/a_0)x \tag{8.9}$$

The new coupled system is then

$$\begin{bmatrix} M & \\ & a_0 \end{bmatrix} \begin{bmatrix} \ddot{x} \\ \ddot{y} \end{bmatrix} + \begin{bmatrix} C & b_1 \\ b_1 & a_1 \end{bmatrix} \begin{bmatrix} \dot{x} \\ \dot{y} \end{bmatrix} + \begin{bmatrix} K + b_1^2/a_0 & b_2 \\ b_2 & a_2 \end{bmatrix} \begin{bmatrix} x \\ y \end{bmatrix} = \begin{bmatrix} 0 \\ 0 \end{bmatrix} \tag{8.10}$$

which has the desired symmetric form.

8.1.2 Realizations

First-order system. Starting with the simplest case, we first try to put the system

$$H(s) = \frac{\beta}{s - \lambda} \tag{8.11}$$

with one real pole λ into the symmetric form of Equation (8.6). Obviously, setting $a_0 = 0$ and $b_1 = 0$, we have

$$\frac{\beta}{s - \lambda} = \frac{a_1\, \beta}{a_1 s - a_1\, \lambda} = \frac{b_2^2}{a_1 s + a_2} \tag{8.12}$$

which has the desired form. The parameters have to satisfy

$$a_2 = -\lambda a_1 \qquad b_2^2 = \beta\, a_1 \tag{8.13}$$

For b_2 to be real, we have to ensure that $\beta\, a_1 \geq 0$. This can always be achieved by choosing the free parameter as $a_1 = \pm 1$. Equation (8.12) corresponds to a system with stiffness and damping but no mass.

A different form can be found by assuming the general form with $a_2 = 0$ and $b_2 = 0$:

$$\frac{(b_1 s)^2}{a_0\, s^2 + a_1 s} - \frac{b_1^2}{a_0} = -\frac{a_1\, b_1^2 s/a_0}{a_0\, s^2 + a_1 s} \tag{8.14}$$

We have to expand the original system by $(a_0\, s)$ as

$$\frac{\beta}{s - \lambda} = \frac{\beta a_0\, s}{a_0\, s^2 - a_0\, \lambda\, s} \tag{8.15}$$

to find the parameters

$$a_1 = -a_0\, \lambda \qquad b_1^2 = a_0\, \beta/\lambda \tag{8.16}$$

The free parameter is chosen as $a_0 = \pm 1$ to ensure that $b_1^2 \geq 0$ and hence that b_1 is real. Equation (8.14) corresponds to a system with mass and damping but no stiffness.

Two first-order systems. Proceeding with simple systems, the next step is to consider the combination of two first-order systems,

$$H(s) = \frac{\beta_1}{s - \lambda_1} + \frac{\beta_2}{s - \lambda_2} \tag{8.17}$$

and to change it to a second-order symmetric system. The following identity is useful:

$$\frac{[e_1\,(s - \lambda_2) + e_2\,(s - \lambda_1)]^2}{a_0\,(s - \lambda_1)(s - \lambda_2)} - \frac{(e_1 + e_2)^2}{a_0} = \frac{e_1^2\,(\lambda_1 - \lambda_2)}{a_0\,(s - \lambda_1)} + \frac{e_2^2\,(\lambda_2 - \lambda_1)}{a_0\,(s - \lambda_2)} \tag{8.18}$$

Comparing the right-hand side of Equation (8.18) with Equation (8.17) we find

$$e_1 = \sqrt{a_0\,\beta_1/(\lambda_1 - \lambda_2)} \qquad e_2 = \sqrt{a_0\,\beta_2/(\lambda_2 - \lambda_1)} \tag{8.19}$$

But comparing the left-hand side of Equation (8.18) with Equation (8.6) yields

$$b_1 = e_1 + e_2 \qquad b_2 = -(e_1 \lambda_2 + e_2 \lambda_1) \tag{8.20}$$

$$a_1 = -a_0\,(\lambda_1 + \lambda_2) \qquad a_2 = a_0\,\lambda_1\,\lambda_2 \tag{8.21}$$

If two first-order systems with real poles are combined, e_1 and e_2, and therefore also b_1 and b_2, are not necessarily real, even if the free parameter can be chosen as $a_0 = \pm 1$. Therefore, first-order systems with real poles will generally have to be implemented in one of the forms Equation (8.12) or (8.14). Note that if Equations (8.20) and (8.21) are used for a first-order system by setting $\beta_2 = 0$ and $\lambda_2 = 0$, the result corresponds to the implementation Equation (8.14).

Complex conjugate poles. The case is different if two first-order systems with complex conjugate eigenvalues are combined:

$$H(s) = \frac{\beta}{s - \lambda} + \frac{\bar{\beta}}{s - \bar{\lambda}} \tag{8.22}$$

Then, with $a_0 = 1$, we obtain the parameters

$$e_1 = \sqrt{\beta/(\lambda - \bar{\lambda})} = e \qquad e_2 = \sqrt{\bar{\beta}/(\bar{\lambda} - \lambda)} = \bar{e} \tag{8.23}$$

and

$$b_1 = e + \bar{e} \qquad b_2 = -(e\bar{\lambda} + \bar{e}\lambda) \tag{8.24}$$

$$a_1 = -(\lambda + \bar{\lambda}) \qquad a_2 = \lambda\bar{\lambda} \tag{8.25}$$

All coefficients are real because all complex numbers appear in sums or products of complex conjugate values. Therefore, two first-order systems with complex conjugate poles are always implemented in the form of Equation (8.6).

8.1.3 Physical models

The symmetric systems derived algebraically can be represented by simple physical models, also called lumped-parameter models. These models are sometimes useful for understanding the physical background, as in the case of visco-elastic material models [Flu75]. For the description of mathematical systems they are of limited use, because mass, damping and stiffness may have negative values, which is not compatible with real mechanical systems. Also, additional internal degrees of freedom have to be introduced and the number of lumped parameters is generally greater than the number of coefficients in the mathematical system. This makes the lumped-parameter models more difficult to understand

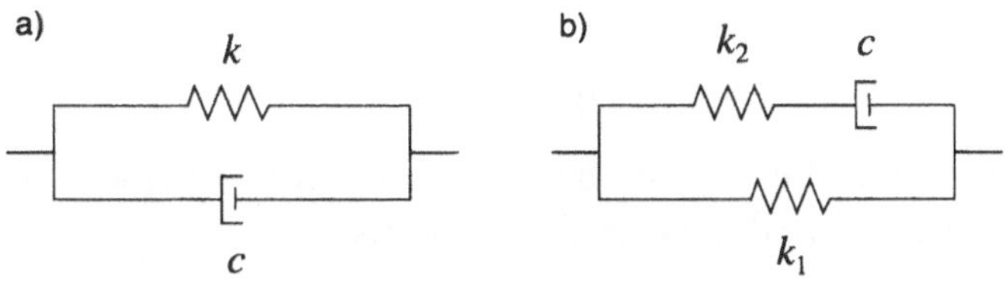

Figure 8.1: Physical models for first-order system: a) flexibility, b) stiffness

and less efficient in the implementation than the mathematical systems. Nevertheless, they have been discussed extensively in the literature [Wol91].

Here, only some of the simple cases are shown. Force-displacement models are used (not pressure-velocity models). The lumped parameters are determined such that the stiffness or the flexibility matches the given transfer function. The model shown in Figure 8.1a) has the flexibility

$$F(s) = \frac{1}{k + cs} \tag{8.26}$$

which has the desired form of Equation (8.12). A model whose stiffness has the same form is shown in Figure 8.1b). One additional degree of freedom and one additional element have to be introduced. The stiffness is

$$S(s) = k_1 + \frac{1}{1/k_2 + 1/(cs)} = \frac{k_1 k_2 + (k_1 + k_2)cs}{k_2 + cs} \tag{8.27}$$

eliminating the s-term in the numerator by setting $k_1 = -k_2 = k$ yields the stiffness

$$S(s) = \frac{k^2}{k - cs} \tag{8.28}$$

Unfortunately, already for this simple model an additional internal degree of freedom has to be introduced which is not present in the mathematical model and the spring constant k_2 has a non-physical (negative) value.

Even more complicated models have to be introduced for the forms of Equations (8.6) and (8.14). An example of the former is shown in [WP92]. Also, here an additional internal degree of freedom is needed. Moreover, these physical models may have negative parameters for mass, damping and stiffness, in which case they are not really "physical". Because of these drawbacks and because there is no simplification, we prefer to work with matrices rather than with these physical models.

8.2 Multivariable systems

8.2.1 Vector formulation

General formulation. A first generalization to the scalar system Equation (8.6) is

$$\mathbf{H}(s) = \frac{(\mathbf{b}_1 s + \mathbf{b}_2)(\mathbf{b}_1 s + \mathbf{b}_2)^{\mathrm{T}}}{a_0\, s^2 + a_1 s + a_2} - \frac{\mathbf{b}_1 \mathbf{b}_1^{\mathrm{T}}}{a_0} \tag{8.29}$$

The outer products of the vectors $\mathbf{b}_1$ and $\mathbf{b}_2$ yield square symmetric matrices. Let us consider the combination analogous to Equation (8.17),

$$\mathbf{H}(s) = \frac{\gamma_1 \boldsymbol{\beta}_1^{\mathrm{T}}}{s - \lambda_1} + \frac{\gamma_2 \boldsymbol{\beta}_2^{\mathrm{T}}}{s - \lambda_2} \tag{8.30}$$

where $\boldsymbol{\beta}$ and $\boldsymbol{\gamma}$ are generally complex vectors but the outer product $\boldsymbol{\gamma}\boldsymbol{\beta}^{\mathrm{T}}$ has to be symmetric, that is,

$$\boldsymbol{\gamma}\boldsymbol{\beta}^{\mathrm{T}} = \boldsymbol{\beta}\boldsymbol{\gamma}^{\mathrm{T}} \tag{8.31}$$

This assumption will always be true if the system is input-output symmetric. Postmultiplying Equation (8.31) by $\boldsymbol{\beta}$ and solving for $\boldsymbol{\gamma}$ yields

$$\boldsymbol{\gamma} = \boldsymbol{\beta}\,\frac{\boldsymbol{\gamma}^{\mathrm{T}}\boldsymbol{\beta}}{\boldsymbol{\beta}^{\mathrm{T}}\boldsymbol{\beta}} = \boldsymbol{\beta} t \tag{8.32}$$

where

$$t = \frac{\boldsymbol{\gamma}^{\mathrm{T}}\boldsymbol{\beta}}{\boldsymbol{\beta}^{\mathrm{T}}\boldsymbol{\beta}} \tag{8.33}$$

Thus assumption Equation (8.31) implies that $\boldsymbol{\beta}$ and $\boldsymbol{\gamma}$ are equal up to a scalar factor t. They can therefore be scaled to be equal. In fact

$$\boldsymbol{\gamma}/\sqrt{t} = \boldsymbol{\beta} t/\sqrt{t} = \boldsymbol{\beta}\sqrt{t} \tag{8.34}$$

The identity Equation (8.18) is generalized to

$$\frac{\Big[\mathbf{e}_1\,(s-\lambda_2) + \mathbf{e}_2\,(s-\lambda_1)\Big]\Big[\mathbf{e}_1\,(s-\lambda_2) + \mathbf{e}_2\,(s-\lambda_1)\Big]^{\mathrm{T}}}{a_0\,(s-\lambda_1)(s-\lambda_2)} - \frac{(\mathbf{e}_1+\mathbf{e}_2)(\mathbf{e}_1+\mathbf{e}_2)^{\mathrm{T}}}{a_0}$$
$$= \frac{\mathbf{e}_1\mathbf{e}_1^{\mathrm{T}}\,(\lambda_1-\lambda_2)}{a_0\,(s-\lambda_1)} + \frac{\mathbf{e}_2\mathbf{e}_2^{\mathrm{T}}\,(\lambda_2-\lambda_1)}{a_0\,(s-\lambda_2)} \tag{8.35}$$

Comparing the right-hand side of Equation (8.35) with Equation (8.30) and substituting Equation (8.32) leads to

$$\frac{\lambda_1-\lambda_2}{a_0}\,\mathbf{e}_1\mathbf{e}_1^{\mathrm{T}} = \boldsymbol{\gamma}_1\boldsymbol{\beta}_1^{\mathrm{T}} = \boldsymbol{\beta}_1\boldsymbol{\beta}_1^{\mathrm{T}} t_1 \tag{8.36}$$

with

$$t_1 = \frac{\boldsymbol{\gamma}_1^{\mathrm{T}}\boldsymbol{\beta}_1}{\boldsymbol{\beta}_1^{\mathrm{T}}\boldsymbol{\beta}_1} \tag{8.37}$$

and an analogous expression for $\mathbf{e}_2$. Therefore, we have

$$\mathbf{e}_1 = \boldsymbol{\beta}_1\sqrt{\frac{a_0\,t_1}{\lambda_1-\lambda_2}} \qquad \mathbf{e}_2 = \boldsymbol{\beta}_2\sqrt{\frac{a_0\,t_2}{\lambda_2-\lambda_1}} \tag{8.38}$$

Comparing the left-hand side of Equation (8.35) with Equation (8.29) yields

$$\mathbf{b}_1 = \mathbf{e}_1 + \mathbf{e}_2 \qquad \mathbf{b}_2 = -(\lambda_2\mathbf{e}_1 + \lambda_1\mathbf{e}_2) \tag{8.39}$$

and, as in the scalar case (Equation 8.21),

$$a_1 = -a_0\,(\lambda_1+\lambda_2) \qquad a_2 = a_0\,\lambda_1\,\lambda_2 \tag{8.40}$$

Real pole. In the case of a single real pole we set $\boldsymbol{\beta}_2 = \boldsymbol{\gamma}_2 = \mathbf{0}$ and $\lambda_2 = 0$ and obtain

$$\mathbf{b}_1 = \mathbf{e}_1 = \boldsymbol{\beta}_1\sqrt{\frac{a_0\,t_1}{\lambda_1}} \qquad \mathbf{b}_2 = \mathbf{0} \tag{8.41}$$

The denominator coefficients are

$$a_0 = \pm 1 \qquad a_1 = -a_0\lambda \qquad a_2 = 0 \tag{8.42}$$

We want to show that $\mathbf{b}_1$ is real if $\gamma_1\boldsymbol{\beta}_1^{\mathrm{T}}$ is real. The latter is satisfied because the system has real input and real output. Observing that $\gamma_1 = \boldsymbol{\beta}_1 t_1$ (Equation 8.32), we can write

$$\frac{a_0}{\lambda_1}\gamma_1\boldsymbol{\beta}_1^{\mathrm{T}} = \frac{a_0\, t_1}{\lambda_1}\boldsymbol{\beta}_1\boldsymbol{\beta}_1^{\mathrm{T}} \tag{8.43}$$

The left-hand side of this expression is real for a real coefficient a_0/λ_1. But that means for the right-hand side that $\boldsymbol{\beta}_1\sqrt{t_1}$ is pure real or pure imaginary. By choosing $a_0 = \pm 1$ we can always ensure that the vector $\mathbf{b}_1 = \boldsymbol{\beta}_1\sqrt{t_1}\sqrt{a_0/\lambda_1}$ is real.

Complex conjugate poles. In the case of complex conjugate poles the transfer function is

$$\mathbf{H}(s) = \frac{\gamma\boldsymbol{\beta}^{\mathrm{T}}}{s-\lambda} + \frac{\bar{\gamma}\bar{\boldsymbol{\beta}}^{\mathrm{T}}}{s-\bar{\lambda}} \tag{8.44}$$

The parameters are with $a_0 = 1$

$$\mathbf{e}_1 = \mathbf{e} = \boldsymbol{\beta}\sqrt{\frac{t}{\lambda - \bar{\lambda}}} \qquad \mathbf{e}_2 = \bar{\mathbf{e}} \tag{8.45}$$

and

$$\mathbf{b}_1 = \mathbf{e} + \bar{\mathbf{e}} \qquad \mathbf{b}_2 = -(\bar{\lambda}\mathbf{e} + \lambda\bar{\mathbf{e}}) \tag{8.46}$$

Both $\mathbf{b}_1$ and $\mathbf{b}_2$ are real in that case. The denominator coefficients are

$$a_0 = 1 \qquad a_1 = -(\lambda + \bar{\lambda}) \qquad a_2 = \lambda\bar{\lambda} \tag{8.47}$$

8.2.2 Matrix formulation

With the results of the vector formulation we proceed to the general matrix formulation for multivariable systems. Unfortunately, a direct symmetrization of the system matrices does not seem to be possible. Even when considering the sign symmetry of the balanced system matrices, the only way to proceed is to diagonalize the system and then use the technique explained in the last subsection for each diagonal term. Therefore, this section is a matrix version of the preceding section. It is notationally more compact, but the actual implementation is more easily performed using the vector formulation.

The first-order system

$$\mathbf{H}(s) = \mathbf{C}(s\mathbf{I} - \mathbf{A})^{-1}\mathbf{B} \tag{8.48}$$

obtained from the approximation has to be transformed to the second-order symmetric system

$$\mathbf{H}(s) = (s\mathbf{B}_1 + \mathbf{B}_2)^{\mathrm{T}}(s^2\mathbf{A}_0 + s\mathbf{A}_1 + \mathbf{A}_2)^{-1}(s\mathbf{B}_1 + \mathbf{B}_2) + s\mathbf{D}_1 + \mathbf{D}_2 \tag{8.49}$$

with $\mathbf{A}_0, \mathbf{A}_1, \mathbf{A}_2, \mathbf{D}_1, \mathbf{D}_2$ symmetric and all matrices real. This expression is more general than the standard mass-damping-stiffness formulation. It considers not only the input of the right-hand side itself but also its first time derivative. Also, since the matrices $\mathbf{B}_1$ and $\mathbf{B}_2$ do not have to be quadratic, the number of internal variables (state variables) and the number of external variables may be different.

It has been shown (p. 133) that the system matrix $\mathbf{A}$ can always be brought into diagonal form

$$\boldsymbol{\Lambda} = \mathbf{T}^{-1}\mathbf{A}\mathbf{T} \tag{8.50}$$

by solving the corresponding eigenvalue problem. The other system matrices are transformed as

$$\hat{\mathbf{B}} = \mathbf{T}^{-1}\mathbf{B} \tag{8.51}$$

$$\hat{\mathbf{C}} = \mathbf{C}\mathbf{T} \tag{8.52}$$

The diagonal system is

$$\hat{\mathbf{C}}(s\mathbf{I}-\mathbf{\Lambda})^{-1}\hat{\mathbf{B}} = \begin{bmatrix} \gamma_1 & \gamma_2 & \cdots \end{bmatrix} \begin{bmatrix} s-\lambda_1 & & \\ & s-\lambda_2 & \\ & & \ddots \end{bmatrix} \begin{bmatrix} \beta_1^{\mathrm{T}} \\ \beta_2^{\mathrm{T}} \\ \vdots \end{bmatrix} \tag{8.53}$$

Therefore, the procedure given before can be applied to each real eigenvalue and to each pair of complex conjugate eigenvalues.

Next the procedure is also explained in matrix form for compact notation. The derivation also shows why the diagonalization is essential.

The first task is to generalize Equation (8.35) to matrix form. We claim that

$$\begin{aligned} &\left[(s\mathbf{I}-\mathbf{\Lambda}_2)\mathbf{E}_1+(s\mathbf{I}-\mathbf{\Lambda}_1)\mathbf{E}_2\right]^{\mathrm{T}}\left[(s\mathbf{I}-\mathbf{\Lambda}_1)(s\mathbf{I}-\mathbf{\Lambda}_2)\right]^{-1}\left[(s\mathbf{I}-\mathbf{\Lambda}_2)\mathbf{E}_1+(s\mathbf{I}-\mathbf{\Lambda}_1)\mathbf{E}_2\right] \\ &\qquad -(\mathbf{E}_1+\mathbf{E}_2)^{\mathrm{T}}(\mathbf{E}_1+\mathbf{E}_2) \\ &= \left[(\mathbf{\Lambda}_1-\mathbf{\Lambda}_2)^{1/2}\mathbf{E}_1\right]^{\mathrm{T}}(s\mathbf{I}-\mathbf{\Lambda}_1)^{-1}\left[(\mathbf{\Lambda}_1-\mathbf{\Lambda}_2)^{1/2}\mathbf{E}_1\right] \\ &\qquad +\left[(\mathbf{\Lambda}_2-\mathbf{\Lambda}_1)^{1/2}\mathbf{E}_2\right]^{\mathrm{T}}(s\mathbf{I}-\mathbf{\Lambda}_2)^{-1}\left[(\mathbf{\Lambda}_2-\mathbf{\Lambda}_1)^{1/2}\mathbf{E}_2\right] \end{aligned} \tag{8.54}$$

The verification is done by first multiplying out the left-hand side. Note that for matrix multiplication two diagonal matrices may be interchanged. Because this fact is used, this derivation is only valid for diagonal matrices $\mathbf{\Lambda}$. The terms involving $\mathbf{E}_1$ are

$$\mathbf{E}_1^{\mathrm{T}}(s\mathbf{I}-\mathbf{\Lambda}_2)(s\mathbf{I}-\mathbf{\Lambda}_1)^{-1}\mathbf{E}_1 - \mathbf{E}_1^{\mathrm{T}}\mathbf{E}_1 \tag{8.55}$$

Analogous terms arise for $\mathbf{E}_2$ with interchanged subscripts 1 and 2, whereas mixed terms involving $\mathbf{E}_1$ and $\mathbf{E}_2$ cancel out. Continuing with the expression involving $\mathbf{E}_1$ we find

$$\begin{aligned} &\mathbf{E}_1^{\mathrm{T}}(s\mathbf{I}-\mathbf{\Lambda}_2)(s\mathbf{I}-\mathbf{\Lambda}_1)^{-1}\mathbf{E}_1 - \mathbf{E}_1^{\mathrm{T}}\mathbf{E}_1 \\ &\quad = \mathbf{E}_1^{\mathrm{T}}(s\mathbf{I}-\mathbf{\Lambda}_2)(s\mathbf{I}-\mathbf{\Lambda}_1)^{-1}\mathbf{E}_1 - \mathbf{E}_1^{\mathrm{T}}(s\mathbf{I}-\mathbf{\Lambda}_1)(s\mathbf{I}-\mathbf{\Lambda}_1)^{-1}\mathbf{E}_1 \\ &\quad = \mathbf{E}_1^{\mathrm{T}}(\mathbf{\Lambda}_1-\mathbf{\Lambda}_2)(s\mathbf{I}-\mathbf{\Lambda}_1)^{-1}\mathbf{E}_1 \\ &\quad = \mathbf{E}_1^{\mathrm{T}}(\mathbf{\Lambda}_1-\mathbf{\Lambda}_2)^{1/2}(s\mathbf{I}-\mathbf{\Lambda}_1)^{-1}(\mathbf{\Lambda}_1-\mathbf{\Lambda}_2)^{1/2}\mathbf{E}_1 \end{aligned} \tag{8.56}$$

which indeed equals the first part of the rihgt-hand side in Equation (8.54). An analogous expression for $\mathbf{E}_2$, which equals the second part, is obtained by interchanging the indexes 1 and 2, and thus Equation (8.54 is verified.

Because $\mathbf{A}$ is a real matrix, its eigenvalues are real or complex conjugate pairs. The first-order diagonal system can be ordered accordingly and is written as

$$\mathbf{H}(s) = \begin{bmatrix} \mathbf{C}_r & \mathbf{C}_c & \bar{\mathbf{C}}_c \end{bmatrix} \begin{bmatrix} s\mathbf{I}-\mathbf{\Lambda}_r & & \\ & s\mathbf{I}-\mathbf{\Lambda}_c & \\ & & s\mathbf{I}-\bar{\mathbf{\Lambda}}_c \end{bmatrix}^{-1} \begin{bmatrix} \mathbf{B}_r \\ \mathbf{B}_c \\ \bar{\mathbf{B}}_c \end{bmatrix} \tag{8.57}$$

where entries with subscript r pertain to the real eigenvalues, the entries with subscript c to complex eigenvalues. Because the first-order system is real and input-output symmetric, the transformation matrix $\mathbf{T}$ can be scaled such that

$$\mathbf{C}_r = \mathbf{B}_r^{\mathrm{T}}\mathbf{\Theta}^{-1} \quad \text{with } \mathbf{B}_r \text{ real and } \mathbf{\Theta} = \mathbf{diag}(\pm 1, \pm 1, \cdots) \tag{8.58}$$

$$\mathbf{C}_c = \mathbf{B}_c^{\mathrm{T}} \quad \text{with } \mathbf{B}_c \text{ complex} \tag{8.59}$$

The proof of this fact is the same as for the vector formulation and is not repeated here.

The partial system corresponding to the real eigenvalues is the first-order symmetric system

$$\mathbf{H}_r(s) = \mathbf{B}_r^{\mathrm{T}} \mathbf{\Theta}^{-1/2} (s\mathbf{I} - \mathbf{\Lambda}_r)^{-1} \mathbf{\Theta}^{-1/2} \mathbf{B}_r \tag{8.60}$$

It could be kept in this form as a stiffness-damping system without mass or it can be transformed using the identity Equation (8.54). Comparing Equation (8.60) with the right-hand side of Equation (8.54) we see that

$$\mathbf{\Lambda}_1 = \mathbf{\Lambda}_r \qquad \mathbf{\Lambda}_2 = \mathbf{0} \tag{8.61}$$

$$\mathbf{E}_1 = \mathbf{\Lambda}_r^{-1/2} \mathbf{\Theta}^{-1/2} \mathbf{B}_r \qquad \mathbf{E}_2 = \mathbf{0} \tag{8.62}$$

With these matrices the left-hand side of Equation (8.54) yields the two expressions

$$(s\mathbf{I} - \mathbf{\Lambda}_2)\mathbf{E}_1 + (s\mathbf{I} - \mathbf{\Lambda}_1)\mathbf{E}_2 = s\mathbf{\Lambda}_r^{-1/2} \mathbf{\Theta}^{-1/2} \mathbf{B}_r \tag{8.63}$$

and

$$(s\mathbf{I} - \mathbf{\Lambda}_1)(s\mathbf{I} - \mathbf{\Lambda}_2) = s^2\mathbf{I} - s\mathbf{\Lambda}_r \tag{8.64}$$

The corresponding second-order system is

$$\mathbf{H}_r(s) = \left[s\mathbf{\Lambda}_r^{-1/2} \mathbf{\Theta}^{-1/2} \mathbf{B}_r\right]^{\mathrm{T}} (s^2\mathbf{I} - s\mathbf{\Lambda}_r)^{-1} \left[s\mathbf{\Lambda}_r^{-1/2} \mathbf{\Theta}^{-1/2} \mathbf{B}_r\right] \tag{8.65}$$

or, slightly modified,

$$\begin{aligned} \mathbf{H}_r(s) &= \left[s(-\mathbf{\Lambda}_r)^{-1/2} \mathbf{B}_r\right]^{\mathrm{T}} (-s^2\mathbf{\Theta} + s\mathbf{\Theta}\mathbf{\Lambda}_r)^{-1} \left[s(-\mathbf{\Lambda}_r)^{-1/2} \mathbf{B}_r\right] \\ &= \mathbf{E}_r^{\mathrm{T}} (-s^2\mathbf{\Theta} + s\mathbf{\Theta}\mathbf{\Lambda}_r)^{-1} \mathbf{E}_r \end{aligned} \tag{8.66}$$

with

$$\mathbf{E}_r = (-\mathbf{\Lambda}_r)^{-1/2} \mathbf{B}_r \tag{8.67}$$

Equation (8.66) represents a mass-damping system without stiffness. The minus sign in $(-\mathbf{\Lambda}_r)^{-1/2}$ is necessary because the poles $\mathbf{\Lambda}_r$ are negative due to the stability condition.

The partial system corresponding to complex conjugate pairs of eigenvalues is

$$\mathbf{H}_c(s) = \mathbf{B}_c^{\mathrm{T}} (s\mathbf{I} - \mathbf{\Lambda}_c)^{-1} \mathbf{B}_c + \bar{\mathbf{B}}_c^{\mathrm{T}} (s\mathbf{I} - \bar{\mathbf{\Lambda}}_c)^{-1} \bar{\mathbf{B}}_c \tag{8.68}$$

Comparing this with the right-hand side of Equation (8.54) we conclude

$$\mathbf{\Lambda}_1 = \mathbf{\Lambda}_c \qquad \mathbf{\Lambda}_2 = \bar{\mathbf{\Lambda}}_c \tag{8.69}$$

$$\mathbf{E}_1 = \mathbf{E}_c = (\mathbf{\Lambda}_c - \bar{\mathbf{\Lambda}}_c)^{-1/2} \mathbf{B}_c \qquad \mathbf{E}_2 = \bar{\mathbf{E}}_c \tag{8.70}$$

With these matrices the left-hand side of Equation (8.54) yields the expressions

$$(s\mathbf{I} - \mathbf{\Lambda}_2)\mathbf{E}_1 + (s\mathbf{I} - \mathbf{\Lambda}_1)\mathbf{E}_2 = s(\mathbf{E}_c + \bar{\mathbf{E}}_c) - (\bar{\mathbf{\Lambda}}_c\mathbf{E}_c + \mathbf{\Lambda}_c\bar{\mathbf{E}}_c) \tag{8.71}$$

and

$$(s\mathbf{I} - \mathbf{\Lambda}_1)(s\mathbf{I} - \mathbf{\Lambda}_2) = s^2\mathbf{I} - s(\mathbf{\Lambda}_c + \bar{\mathbf{\Lambda}}_c) + \mathbf{\Lambda}_c\bar{\mathbf{\Lambda}}_c \tag{8.72}$$

The corresponding second-order system is

$$\mathbf{H}_c(s) = \left[s(\mathbf{E}_c + \bar{\mathbf{E}}_c) - (\bar{\mathbf{\Lambda}}_c\mathbf{E}_c + \mathbf{\Lambda}_c\bar{\mathbf{E}}_c)\right]^{\mathrm{T}} \left[s^2\mathbf{I} - s(\mathbf{\Lambda}_c + \bar{\mathbf{\Lambda}}_c) + \mathbf{\Lambda}_c\bar{\mathbf{\Lambda}}_c\right]^{-1} \left[\cdots\right] \tag{8.73}$$

where the last brackets contain the same expression as the first brackets. Note that the resulting matrices are real because they consist of sums or products of complex conjugate pairs.

The result obtained in Equations (8.66), (8.71) and (8.72) can be put into the matrices of Equation (8.49) as follows

$$\mathbf{A}_0 = \begin{bmatrix} -\mathbf{\Theta} & \\ & \mathbf{I} \end{bmatrix} \tag{8.74}$$

$$\mathbf{A}_1 = \begin{bmatrix} \mathbf{\Theta}\mathbf{\Lambda}_r & \\ & -(\mathbf{\Lambda}_c + \bar{\mathbf{\Lambda}}_c) \end{bmatrix} \tag{8.75}$$

$$\mathbf{A}_2 = \begin{bmatrix} \mathbf{0} & \\ & \mathbf{\Lambda}_c\bar{\mathbf{\Lambda}}_c \end{bmatrix} \tag{8.76}$$

$$\mathbf{B}_1 = \begin{bmatrix} \mathbf{E}_r \\ \mathbf{E}_c + \bar{\mathbf{E}}_c \end{bmatrix} \tag{8.77}$$

$$\mathbf{B}_2 = \begin{bmatrix} \mathbf{0} \\ -(\bar{\mathbf{\Lambda}}_c\mathbf{E}_c + \mathbf{\Lambda}_c\bar{\mathbf{E}}_c) \end{bmatrix} \tag{8.78}$$

$$\mathbf{D}_2 = \mathbf{D} - \mathbf{B}_1^{\mathrm{T}}\mathbf{B}_1 \tag{8.79}$$

with (Equations 8.67 and 8.70)

$$\mathbf{E}_r = (-\mathbf{\Lambda}_r)^{-1/2}\,\mathbf{B}_r \tag{8.80}$$

$$\mathbf{E}_c = (\mathbf{\Lambda}_c - \bar{\mathbf{\Lambda}}_c)^{-1/2}\,\mathbf{B}_c \tag{8.81}$$

8.3 Example

The system matrices obtained in the two-mass example, Equation (7.99), are transformed to a second-order symmetric system. First, the system is brought to diagonal form:

$$\mathbf{\Lambda} = \begin{bmatrix} -0.0608 + 1.580i & \\ & -0.0608 + 1.580i \end{bmatrix} = \begin{bmatrix} \lambda & \\ & \bar{\lambda} \end{bmatrix} \tag{8.82}$$

$$\hat{\mathbf{B}} = \begin{bmatrix} 0.306 - 0.302i & 0.376 - 0.347i \\ 0.306 + 0.302i & 0.376 + 0.347i \end{bmatrix} = \begin{bmatrix} \boldsymbol{\beta}^{\mathrm{T}} \\ \bar{\boldsymbol{\beta}}^{\mathrm{T}} \end{bmatrix} \tag{8.83}$$

$$\hat{\mathbf{C}} = \begin{bmatrix} 0.294 - 0.313i & 0.294 + 0.313i \\ 0.361 - 0.361i & 0.361 + 0.361i \end{bmatrix} = \begin{bmatrix} \boldsymbol{\gamma} & \bar{\boldsymbol{\gamma}} \end{bmatrix} \tag{8.84}$$

$$\mathbf{D} = \begin{bmatrix} 0.018 & -0.027 \\ -0.027 & 0.038 \end{bmatrix} \tag{8.85}$$

The matrix $\mathbf{D}$ remains unchanged by the diagonal transformation. The two diagonal terms of the matrix $\mathbf{\Lambda}$ are a complex conjugate pair. The matrices $\hat{\mathbf{B}}$ and $\hat{\mathbf{C}}$ are not unique but depend on the scaling of the eigenvectors in the diagonal transformation. Here, the eigenvectors have been scaled to unit length. The matrices $\hat{\mathbf{B}}$ and $\hat{\mathbf{C}}$ can be transformed further by a diagonal transformation matrix $\mathbf{T}$ to the form $\hat{\mathbf{B}} = \hat{\mathbf{C}}^{\mathrm{T}}$. This is possible because for input-output symmetric systems, corresponding rows of $\hat{\mathbf{B}}$ and columns of $\hat{\mathbf{C}}$ are proportional to each other (Equation 8.32). The diagonal transformation matrix does not alter the matrix $\mathbf{\Lambda}$. The scaling factor according to Equation (8.33) is

$$t = \frac{\boldsymbol{\gamma}^{\mathrm{T}}\boldsymbol{\beta}}{\boldsymbol{\beta}^{\mathrm{T}}\boldsymbol{\beta}} \tag{8.86}$$

and the (inverse) transformation matrix is

$$\mathbf{T}^{-1} = \begin{bmatrix} \sqrt{t} & \\ & \sqrt{\bar{t}} \end{bmatrix} = \begin{bmatrix} 0.999 - 0.020i & \\ & 0.999 + 0.020i \end{bmatrix} \tag{8.87}$$

With this additional transformation we get

$$\hat{\mathbf{B}}\,\mathbf{T}^{-1} = \mathbf{T}^{\mathrm{T}}\,\hat{\mathbf{C}}^{\mathrm{T}} = \begin{bmatrix} 0.300 - 0.308i & 0.369 - 0.354i \\ 0.300 + 0.308i & 0.369 + 0.354i \end{bmatrix} \tag{8.88}$$

We calculate the temporary vector (Equation 8.45)

$$\mathbf{e} = \boldsymbol{\beta}\sqrt{\frac{t}{\lambda - \bar{\lambda}}} = \begin{bmatrix} -0.003 - 0.242i & 0.006 - 0.288i \end{bmatrix} \tag{8.89}$$

and with this the two vectors (Equation 8.46)

$$\mathbf{b}_1 = \mathbf{e} + \bar{\mathbf{e}} = \begin{bmatrix} -0.006 & 0.012 \end{bmatrix} \tag{8.90}$$

and

$$\mathbf{b}_2 = -(\bar{\lambda}\mathbf{e} + \lambda\bar{\mathbf{e}}) = \begin{bmatrix} 0.763 & 0.909 \end{bmatrix} \tag{8.91}$$

The $\mathbf{A}$ matrices are calculated from the eigenvalues $\boldsymbol{\Lambda}$ as

$$\mathbf{A}_0 = 1 \qquad \mathbf{A}_1 = -(\lambda + \bar{\lambda}) = 0.1216 \qquad \mathbf{A}_2 = \lambda\bar{\lambda} = 2.499 \tag{8.92}$$

Finally the matrix $\mathbf{D}$ is updated as

$$\mathbf{D}_2 = \mathbf{D} - \mathbf{b}_1\,\mathbf{b}_1^{\mathrm{T}} = \begin{bmatrix} 0.018 & -0.026 \\ -0.026 & 0.037 \end{bmatrix} \tag{8.93}$$

The matrix $\mathbf{D}_1$ corresponds to the direct damping term and is not used here.

Chapter 9

Fluid Model and Analytical Solutions

So far simple model problems were used to investigate the numerical methods. In this chapter we now introduce some larger, more realistic problems which are used later on as verification problems for the proposed numerical methods and the computer program. In the first part the mathematical fluid model is presented. This part also serves as background for the next chapter on finite element models. In the second part three problems with simple geometry are analyzed: The two-dimensional reservoir model, the rectangular and the semi-circular channel. For all problems reservoir bottom absorption is included and the dam is assumed to be rigid. For the two-dimensional problem also some commonly-used simplified models and boundary conditions are analyzed. The comparison of the analytical with the numerical results follows in Chapters 10 and 11.

9.1 Fluid model

This section describes the differential equation and the boundary conditions for the fluid. The fluid is assumed to be compressible, irrotational with small velocity amplitudes and no viscosity effects. Then the behavior of the fluid can be described by the velocity potential φ. The governing differential equation is the wave equation

$$\Delta\varphi = \frac{1}{c^2}\ddot{\varphi} \qquad \text{in } \Omega \tag{9.1}$$

where c is the acoustic-wave velocity and Ω the interior of the fluid domain.

The physical variables are the velocity vector

$$\mathbf{v} = \mathbf{grad}\varphi \tag{9.2}$$

and the dynamic pressure

$$p = -\rho\dot{\varphi} \tag{9.3}$$

where ρ is the mass density.

The boundary conditions are the following: At the free surface, the pressure vanishes. The velocity potential is therefore an arbitrary constant, chosen as zero

$$\varphi = 0 \qquad \text{on } \Gamma_p \tag{9.4}$$

where Γ_p is the boundary with prescribed pressure. At the interface to the dam the normal velocity is prescribed.

$$\frac{\partial\varphi}{\partial n} = v_n \qquad \text{on } \Gamma_v \tag{9.5}$$

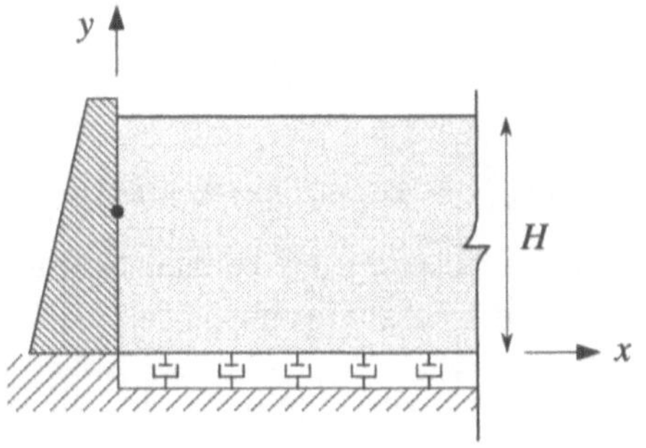

Figure 9.1: 2-D fluid problem: geometry

where Γ_v is the boundary with prescribed velocity and $\partial\varphi/\partial n$ is the normal derivative with the normal pointing outward of the fluid.

At the interface to the foundation also the normal velocity is prescribed. To take into account the flexibility of the foundation in a simple manner, an absorptive foundation model [FC83]. is used. The corresponding boundary condition is

$$\frac{\partial\varphi}{\partial n} + q\dot{\varphi} = v_n \qquad \text{on } \Gamma_v \tag{9.6}$$

The value q is determined considering the one-dimensional problem of a plane, normal incident wave passing from the fluid to the foundation. The value of q is determined by

$$q = \frac{\rho}{c_s \rho_s} \tag{9.7}$$

with the compression-wave velocity of the foundation c_s and the mass density of the foundation ρ_s. The portion of the amplitude reflected back to the the fluid is given by the reflection coefficient α as

$$\alpha = \frac{1 - cq}{1 + cq} \tag{9.8}$$

Numerical values for water are $c = 1440$ m/s^2 and $\rho = 1000$ kg/m^3. For a stiff bedrock typical values are $c_s = 5000$ m/s^2 and $\rho_s = 2600$ kg/m^3 resulting in a reflection coefficient of $\alpha = 0.80$. This shows that even a stiff rock is by no means rigid in the present context. The model should therefore include the effect of the absorptive foundation.

9.2 Analytical solutions for 2-D fluid problem

In this section we introduce the model of a two-dimensional reservoir of constant depth. This example is relatively simple, yet includes all the features of the more complicated three-dimensional models. The first solutions found in the literature assume a rigid reservoir bottom [Wes33, Kot59, Cho67]. Bottom absorption has been introduced by Fenves and Chopra [FC83].

The model with height H is shown in Figure 9.1. The dot at $y = 0.6H$ indicates the point where results are shown subsequently. The wave equation is solved in the frequency domain by separation of variables.

$$\varphi = \Phi(y)\, e^{i(\omega t - kx)} \tag{9.9}$$

Substituted into the wave equation and taking $\Delta\Phi = \Phi''$ yields

$$\Phi'' + \lambda^2 \Phi = 0 \tag{9.10}$$

with

$$\lambda^2 = \frac{\omega^2}{c^2} - k^2 \tag{9.11}$$

The general solution is

$$\Phi(y) = A\cos\lambda y + B\sin\lambda y \tag{9.12}$$

The constants A and B have to be determined by the boundary conditions.

The boundary condition for the free surface is

$$\varphi = 0 \qquad \text{at } y = H \tag{9.13}$$

In terms of the variable Φ the boundary condition is

$$\Phi(H) = A\cos\lambda H + B\sin\lambda H = 0 \tag{9.14}$$

which determines B as

$$B = -A\cot\lambda H \tag{9.15}$$

The general solution satisfying the free surface boundary condition is thus

$$\Phi(y) = A\,(\cos\lambda y - \cot\lambda H\ \sin\lambda y) \tag{9.16}$$

The boundary condition at the foundation-reservoir interface with the outward normal pointing to the negative y-direction is

$$-\frac{\partial\varphi}{\partial y} + i\omega q\varphi = -v_y(x,t) \qquad \text{at } y = 0 \tag{9.17}$$

Similarly, the boundary condition at the dam is

$$-\frac{\partial\varphi}{\partial x} = -v_x(y,t) \qquad \text{at } y = 0 \tag{9.18}$$

Two cases have now to be distinguished. For a horizontal upstream excitation, the vertical velocity $v_y(x,t) = 0$ and the boundary conditions for Φ are homogeneous, leading to an eigenvalue problem. For a vertical excitation the boundary conditions are inhomogeneous, leading to a boundary value problem.

Upstream excitation. In the case of an upstream excitation, $v_y(x,t) = 0$ and the boundary condition at the foundation level (Equation 9.17) becomes, for the variable Φ,

$$\Phi' = i\omega q\Phi \qquad \text{at } y = 0 \tag{9.19}$$

First we consider the case with no bottom absorption, that is $q = 0$. Applying the boundary condition to the general solution Equation (9.16) which already satisfies the free surface boundary condition yields

$$\Phi'(y) = -A\lambda\,(\sin\lambda y + \cot\lambda H\ \cos\lambda y) = 0 \qquad \text{at } y = 0 \tag{9.20}$$

or

$$-A\lambda\cot\lambda H = 0 \tag{9.21}$$

This imposes the following condition on λ

$$\cos\lambda H = 0 \tag{9.22}$$

which determines the eigenvalues

$$\lambda_n = \frac{\pi}{2H}, \frac{3\pi}{2H}, \ldots = \frac{2n-1}{2H}\pi \tag{9.23}$$

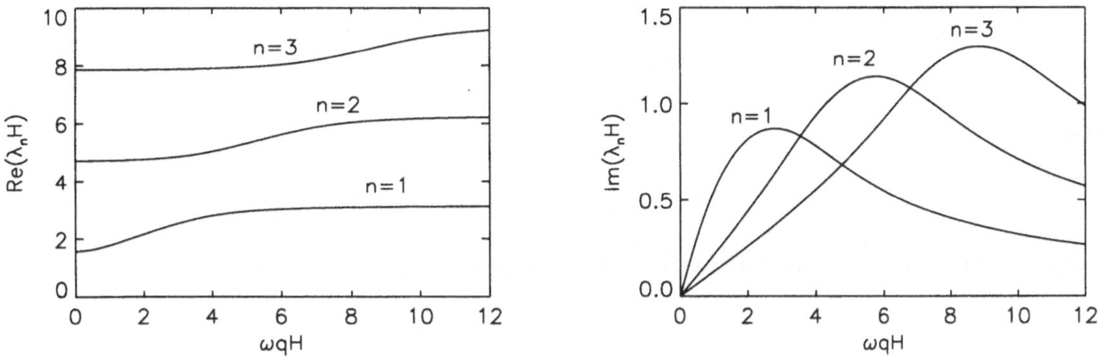

Figure 9.2: 2-D fluid problem: frequency-dependent eigenvalues

If bottom absorption is included, the condition for the eigenvalues is

$$-A\lambda \cot \lambda H = Ai\omega q \tag{9.24}$$

or

$$\cot \lambda H = -\frac{i\omega q}{\lambda} = -i\frac{\omega qH}{\lambda H} \tag{9.25}$$

The roots of this transcendental equation yield the frequency-dependent, complex eigenvalues λ_n. From it it is clear that the problem only involves the dimensionless quantities ωqH and λH. Real and imaginary parts of λH are plotted in Figure 9.2. Note that, if foundation absorption is considered, the eigenvalues and the eigenvectors are frequency-dependent. For $\omega = 0$ or $q = 0$ the equation reduces to $\cos \lambda H = 0$.

Equation (9.25) has one solution λ_n for each mode. With the arbitrary scaling $A = 1$ the corresponding eigenvectors are

$$\Phi_n = \cos \lambda_n y - \cot \lambda_n H \ \sin \lambda_n y \tag{9.26}$$

The differential equation Equation (9.10) with the boundary conditions Equations (9.14) and (9.19) constitutes a Sturm-Liouville problem [NU88, p. 299] and hence the eigenvectors are orthogonal. This is shown in the following for the problem at hand but can be derived for the general case.

Consider the integrals

$$\int_0^H \Phi_m \left(\Phi_n'' + \lambda_n^2 \Phi_n\right) dy = 0 \tag{9.27}$$

$$\int_0^H \Phi_n \left(\Phi_m'' + \lambda_m^2 \Phi_m\right) dy = 0 \tag{9.28}$$

They are zero because the expressions in the parentheses are solutions of the eigenvalue problem. The first part of the first integral can be integrated by parts twice.

$$\int_0^H \Phi_m \Phi_n'' \, dy = \Phi_m \Phi_n' \big|_0^H - \Phi_m' \Phi_n \big|_0^H + \int_0^H \Phi_m'' \Phi_n \, dy \tag{9.29}$$

Now we take the difference between the integral Equation (9.27) and the integral Equation (9.28) and make use of the partial integration in Equation (9.29). This yields

$$\int_0^H (\lambda_n^2 - \lambda_m^2)\, \Phi_n \Phi_m \, dy = \Phi_m' \Phi_n \big|_0^H - \Phi_m \Phi_n' \big|_0^H \tag{9.30}$$

The right-hand side is zero because of the boundary conditions Equation (9.14)

$$\Phi_m|_{y=H} = \Phi_n|_{y=H} = 0 \tag{9.31}$$

and Equation (9.19)

$$\Phi'_m \Phi_n|_{y=0} - \Phi_m \Phi'_n|_{y=0} = i\omega q \left[\Phi_m \Phi_n - \Phi_n \Phi_m\right]_{y=0} = 0 \tag{9.32}$$

With the right-hand side of Equation (9.30) equal to zero we have

$$\int_0^H (\lambda_n^2 - \lambda_m^2)\, \Phi_n \Phi_m \, dy = 0 \tag{9.33}$$

For any $\lambda_n \neq \lambda_m$ the orthogonality follows

$$\int_0^H \Phi_n \Phi_m \, dy = 0 \tag{9.34}$$

The complete solution for the velocity potential is given by a linear combination of the eigenvectors

$$\varphi = e^{i\omega t} \sum_{n=1}^{\infty} A_n e^{-ik_n x} \Phi_n(y) \tag{9.35}$$

It is determined up to the coefficients A_n. These have to be found from the velocity boundary condition at the dam. For a time-harmonic velocity

$$v_x(y,t) = \hat{v}_x(y)\, e^{i\omega t} \tag{9.36}$$

the boundary condition Equation (9.18) is

$$\hat{v}_x(y) = -\sum_{n=1}^{\infty} ik_n A_n e^{-ik_n x} \Phi_n(y) \tag{9.37}$$

Because the problem is invariant with respect to the x-coordinate, the derivation is somewhat simplified by choosing $x = 0$ at the upstream face of the dam. Then

$$-\sum_{m=1}^{\infty} ik_m A_m \Phi_m(y) = \hat{v}_x(y) \tag{9.38}$$

Multiplying this equation by Φ_n and integrating over the height determines A_n. Because of the orthogonality of the eigenvectors only one term of the summation remains.

$$-ik_n A_n \int_0^H \Phi_n^2(y)\, dy = \int_0^H \hat{v}_x(y)\, \Phi_n(y)\, dy \tag{9.39}$$

Defining the modal velocity as

$$\hat{V}_n = \frac{\int_0^H \hat{v}_x(y)\, \Phi_n(y)\, dy}{\int_0^H \Phi_n^2(y)\, dy} \tag{9.40}$$

the amplitude A_n of each mode is given by

$$A_n = -\frac{\hat{V}_n}{ik_n} \tag{9.41}$$

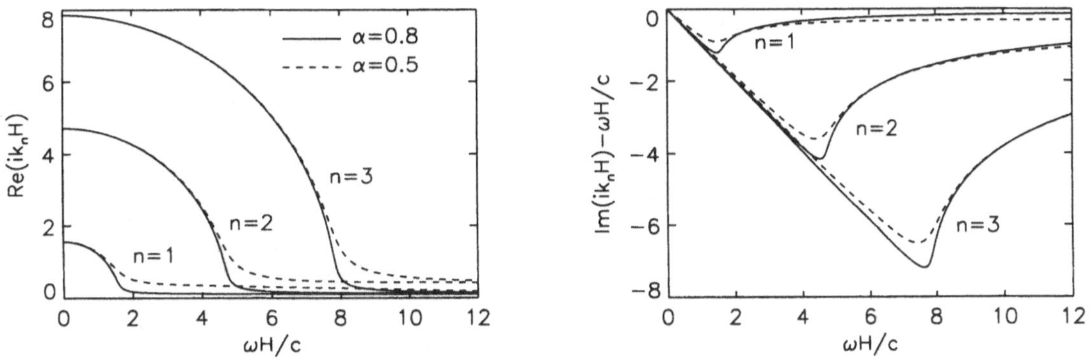

Figure 9.3: 2-D fluid problem: impedance of modes

The value ik_n is determined by Equation (9.11) as

$$ik_n = \sqrt{\lambda_n^2 - \frac{\omega^2}{c^2}} \tag{9.42}$$

As already discussed in Chapter 2, the sign of the square root is determined by the radiation condition. The expression

$$e^{i(\omega t - kx)} = e^{i(\omega t - \Im(ik)x)}\, e^{-\Re(ik)x} \tag{9.43}$$

has to be an outgoing decreasing wave, which is the case for $\Re(ik) > 0$ and $\Im(ik) > 0$. For $\omega < \lambda c$, the value ik is close to real and the waves decay with distance. For $\omega > \lambda c$, the value ik is almost pure imaginary, resulting in pressure waves propagating to the right. For high frequencies the function behaves as $ik_n = i\omega/c$. In order to have an expression which does not go to infinity, either ik_n/ω or $ik_n - i\omega/c$ is usually plotted for the imaginary part. We prefer the latter because this is the form used later for the time-domain formulation. The real and imaginary parts of this value are plotted in Figure 9.3 as a function of frequency for bottom-reflection coefficients $\alpha = 0.8$ and 0.5. The dimensionless quantities ik_nH and $\omega H/c$ are shown.

The term ik_n may be viewed as a "dynamic stiffness" in analogy to solid problems. In physical quantities, however, it looks more like a flexibility because with the force-like $p = -\rho\dot{\varphi}$ and the displacement-like $a = \dot{v}$, the relationship is $a = (ik_n/\rho)p$. To avoid confusion we use the more general term 'impedance', restricting the term 'dynamic stiffness' to solid problems.

Assuming now a rigid dam, the velocity at the dam is

$$v_x(y,t) = \hat{v}_x e^{i\omega t} \tag{9.44}$$

where $\hat{v}_x$ is now constant over the height y. Then the modal velocity is

$$\hat{V}_n = \hat{v}_x \frac{\int_0^H \Phi_n(y)\, dy}{\int_0^H \Phi_n^2(y)\, dy} = \hat{v}_x \frac{I_1}{I_2} \tag{9.45}$$

where

$$\begin{aligned} I_1 &= \int_0^H \Phi_n\, dy = \int_0^H (\cos\lambda_n y + i\omega q/\lambda_n \sin\lambda_n y)\, dy \\ &= H\left(1 - \frac{\omega^2 q^2}{\lambda_n^2}\right)\frac{\sin\lambda_n H}{\lambda_n H} + \frac{i\omega q}{\lambda_n^2} \end{aligned} \tag{9.46}$$

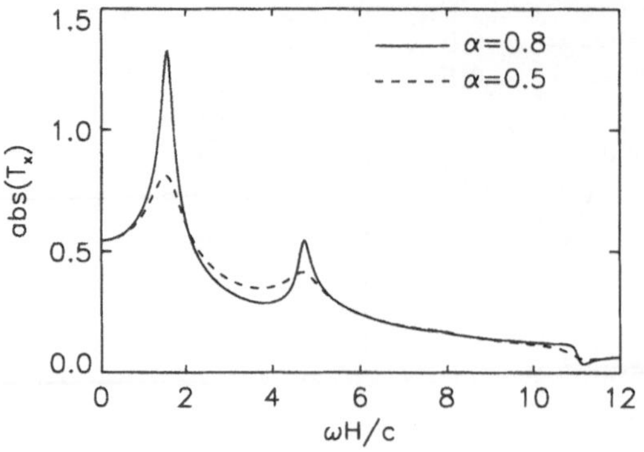

Figure 9.4: 2-D fluid problem: pressure transfer function for horizontal excitation

and

$$\begin{aligned} I_2 &= \int_0^H \Phi_n^2\,dy = \int_0^H (\cos\lambda_n y + i\omega q/\lambda_n\,\sin\lambda_n y)^2\,dy \\ &= \frac{H}{2}\left(1-\frac{\omega^2 q^2}{\lambda_n^2}\right)\left(1+\frac{\sin^2\lambda_n H}{\lambda_n H}\frac{i\omega q}{\lambda_n}\right) \end{aligned} \tag{9.47}$$

The velocity potential at the dam can be calculated as

$$\varphi(y) = e^{i\omega t}\sum_{n=1}^{\infty} A_n\Phi_n(y) = -\left(\sum_{n=1}^{\infty}\frac{1}{ik_n}\frac{I_1}{I_2}\Phi_n(y)\right)\hat{v}_x e^{i\omega t} \tag{9.48}$$

This relationship can be expressed in the form of a transfer function as

$$T_x(\omega) = \frac{\hat{p}}{H\rho\hat{a}_x} = -\frac{\hat{\varphi}}{H\hat{v}_x} = \sum_{n=1}^{\infty}\frac{1}{ik_n H}\frac{I_1}{I_2}\Phi_n(y) \tag{9.49}$$

where the variables $\hat{\varphi}, \hat{p}, \hat{a}_x$ denote the complex amplitudes of the velocity potential, the pressure and the horizontal acceleration of the dam, respectively. Figure 9.4 shows the amplitude of the transfer function taking 6 modes into account. The response is calculated at $y = 0.6H$ as indicated in Figure 9.1. The peaks arise approximately at the eigenfrequencies of the undamped system $\omega H/c = \lambda_n H = \pi/2,\ 3\pi/2,\ 5\pi/2,\ \ldots$.

Although the pressure at the dam is the main issue of this analysis, it is also very instructive to look at the pressure distribution and the displacement of the fluid in the reservoir itself. Similar to the transfer function, the non-dimensional pressure in the reservoir is

$$\frac{p(x,y,t)}{H\rho\hat{a}_x} = \left(\sum_{n=1}^{\infty}\frac{1}{ik_n H}\frac{I_1}{I_2}\Phi_n(y)\,e^{-ik_n x}\right)e^{i\omega t} \tag{9.50}$$

The displacements are found by integrating the velocity, which is equal to the gradient of the velocity potential. The displacement in the x-direction is

$$u_x(x,y,t) = \frac{1}{i\omega}\frac{\partial\varphi}{\partial x} = \left(\sum_{n=1}^{\infty}\frac{I_1}{I_2}\Phi_n(y)\,e^{-ik_n x}\right)\frac{\hat{a}_x}{(i\omega)^2}e^{i\omega t} \tag{9.51}$$

and the one in the y-direction is

$$u_y(x,y,t) = \frac{1}{i\omega}\frac{\partial\varphi}{\partial y} = -\left(\sum_{n=1}^{\infty}\frac{1}{ik_n}\frac{I_1}{I_2}\Phi_n'(y)\,e^{-ik_n x}\right)\frac{\hat{a}_x}{(i\omega)^2}e^{i\omega t} \tag{9.52}$$

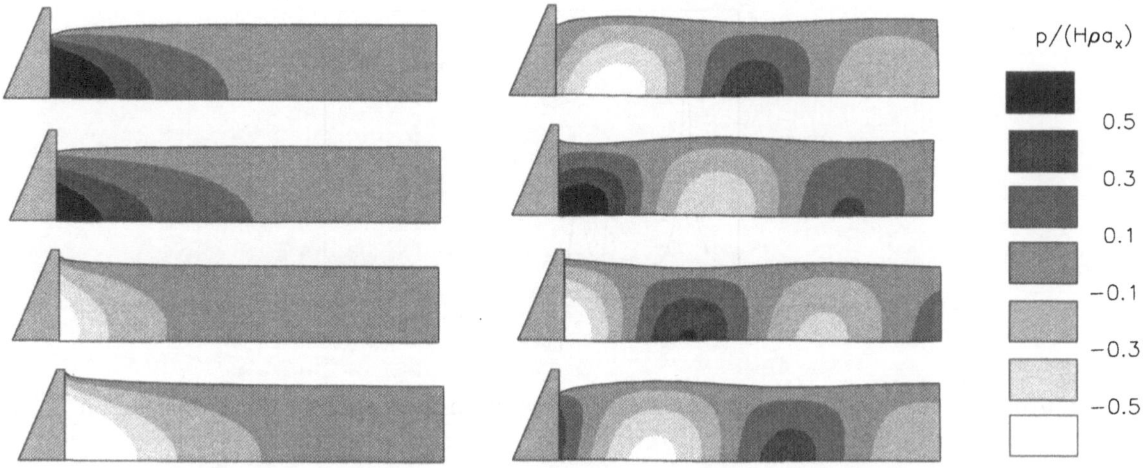

Figure 9.5: 2-D fluid problem: pressure distribution in the reservoir. Left: $\omega = 0.8\,\omega_1$, decaying waves (displacements ×1640). Right: $\omega = 1.5\,\omega_1$, traveling waves (displacements ×5760).

The pressure distribution is shown in Figure 9.5 as contour maps in the displaced reservoir. The horizontal acceleration of the dam is $\hat{a}_x = 1$ m/s^2, the reservoir height is $H = 100$ m and the displacements are exaggerated by the factors indicated in the caption. To visualize the time behavior, snapshots with a time increment of $\Delta t = 0.25\,T_1$ are plotted, where T_1 is the fundamental period $T_1 = 4H/c$. (These figures are identical to the right sequences in Figures 1.2 and 1.3 in the introduction.) To show the difference between decaying and propagating waves, the pressure distribution is evaluated at two frequencies, one below and one above the cut-off frequency ω_1. The cut-off frequency corresponds to the fundamental frequency and is given by $\omega_1 H/c = \pi/2$. On the left in Figure 9.5 the frequency is $\omega = 0.8\,\omega_1$. As expected for decaying waves, the pressure is basically in phase and is restricted to a region close to the dam. On the right the frequency is $\omega = 1.5\,\omega_1$. The waves travel to the right, decaying slightly due to the bottom absorption ($\alpha = 0.8$).

Vertical excitation. For a vertical excitation, the prescribed velocity is assumed to be independent of x. Then, $k = 0$ and $\lambda = \omega/c$. With the foundation velocity $v_y = \hat{v}_y e^{i\omega t}$, the boundary condition for the bottom (Equation 9.17) becomes

$$\Phi' = i\omega q\Phi + \hat{v}_y \qquad \text{at } y = 0 \tag{9.53}$$

This boundary condition is applied to the general solution for Φ which already satisfies the free surface boundary condition (Equation 9.16) to get

$$-A\lambda\,(\sin\lambda y + \cot\lambda H\ \cos\lambda y) = i\omega q A\,(\cos\lambda y - \cot\lambda H\ \sin\lambda y) + \hat{v}_y \qquad \text{at } y = 0 \tag{9.54}$$

or

$$-A\lambda\cot\lambda H = i\omega q A + \hat{v}_y \tag{9.55}$$

This yields for A

$$A = \frac{-\hat{v}_y}{\lambda\cot\lambda H + i\omega q} \tag{9.56}$$

The boundary condition at the dam, that is, $\partial\varphi/\partial x = -ik\varphi = 0$, is automatically satisfied because $k = 0$. The velocity potential at the dam is

$$\varphi(y) = -\frac{\cos\lambda y - \cot\lambda H\ \sin\lambda y}{\ldots}\,\hat{v}_y e^{i\omega t} \tag{9.57}$$

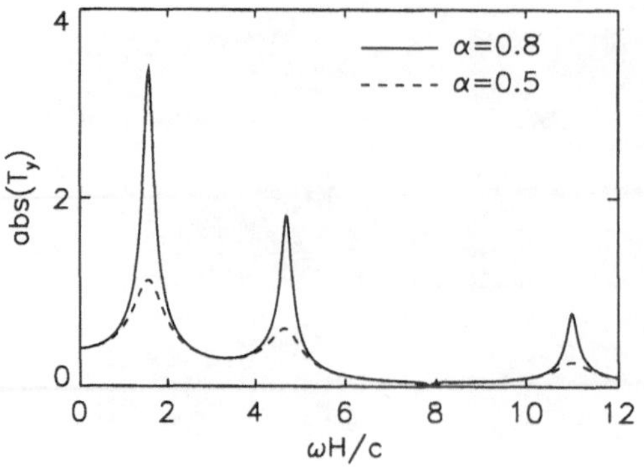

Figure 9.6: 2-D fluid problem: pressure transfer function for vertical excitation

It is interesting to investigate the limit as $\omega \to 0$. Then also $\lambda \to 0$ and $\lambda \cot \lambda H = 1/H$ and $\cot \lambda H \ \sin \lambda y = y/H$. The pressure is $\hat{p} = \rho i \omega \hat{v}_y H(1 - y/H)$. With the acceleration $\hat{a}_y = i\omega \hat{v}_y$ the pressure has the hydrostatic pressure distribution

$$\hat{p}(y) = \hat{a}_y \rho (H - y) \tag{9.58}$$

The amplitude of the transfer function

$$T_y(\omega) = \frac{\hat{p}}{H\rho\hat{a}_y} = -\frac{\hat{\varphi}}{H\hat{v}_y} = \frac{\cos \lambda y - \cot \lambda H \ \sin \lambda y}{\lambda H \cot \lambda H + i\omega q H} \tag{9.59}$$

is plotted in Figure 9.6. As for the upstream excitation, the peaks arise at the eigenfrequencies of the undamped system $\omega H/c = \pi/2,\ 3\pi/2,\ \ldots$. For the limiting value $\omega \to 0$ also $\lambda \to 0$ and the transfer function becomes

$$T_y(0) = 1 - y/H \tag{9.60}$$

9.3 Analysis of simplified models

In this section some commonly used simplified models and boundary conditions are analyzed for the two-dimensional fluid problem: The fixed boundary, the viscous boundary and the incompressible fluid model. Although these models are designed for time-domain applications, the analysis of them is best done in the frequency domain. All examples are analyzed with a bottom-reflection coefficient of $\alpha = 0.8$.

Figure 9.7 shows the general setup. The artificial boundary with the various boundary conditions is placed at a distance L. The pressure at $y = 0.6H$, indicated by a dot, is considered for the transfer functions.

9.3.1 Upstream excitation

To treat an artificial boundary at a distance L from the dam, we have to include a reflected wave in the x-direction. Analogously to Equation (9.35), the general form of the velocity potential including a reflected wave is

$$\varphi = e^{i\omega t} \sum_{n=1}^{\infty} A_n' \left(e^{-ik_n x} + R_n e^{ik_n x} \right) \Phi_n(y) \tag{9.61}$$

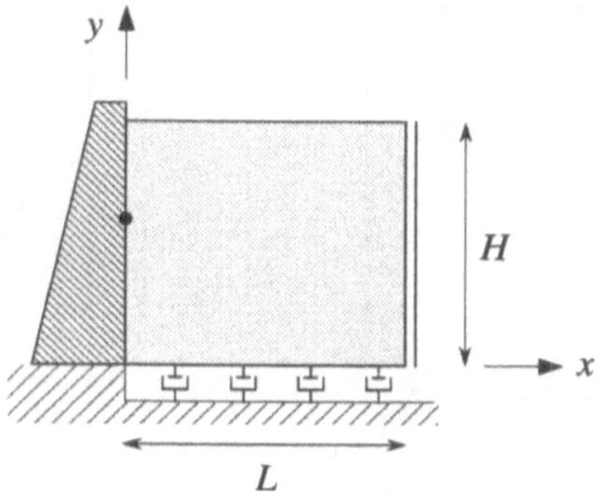

Figure 9.7: 2-D fluid problem: nearfield with artificial boundary

This form already takes care of the boundary condition at the free surface and the bottom of the reservoir. The boundary condition at the dam ($x = 0$) requires

$$-\sum_{m=1}^{\infty} ik_m(1 - R_m)A'_m \Phi_m(y) = \hat{v}_x(y) \tag{9.62}$$

Premultiplying by Φ_n and integrating over the height yields

$$-ik_n(1 - R_n)A'_n \int_0^H \Phi_n^2(y)\,dy = \int_0^H \hat{v}_x(y)\,\Phi_n(y)\,dy \tag{9.63}$$

The amplitude for each mode is

$$A'_n = -\frac{\hat{V}_n}{ik_n(1 - R_n)} = \frac{A_n}{1 - R_n} \tag{9.64}$$

where $\hat{V}_n$ is the modal velocity defined in Equation (9.40) and A_n is the amplitude of the case with the rigorous boundary condition.

For a rigid dam the modal velocity is $\hat{V}_n = \hat{v}_x I_1/I_2$ as described in Equations (9.45, 9.46 and 9.47). The velocity potential at the dam is then

$$\varphi(y) = e^{i\omega t} \sum_{n=1}^{\infty} \frac{1 + R_n}{1 - R_n} A_n \Phi_n(y) = -\hat{v}_x e^{i\omega t} \sum_{n=1}^{\infty} \frac{1 + R_n}{1 - R_n} \frac{1}{ik_n} \frac{I_1}{I_2} \Phi_n(y) \tag{9.65}$$

and the transfer function is

$$T_x(\omega) = \frac{\hat{p}}{H\rho\hat{a}_x} = \sum_{n=1}^{\infty} \frac{1 + R_n}{1 - R_n} \frac{1}{ik_n H} \frac{I_1}{I_2} \Phi_n(y) \tag{9.66}$$

The coefficient R_n depends on the boundary condition at the artificial boundary and is derived next.

Fixed boundary condition. Assuming a fixed boundary at $x = L$, yields the boundary condition

$$\frac{\partial \varphi}{\partial x} = 0 \qquad \text{at } x = L \tag{9.67}$$

Using the velocity potential Equation (9.61) the boundary condition is

$$\sum_{n=1}^{\infty} -ik_n\, A'_n \left(e^{-ik_n L} - R_n e^{ik_n L}\right) \Phi_n(y) = 0 \tag{9.68}$$

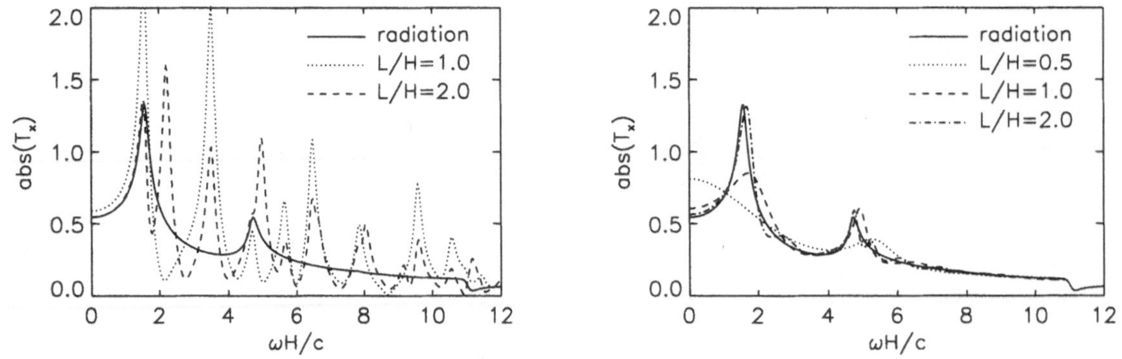

Figure 9.8: 2-D fluid problem: pressure transfer function for horizontal excitation. Left: fixed boundary. Right: viscous boundary.

The expression in the parentheses must vanish yielding

$$R_n = e^{-2ik_nL} \tag{9.69}$$

The amplitude of the transfer function (Equation 9.66) is plotted for different values L/H in Figure 9.8 on the left. In general, the fixed boundary condition leads to completely wrong results. However, for frequencies below the first eigenfrequency the curves compare quite well with the exact model ('radiation'). Especially, the results are much better than the ones obtained by an incompressible-fluid model shown later on. In this frequency range no radiation of wave energy occurs, yet the compressibility of the nearfield has some effect.

Viscous boundary. For the viscous boundary the condition is

$$\frac{\partial \varphi}{\partial x} + \frac{1}{c}\dot{\varphi} = 0 \qquad \text{at } x = L \tag{9.70}$$

Applied to the velocity potential the viscous boundary yields

$$\sum_{n=1}^{\infty} A'_n \left[-ik_n\left(e^{-ik_nL} - R_n e^{ik_nL}\right) + i\omega/c\left(e^{-ik_nL} + R_n e^{ik_nL}\right)\right]\Phi_n(y) = 0 \tag{9.71}$$

The bracketed expression must vanish. Therefore

$$R_n = \frac{ik_n - i\omega/c}{ik_n + i\omega/c}e^{-2ik_nL} \tag{9.72}$$

The coefficient R_n is identical to the reflection coefficient determined earlier in Equation (3.18). On the right of Figure 9.8, viscous boundaries at various distances are compared with the exact radiation condition. The behavior is generally better than for the fixed boundary. Acceptable results are, however, only obtained for a reservoir length of at least $L = 2H$. This would lead to quite a large model when using finite elements. The actual length necessary depends on the foundation absorption.

Incompressible fluid. The incompressibility assumption is a formulation that completely avoids the problem of radiation at the cost of neglecting some important physical aspects of the compressible model. No waves occur and therefore no energy is radiated to the farfield. The response is frequency-independent and there are no resonance phenomena.

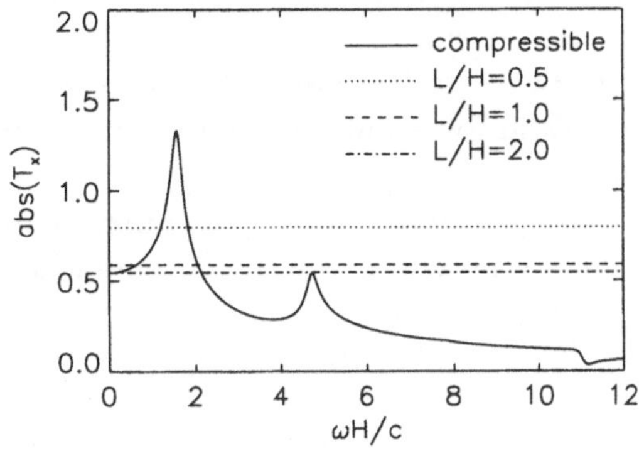

Figure 9.9: 2-D fluid problem: pressure transfer function for horizontal excitation, incompressible fluid

For the incompressible case the differential equation is

$$\Delta\varphi = 0 \qquad \text{in } \Omega \tag{9.73}$$

The boundary conditions at the free surface, at the dam and at the reservoir bottom are the same as in the compressible case (Equations 9.4, 9.5 and 9.6). For the reservoir bottom, no absorption is considered ($q = 0$). Although the mathematical formulation would still be valid, the notion of reflected waves has no meaning because no waves exist in the reservoir. It can easily be seen that the differential equation and all boundary conditions are the same as for the compressible case with $\omega = 0$. Also, the problem is clearly frequency-independent. Therefore, a practical way to calculate the incompressible behavior is to take the zero-frequency solution of the compressible case for all frequencies.

The zero-frequency solution of the compressible case yields

$$ik_n = \lambda_n = \frac{2n-1}{2H}\pi \tag{9.74}$$

For $\omega = 0$ the reflection coefficient for both the fixed (Equation 9.69) and the viscous boundary (Equation 9.72) lead to

$$R_n = e^{-2ik_nL} = e^{-(2n-1)\pi L/H} \tag{9.75}$$

The corresponding transfer function for different values of L/H is plotted in Figure 9.9. The artificial boundary has to be placed at a distance comparable to the height of the reservoir to reach the zero-frequency pressure of the compressible model. Reasonable results are only obtained for frequencies far below the fundamental frequency.

9.3.2 Vertical excitation

Fixed boundary condition. For the infinite reservoir the vertical excitation was assumed to be independent of x and therefore we had $ik = 0$. The horizontal (x-direction) velocity is

$$\frac{\partial\varphi}{\partial x} = 0 \tag{9.76}$$

at any cross-section of the channel. This corresponds to a finite reservoir with a rigid dam and a fixed boundary for any length of the reservoir. So there is no difference between a finite and a semi-infinite reservoir in this case.

Incompressible fluid. The incompressible case is derived as the zero-frequency solution of the compressible case. For an infinite reservoir length, the transfer function at $\omega = 0$ is (Equation 9.60) $T_y = 1 - y/H$. This solution corresponding to the hydrostatic pressure distribution is also valid for a finite reservoir with a fixed boundary as has been shown in the last paragraph.

Viscous boundary. The viscous boundary requires a more complicated analysis. The general form of the velocity potential is written somewhat differently than before as

$$\varphi = \Phi(y) \cos kx \, e^{i\omega t} \tag{9.77}$$

where Φ is the solution satisfying the free surface boundary condition (Equation 9.16)

$$\Phi(y) = A (\cos \lambda y - \cot \lambda H \ \sin \lambda y) \tag{9.78}$$

In the x-direction $\sin kx$ and $\cos kx$ have been used instead of e^{-ikx} and e^{ikx} but the term $\sin kx$ has been dropped because it does not satisfy the boundary condition $\partial\varphi/\partial x = 0$ at the dam. So the above form satisfies the boundary conditions at the free surface and at the dam. For a viscous boundary, the boundary condition at the artificial boundary is

$$\frac{\partial \varphi}{\partial x} + \frac{1}{c}\dot{\varphi} = 0 \qquad \text{at } x = L \tag{9.79}$$

With the partial derivative

$$\frac{\partial \varphi}{\partial x} = -\Phi(y)\, k \sin kx \, e^{i\omega t} \tag{9.80}$$

the viscous boundary reads

$$-k \sin kx + \frac{i\omega}{c} \cos kx = 0 \qquad \text{at } x = L \tag{9.81}$$

or

$$\tan k_n L = \frac{i\omega L}{c k_n L} \tag{9.82}$$

This equation determines the eigenvalues k_n in the x-direction. Note, that the problem analyzed here leads to an eigenvalue problem in the horizontal direction and a boundary value problem in the vertical direction. This is in contrast to the problem of an semi-infinite reservoir with an upstream excitation where the eigenvalue problem is in the vertical direction and the boundary value problem in the horizontal direction. The corresponding λ_n are given by Equation (9.11) as

$$\lambda_n = \sqrt{\frac{\omega^2}{c^2} - k_n^2} \tag{9.83}$$

The sign of the square root is of no importance here, because Φ is an even function in λ. The expression for φ is now

$$\varphi = e^{i\omega t} \sum_{n=1}^{\infty} A_n (\cos \lambda_n y - \cot \lambda_n H \ \sin \lambda_n y) \cos k_n x \tag{9.84}$$

It remains to consider the boundary condition at the reservoir-foundation interface (Equation 9.17)

$$-\frac{\partial \varphi}{\partial y} + i\omega q \varphi = -v_y(x,t) \qquad \text{at } y = 0 \tag{9.85}$$

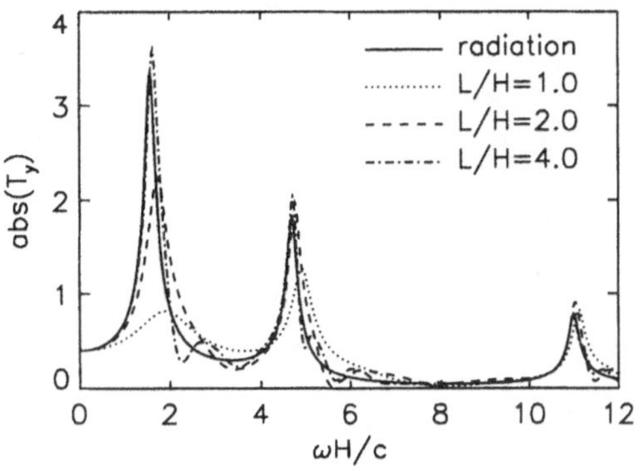

Figure 9.10: 2-D fluid problem: pressure transfer function for vertical excitation, viscous boundary

This boundary condition is now used to determine the factors A_n. The partial derivative is

$$-\frac{\partial \varphi}{\partial y} = e^{i\omega t} \sum_{n=1}^{\infty} A_n\, \lambda_n\, (\sin \lambda_n y + \cot \lambda_n H\, \cos \lambda_n y) \cos k_n x = -\hat{v}_y \tag{9.86}$$

and the boundary condition is

$$\sum_{n=1}^{\infty} A_n\, (\lambda_n \cot \lambda_n H + i\omega q) \cos k_n x = -\hat{v}_y \tag{9.87}$$

Premultiplying by $\cos k_m x$ and integrating over the length L yields

$$A_m\, (\lambda_m \cot \lambda_m H + i\omega q) \int_0^L \cos^2 k_m x\, dx = -\hat{v}_y \int_0^L \cos k_m x\, dx \tag{9.88}$$

or with the index variable m changed back to n

$$\begin{aligned} A_n &= -\frac{\hat{v}_y}{\lambda_n \cot \lambda_n H + i\omega q}\, \frac{\int_0^L \cos k_n x\, dx}{\int_0^L \cos^2 k_n x\, dx} \\ &= -\frac{\hat{v}_y}{\lambda_n \cot \lambda_n H + i\omega q}\, \frac{2\, \sin k_n L}{\cos k_n L\, \sin k_n L + k_n L} \end{aligned} \tag{9.89}$$

The velocity potential at the dam ($x = 0$) is

$$\varphi = e^{i\omega t} \sum_{n=1}^{\infty} A_n\, (\cos \lambda_n y - \cot \lambda_n H\, \sin \lambda_n y) \tag{9.90}$$

which leads to the transfer function

$$T_y(\omega) = \frac{\hat{p}}{H\rho \hat{a}_y} = -\frac{\hat{\varphi}}{H\hat{v}_y} = \sum_{n=1}^{\infty} \frac{\cos \lambda_n y - \cot \lambda_n H\, \sin \lambda_n y}{\lambda_n H \cot \lambda_n H + i\omega q H}\, \frac{2 \sin k_n L}{\cos k_n L\, \sin k_n L + k_n L} \tag{9.91}$$

Its amplitude is shown in Figure 9.10 for various lengths of the reservoir. To get reasonable results, the length of the reservoir has to be four times its height, which is even more than for the horizontal excitation. Again, the necessary length depends on the foundation absorption.

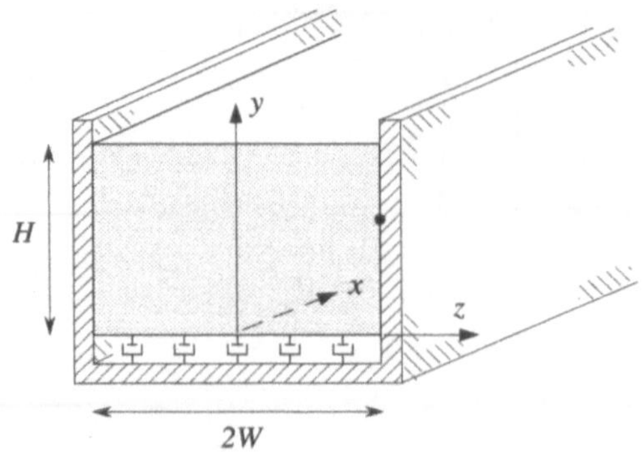

Figure 9.11: Rectangular channel: geometry

9.4 Analytical solutions for 3-D fluid problems

Two three-dimensional models, the rectangular and the semi-circular channel, are considered in this section. Problems with these and other geometries are investigated in [WS49, Shu87] for the case with no bottom absorption.

9.4.1 Rectangular channel

The simplest three-dimensional example is the rectangular channel shown in Figure 9.11. The bottom foundation is absorptive whereas the side walls are rigid with no absorption. This problem is somewhat simpler than the more general case with absorptive side walls and can be treated in a quick manner because for the upstream and the vertical excitation the results are identical to the two-dimensional case. The interesting new aspect which is treated in more detail is the cross-stream excitation.

Again, the method of separation of variables is used. The velocity potential is written as

$$\varphi = \Phi_y(y)\, \Phi_z(z)\, e^{i(\omega t - kx)} \tag{9.92}$$

Substituted into the wave equation two differential equations result:

$$\Phi_y'' + \lambda^2 \Phi_y = 0 \tag{9.93}$$

$$\Phi_z'' + \mu^2 \Phi_z = 0 \tag{9.94}$$

and λ and μ are related by

$$\lambda^2 + \mu^2 = \frac{\omega^2}{c^2} - k^2 \tag{9.95}$$

The solutions are

$$\Phi_y = A \cos \lambda y + B \sin \lambda y \tag{9.96}$$

$$\Phi_z = C \cos \mu z + D \sin \mu z \tag{9.97}$$

Upstream excitation. For an upstream excitation the boundary conditions at the free surface and at the bottom are

$$\Phi_y = 0 \qquad \text{at } y = H \tag{9.98}$$

$$\Phi_y' = i\omega q \Phi_y \qquad \text{at } y = 0 \tag{9.99}$$

For the side walls we have

$$\Phi_z' = 0 \qquad \text{at } z = -W \text{ and } z = W \tag{9.100}$$

and the prescribed motion of the dam is enforced by the boundary condition

$$-ik\Phi_y\Phi_z = \hat{v}_x \qquad \text{at } x = 0 \tag{9.101}$$

For a rigid dam, the excitation is constant over the cross-section. Obviously, Φ_z can then be assumed to be constant, since a constant function satisfies the boundary condition Equation (9.100). This results in $\mu = 0$ and, since a general constant is included in Φ_y, we take $\Phi_z = 1$. Clearly, the differential equation (Equation 9.93) and the boundary conditions (Equations 9.98 and 9.99) for Φ_y are identical to the two-dimensional case. Also, the expression for k (Equation 9.95) matches and therefore the results of the two-dimensional case apply. (This result is only true because the side walls were assumed to have no absorption.)

Vertical excitation. For a vertical excitation, Φ_z may again be assumed to be constant. The boundary conditions for Φ_y are again identical to the ones of the two-dimensional model with vertical excitation and the two-dimensional results apply also for the three-dimensional case.

Cross-stream excitation. For a cross-stream excitation the situation is different. The boundary conditions for the free surface and the bottom wall are

$$\Phi_y = 0 \qquad \text{at } y = H \tag{9.102}$$

and

$$\Phi'_y = i\omega q\Phi_y \qquad \text{at } y = 0 \tag{9.103}$$

which are the same as for the two-dimensional model. For the side walls, considering a uniform input motion, we have

$$\frac{\partial\varphi}{\partial z} = \hat{v}_z e^{i\omega t} \qquad \text{at } z = W \tag{9.104}$$

and

$$-\frac{\partial\varphi}{\partial z} = -\hat{v}_z e^{i\omega t} \qquad \text{at } z = -W \tag{9.105}$$

In terms of the variables Φ_y and Φ_z this is

$$\Phi_y\Phi'_z = \hat{v}_z \qquad \text{at } z = -W \text{ and } z = W \tag{9.106}$$

For the last condition (Equation 9.106) we need the derivative

$$\Phi'_z = -C\mu\sin\mu z + D\mu\cos\mu z \tag{9.107}$$

Clearly, only the second term satisfies $\Phi'_z(W) = \Phi'_z(-W)$ and hence $C = 0$. For the cross-stream excitation the motion is thus anti-symmetric. Because the general constant is included in Φ_y we again take $D = 1$. Boundary conditions (9.102) and (9.103) are, as in the two-dimensional case, satisfied by Equation (9.26). Therefore, we have $\Phi_y = \Phi_n(y)$ of the two-dimensional case. The constant distribution over the height is achieved by a linear combination of all eigenvectors. Equation (9.106) is now

$$\sum_{n=1}^{\infty} A_n\mu_n\Phi_n(y)\cos\mu_n W = \hat{v}_z \tag{9.108}$$

Because $\hat{v}_z$ is independent of x, it follows that $k = 0$ and with Equation (9.95)

$$\mu_n^2 = \frac{\omega^2}{c^2} - \lambda_n^2 \tag{9.109}$$

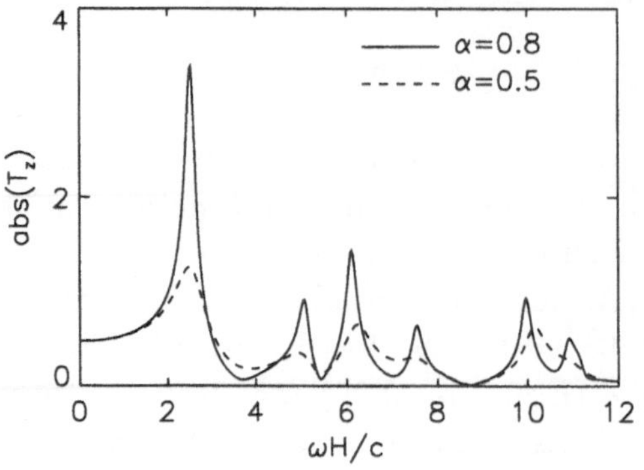

Figure 9.12: Rectangular channel: pressure transfer function for cross-stream excitation

Equation (9.108) is multiplied by $\Phi_n(y)$ and integrated over the height. By virtue of the orthogonality of Φ_n only one term of the summation remains. It follows

$$\mu_n A_n \cos \mu_n W \int_0^H \Phi_n^2(y)\, dy = \hat{v}_z \int_0^H \Phi_n(y)\, dy \tag{9.110}$$

With the integrals Equation (9.46) and (9.47) A_n becomes

$$A_n = \frac{\hat{v}_z}{\mu_n \cos \mu_n W} \frac{I_1}{I_2} \tag{9.111}$$

The velocity potential at the dam is

$$\varphi(y,z) = e^{i\omega t} \sum_{n=1}^{\infty} A_n \Phi_n(y) \sin \mu_n z = \hat{v}_z e^{i\omega t} \sum_{n=1}^{\infty} \frac{I_1}{I_2} \Phi_n(y) \frac{\sin \mu_n z}{\mu_n \cos \mu_n W} \tag{9.112}$$

The corresponding transfer function is

$$T_z(\omega) = \frac{\hat{p}}{H\rho\hat{a}_z} = -\frac{\hat{\varphi}}{H\hat{v}_z} = -\sum_{n=1}^{\infty} \frac{I_1}{I_2} \Phi_n(y) \frac{\sin \mu_n z}{\mu_n H \cos \mu_n W} \tag{9.113}$$

This transfer function is evaluated for the point $y = 0.6H$ and $z = W$ of a rectangular cross-section with $W = 0.8H$ as shown in Figure 9.11. Note that a point on the center line $z = 0$ would have zero pressure because of symmetry. The plot of the transfer function is shown Figure 9.12.

9.4.2 Semi-circular channel

Another three-dimensional example which can be solved analytically is the reservoir with semi-circular cross-section as shown in Figure 9.13. Cylindrical coordinates x, r, θ are used. The two dots at $\theta = 0$ and $\theta = \pi/4$ indicate the points where the results are evaluated. The solution is taken from [SW92] with a slightly different notation. More details are given in the reference.

The Laplace operator in cylindrical coordinates is

$$\Delta\varphi = \frac{\partial^2\varphi}{\partial x^2} + \frac{1}{r}\frac{\partial\varphi}{\partial r} + \frac{\partial^2\varphi}{\partial r^2} + \frac{1}{r^2}\frac{\partial^2\varphi}{\partial\theta^2} \tag{9.114}$$

and the components of the gradient are

$$v_x = \frac{\partial\varphi}{\partial x} \qquad v_r = \frac{\partial\varphi}{\partial r} \qquad v_\theta = \frac{1}{r}\frac{\partial\varphi}{\partial\theta} \tag{9.115}$$

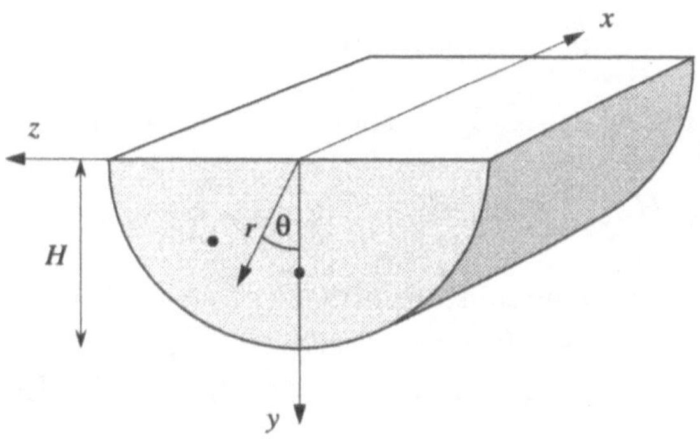

Figure 9.13: Semi-circular channel: geometry

With the separation of variables

$$\varphi = \Phi_r(r)\,\Phi_\theta(\theta)\,e^{i(\omega t - kx)} \tag{9.116}$$

the wave equation leads to the two differential equations

$$\Phi_\theta'' + \nu^2 \Phi_\theta = 0 \tag{9.117}$$

$$r^2 \Phi_r'' + r\Phi_r' + (\lambda^2 r^2 - \nu^2)\Phi_r = 0 \tag{9.118}$$

where

$$\lambda^2 = \frac{\omega^2}{c^2} - k^2 \tag{9.119}$$

The solution of the first differential equation is

$$\Phi_\theta(\theta) = A\cos\nu\theta + B\sin\nu\theta \tag{9.120}$$

The second equation is the Bessel differential equation with the solution

$$\Phi_r(r) = J_\nu(\lambda r) \tag{9.121}$$

J_ν is the Bessel function of the first kind and ν^{th} order. The Bessel function of the second kind Y_ν is also a solution but is excluded because it is singular at the origin.

Upstream excitation. For an upstream excitation the problem is symmetric about $\theta = 0$. Therefore, only cosine terms appear in Equation (9.120) and $B = 0$. The boundary condition for the free surface is

$$\Phi_\theta = 0 \qquad \text{at } \theta = \pm\pi/2 \tag{9.122}$$

or

$$\cos\nu\pi/2 = 0 \tag{9.123}$$

which is satisfied for

$$\nu_n = 2n - 1, \qquad n = 1, 2, \ldots \tag{9.124}$$

At the reservoir-foundation interface the boundary condition with the outward normal pointing to the r-direction is

$$\frac{\partial\varphi}{\partial r} + q\dot{\varphi} = 0 \qquad \text{at } r = H \tag{9.125}$$

or for the variable Φ_r

$$\Phi_r' + i\omega q \Phi_r = 0 \qquad \text{at } r = H \tag{9.126}$$

Substituting the Bessel functions leads to the equation for the eigenvalues

$$\lambda_{mn} H J_{2n-1}'(\lambda_{mn} H) + i\omega q H J_{2n-1}(\lambda_{mn} H) = 0 \tag{9.127}$$

The eigenvalues λ_{mn} have to be evaluated numerically which is not a trivial task, because the argument $\lambda_{mn} H$ of the Bessel function becomes complex in the case of $q \neq 0$. From the eigenvalues λ_{mn} the wave numbers k_{mn} are determined by Equation (9.119) as

$$ik_{mn} = \sqrt{\lambda_{mn}^2 - \frac{\omega^2}{c^2}} \tag{9.128}$$

As for the two-dimensional example, the radiation boundary condition has to be satisfied by taking the proper sign for the square root such that $\Re(ik) > 0$ and $\Im(ik) > 0$. The total solution is obtained by superposition of all modes as

$$\varphi = e^{i\omega t} \sum_{m,n=1}^{\infty} A_{mn}\, e^{-ik_{mn} x}\, J_{2n-1}(\lambda_{mn} r)\, \cos{(2n-1)\theta} \tag{9.129}$$

The velocity in the x-direction is

$$v_x = -e^{i\omega t} \sum_{m,n=1}^{\infty} ik_{mn}\, A_{mn}\, e^{-ik_{mn} x}\, J_{2n-1}(\lambda_{mn} r)\, \cos{(2n-1)\theta} \tag{9.130}$$

For a rigid dam at $x = 0$, the velocity must equal the prescribed one, that is $v_x = \hat{v}_x e^{i\omega t}$. The condition for A_{mn} is therefore

$$-\sum_{m,n=1}^{\infty} ik_{mn}\, A_{mn}\, J_{2n-1}(\lambda_{mn} r)\, \cos{(2n-1)\theta} = \hat{v}_x \tag{9.131}$$

This equation is multiplied by $J_{2j-1}(\lambda_{ij} r) \cos{(2j-1)\theta}$ and integrated over the cross-section of the channel. Due to the orthogonality of the eigenvectors only one term of the summation remains. After renaming the indices back to m and n we get

$$\begin{aligned} -ik_{mn}\, A_{mn} \int_0^H \int_{-\pi/2}^{\pi/2} \Big[J_{2n-1}(\lambda_{mn} r)\, \cos{(2n-1)\theta}\Big]^2\, r\, d\theta\, dr & \\ = \hat{v}_x \int_0^H \int_{-\pi/2}^{\pi/2} \Big[J_{2n-1}(\lambda_{mn} r)\, \cos{(2n-1)\theta}\Big]\, r\, d\theta\, dr & \end{aligned} \tag{9.132}$$

The coefficients A_{mn} are then

$$A_{mn} = -\frac{\hat{v}_x}{ik_{mn}} \frac{I_3}{I_4} \tag{9.133}$$

with the integrals

$$\begin{aligned} I_3 &= \int_0^H \int_{-\pi/2}^{\pi/2} \Big[J_{2n-1}(\lambda_{mn} r) \cos{(2n-1)\theta}\Big]\, r\, d\theta\, dr \\ &= \frac{2}{2n-1} \sin\frac{(2n-1)\pi}{2} \int_0^H r\, J_{2n-1}(\lambda_{mn} r)\, dr \end{aligned} \tag{9.134}$$

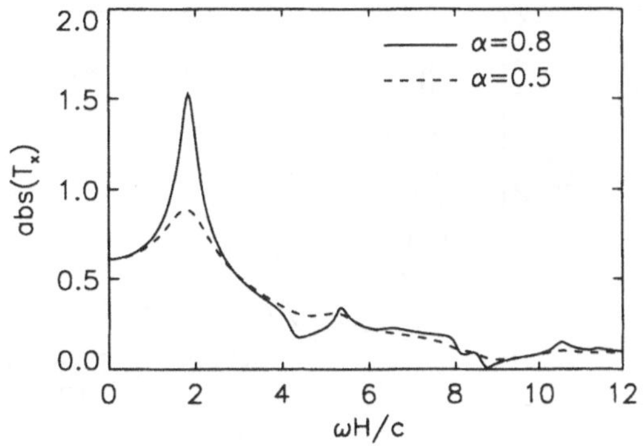

Figure 9.14: Semi-circular channel: pressure transfer function for horizontal excitation

and

$$I_4 = \int_0^H \int_{-\pi/2}^{\pi/2} \Big[J_{2n-1}(\lambda_{mn} r) \cos{(2n-1)\theta}\Big]^2 \, r \, d\theta \, dr$$
$$= \frac{\pi}{2} \int_0^H r J_{2n-1}^2(\lambda_{mn} r) \, dr \tag{9.135}$$

The velocity potential at the dam is

$$\varphi(r,\theta) = -\left(\sum_{m,n=1}^{\infty} \frac{1}{ik_{mn}} \frac{I_3}{I_4} J_{2n-1}(\lambda_{mn} r) \, \cos{(2n-1)\theta} \right) \hat{v}_x \, e^{i\omega t} \tag{9.136}$$

and the transfer function is

$$T_x(\omega) = \frac{\hat{p}}{H \rho \hat{a}_x} = -\frac{\hat{\varphi}}{H \hat{v}_x} = \sum_{m,n=1}^{\infty} \frac{1}{ik_{mn} H} \frac{I_3}{I_4} J_{2n-1}(\lambda_{mn} r) \, \cos{(2n-1)\theta} \tag{9.137}$$

The transfer function for the point $(r = 0.6H, \theta = 0)$ is shown in Figure 9.14.

Vertical excitation. The vertical excitation is assumed to be uniform along the x-direction, that is $k = 0$ and $\lambda = \omega/c$. The boundary condition for the free surface is the same as for the upstream excitation (Equation 9.122). The prescribed motion at $r = H$ is

$$v_r = -\hat{v}_y \, e^{i\omega t} \cos\theta \tag{9.138}$$

and the boundary condition at the dam-reservoir interface is

$$\frac{\partial \varphi}{\partial r} + i\omega q \varphi = v_r \qquad \text{at } r = H \tag{9.139}$$

Substituting $\varphi = \Phi_r \, \Phi_\theta \, e^{i\omega t}$ and taking the prescribed motion $v_r = -\hat{v}_y \, e^{i\omega t} \cos\theta$ at $r = H$ we get

$$\Phi_r' \, \Phi_\theta + i\omega q \Phi_r \, \Phi_\theta + \hat{v}_y \cos\theta = 0 \qquad \text{at } r = H \tag{9.140}$$

Comparing with the general form of Φ_θ (Equation 9.120), it is clear that

$$\Phi_\theta = A \cos\theta \tag{9.141}$$

This implies that $\nu = 1$ and therefore the radial function is

$$\Phi_r = J_1(\lambda r) \tag{9.142}$$

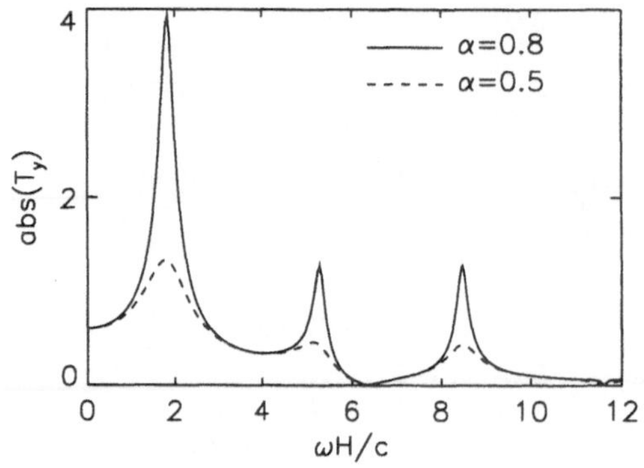

Figure 9.15: Semi-circular channel: pressure transfer function for vertical excitation

The boundary condition for the vertical velocity reads now

$$A\lambda J_1'(\lambda H)\cos\theta + i\omega q A J_1(\lambda H)\cos\theta + \hat{v}_y\cos\theta = 0 \tag{9.143}$$

which yields for A

$$A = \frac{-\hat{v}_y}{\lambda J_1'(\lambda H) + i\omega q J_1(\lambda H)} \tag{9.144}$$

Due to the symmetry of the excitation, the solution is symmetric about $\theta = 0$. The velocity potential at the dam is now

$$\varphi(r,\theta) = \frac{-J_1(\lambda r)\cos\theta}{\lambda J_1'(\lambda H) + i\omega q J_1(\lambda H)}\hat{v}_y e^{i\omega t} \tag{9.145}$$

The transfer function is

$$T_y(\omega) = \frac{\hat{p}}{H\rho\hat{a}_y} = -\frac{\hat{\varphi}}{H\hat{v}_y} = \frac{J_1(\lambda r)\cos\theta}{\lambda H J_1'(\lambda H) + i\omega q H J_1(\lambda H)} \tag{9.146}$$

For the point $(r = 0.6, \theta = 0)$, the amplitude of the transfer function is plotted in Figure 9.15.

As for the two-dimensional case, the hydrostatic pressure is recovered by letting $\omega \to 0$. Then also $\lambda \to 0$ and $i\omega q J_1(\lambda H) \to 0$ and $J_1(\lambda r)/J_1'(\lambda H) \to \lambda r J_1(\lambda r)/J_1'(\lambda H) \to r$, and we are left with the hydrostatic pressure distribution

$$T_y(0) = \frac{r}{H}\cos\theta \tag{9.147}$$

Cross-stream excitation. The cross-stream excitation is also assumed to be uniform along the x-axis. It is defined by

$$v_r = \hat{v}_z\, e^{i\omega t}\sin\theta \tag{9.148}$$

The boundary condition at the reservoir-dam interface is therefore

$$\Phi_r'\,\Phi_\theta + i\omega q\Phi_r\,\Phi_\theta - \hat{v}_z\sin\theta = 0 \tag{9.149}$$

Because the excitation is an odd function, Φ_θ also has to be odd:

$$\Phi_\theta = A_n \sin\nu_n\theta \tag{9.150}$$

Further it has to satisfy the boundary condition at the free surface. Therefore,

$$\sin(\nu_n\pi/2) = 0 \tag{9.151}$$

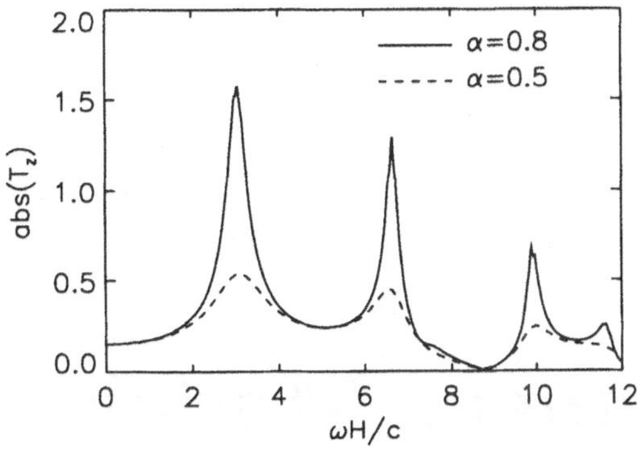

Figure 9.16: Semi-circular channel: pressure transfer function for cross-stream excitation

which is true for

$$\nu_n = 2n, \qquad n = 1, 2, \ldots \tag{9.152}$$

The boundary condition for the cross-stream excitation becomes

$$\sum_{n=1}^{\infty} A_n \lambda J'_{2n}(\lambda H) \sin 2n\theta + i\omega q \sum_{n=1}^{\infty} A_n J_{2n}(\lambda H) \sin 2n\theta - \hat{v}_y \sin\theta = 0 \tag{9.153}$$

Because the velocity potential is expressed in terms of $\sin 2n\theta$, the prescribed velocity has to be expanded into the same form. The expansion of $\sin\theta$ for $-\pi/2 < \theta < \pi/2$ is

$$\sin\theta = \frac{8}{\pi} \sum_{n=1}^{\infty} (-1)^n \frac{n}{1-4n^2} \sin 2n\theta \tag{9.154}$$

Comparing with Equation (9.153) yields for A_n

$$A_n = \frac{-\hat{v}_z(-1)^n 8n}{(4n^2-1)\pi\Big(\lambda J'_{2n}(\lambda H) + i\omega q J_{2n}(\lambda H)\Big)} \tag{9.155}$$

and with this coefficient the velocity potential at the dam is

$$\varphi(r,\theta) = \left(\sum_{n=1}^{\infty} A_n J_{2n}(\lambda r) \sin 2n\theta\right) \hat{v}_z\, e^{i\omega t} \tag{9.156}$$

The amplitude of the transfer function

$$T_z(\omega) = \frac{\hat{p}}{H\rho\hat{a}_z} = -\frac{\hat{\varphi}}{H\hat{v}_z} = \sum_{n=1}^{\infty} \frac{(-1)^n 8n\, J_{2n}(\lambda r) \sin 2n\theta}{(4n^2-1)\pi\Big(\lambda H J'_{2n}(\lambda H) + i\omega q H J_{2n}(\lambda H)\Big)} \tag{9.157}$$

at the point $(r = 0.6H, \theta = -\pi/4)$ is shown in Figure 9.16.

Chapter 10

Finite Element Models

The finite element method is nowadays the standard method for solving partial differential equations, especially in continuum mechanics. In this chapter the fluid, solid and interface elements are derived. The elements are employed for the discretization of the nearfield, that is the dam, the nearfield of the foundation and the nearfield of the reservoir. For the farfield, a semi-analytical approach is taken. The cross-section of the infinite channel is discretized by finite elements, whereas the infinite direction is described analytically. The farfield solution is developed in the frequency domain and the coupling to the nearfield is discussed. In the last part, the frequency-domain analysis is applied to an arch dam. Besides the rigorous solution also some simplified models are analyzed for comparison.

The first step in developing a finite element for a given differential equation is to find the weak form of the problem. The weak form is equivalent to the differential equation and has the from of an integral equation containing the field variables and their partial spatial derivatives. The second step is the discretization, that is, expressing the field variables defined on the continuum as a function of a finite number of parameters. In the finite element method the shape functions are used to describe the variables within an element as a function of the values at the nodes. The third step is to use the discretized problem in the weak form (integral equation) thereby obtaining a set of algebraic equations for the nodal values. The method is standard and is described in many text books [And93, BW76, Hug87, Zie77]. Here, we follow closely the book of Hughes. The different steps are explained in the context of the fluid element. The fluid element is easier to formulate than the solid element because the field variable is a scalar field. The other reason for taking the fluid element is that this element is not as well known as the solid element.

All elements used are of the isoparametric type. Isoparametric elements are very flexible to adapt to any geometry, completeness is guaranteed by construction and they have a standard implementation. The dam is modeled by isoparametric bricks not by shell elements for tow reasons: Firstly, the the dam is not the main interest of this investigation. Therefore, the implementational effort for this part is kept to a minimum. Secondly, isoparametric elements are compatible with isoparametric joint elements which will be used in the future to model the opening and closing of the contraction joints.

10.1 General formulation for fluid

10.1.1 Weak form

Two different types of formulation for the fluid exist in the literature. One is the displacement formulation the other is the potential formulation. The displacement formulation uses directly the displacements as field variables, whereas the potential formulation uses the displacement potential, the velocity potential or the pressure.

The displacement formulation has the advantage that it is analogous to the solid element and can easily be coupled to the latter. However, since the fluid has no shear stiffness, the element has to be stabilized to avoid the so-called hourglass modes [Wil83].

The velocity-potential formulation does not encounter the problem of hourglass modes. It leads to less degrees of freedom because the nodal variable is a scalar. The only difficulty is the coupling to the displacement degrees of freedom. However, an interface element which takes care of this coupling follows immediately from the weak formulation. Moreover, the coupled system is symmetrical. The pressure formulation which is also often used [HC83] is basically identical to the velocity-potential formulation but leads to a non-symmetric coupled system. Another similar formulation is the combination of the displacement potential and the pressure [PA89]. This leads to a symmetric formulation but uses two degrees of freedom per node. The advantages are that the pressure field is continuous even for discontinuous material properties and that the static case can easily be included.

In the following, only the velocity potential formulation is treated [OB85a]. The differential equations and the boundary conditions have been stated in Chapter 9 and are repeated here for convenience. The governing differential equation is the acoustic wave equation

$$\Delta\varphi = \frac{1}{c^2}\ddot{\varphi} \qquad \text{in } \Omega \tag{10.1}$$

where φ is the velocity potential. The related physical variables are the velocity

$$\mathbf{v} = \mathbf{grad}\varphi \tag{10.2}$$

and the dynamic pressure

$$p = -\rho\dot{\varphi} \tag{10.3}$$

For the boundary conditions one has to distinguish between essential and natural boundary conditions. Essential boundary conditions involve the field variables themselves, natural boundary conditions involve spatial derivatives. The essential boundary condition is the free surface condition

$$\varphi = 0 \qquad \text{on } \Gamma_p \tag{10.4}$$

The flexibility of the foundation is taken into account by the absorptive foundation model. The model is based on one-dimensional wave propagation. It is assumed that in the normal direction only a part α of the wave is reflected back to the reservoir. The corresponding natural boundary condition is

$$\frac{\partial\varphi}{\partial n} + q\dot{\varphi} = v_n \qquad \text{on } \Gamma_v \tag{10.5}$$

where v_n is the prescribed normal velocity and q is a parameter related to the reflection coefficient α by $\alpha = (1 - cq)/(1 + cq)$.

To find the weak formulation, the differential equation and the natural boundary conditions are multiplied by a weighting function ψ and integrated over the respective domains.

$$-\int_\Omega \psi\left(\Delta\varphi - \frac{1}{c^2}\ddot{\varphi}\right) d\Omega + \int_{\Gamma_v} \psi\left(\frac{\partial\varphi}{\partial n} + q\dot{\varphi} - v_n\right) d\Gamma = 0 \tag{10.6}$$

Using Green's theorem

$$\int_\Omega \psi\,\Delta\varphi\, d\Omega = \int_\Gamma \psi\frac{\partial\varphi}{\partial n} d\Gamma - \int_\Omega \left(\frac{\partial\psi}{\partial x}\frac{\partial\varphi}{\partial x} + \cdots\right) d\Omega \tag{10.7}$$

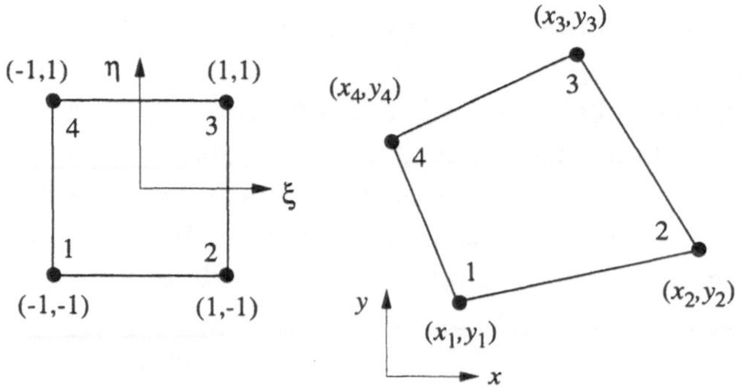

Figure 10.1: Four-node quadrilateral element

with $\Gamma = \Gamma_v \cup \Gamma_p$, the weak form becomes

$$\int_\Omega \left(\frac{\partial \psi}{\partial x} \frac{\partial \varphi}{\partial x} + \cdots \right) d\Omega + \int_\Omega \frac{1}{c^2} \psi \ddot{\varphi} \, d\Omega + \int_{\Gamma_v} \psi \left(q\dot{\varphi} - v_n \right) d\Gamma - \int_{\Gamma_p} \psi \frac{\partial \varphi}{\partial n} \, d\Gamma = 0 \qquad (10.8)$$

Now the weighting function is taken to satisfy the essential boundary conditions, that is

$$\psi = 0 \qquad \text{on } \Gamma_p \qquad (10.9)$$

Then the last term in Equation (10.8) vanishes and, after multiplying by ρ, we end up with the final form

$$\rho \int_\Omega \left(\frac{\partial \psi}{\partial x} \frac{\partial \varphi}{\partial x} + \cdots \right) d\Omega + \frac{\rho}{c^2} \int_\Omega \psi \ddot{\varphi} \, d\Omega + q\rho \int_{\Gamma_v} \psi \dot{\varphi} \, d\Gamma - \rho \int_{\Gamma_v} \psi v_n \, d\Gamma = 0 \qquad (10.10)$$

10.1.2 Shape functions

The region Ω is now divided into elements. The coordinates, the field variables and the weighting functions are interpolated within each element from its nodal values. For the isoparametric elements all these parameters are interpolated using the same shape functions.

It is customary to use a so-called parent element to define the shape functions. The parent element is an element with local coordinates in the range -1 to $+1$, that is, either a line with endpoints $\xi = \pm 1$, a square with the edges $\xi = \pm 1$ and $\eta = \pm 1$ or a cube with surfaces $\xi = \pm 1$, $\eta = \pm 1$ and $\zeta = \pm 1$. A simple example is a four-node quadrilateral shown in Figure 10.1. The mapping from the parent element's local ξ, η-coordinates to the element's global x, y-coordinates is done using the shape functions. For instance, the x-coordinate for the four-node quadrilateral element is given by

$$x(\xi, \eta) = \mathbf{N}^e \mathbf{x}^e = \begin{bmatrix} N_1(\xi,\eta) & N_2(\xi,\eta) & N_3(\xi,\eta) & N_4(\xi,\eta) \end{bmatrix} \begin{bmatrix} x_1 \\ x_2 \\ x_3 \\ x_4 \end{bmatrix} \qquad (10.11)$$

where $\mathbf{N}^e$ is the matrix of element shape functions and $\mathbf{x}^e$ the vector of nodal coordinates of the element using local node numbering.

For the four-node quadrilateral the shape functions are

$$\begin{aligned} N_1(\xi,\eta) &= 1/4\,(1-\xi)(1-\eta) & (10.12) \\ N_2(\xi,\eta) &= 1/4\,(1+\xi)(1-\eta) & (10.13) \\ N_3(\xi,\eta) &= 1/4\,(1+\xi)(1+\eta) & (10.14) \\ N_4(\xi,\eta) &= 1/4\,(1-\xi)(1+\eta) & (10.15) \end{aligned}$$

Note that all shape functions satisfy

$$\sum_{j=1}^{N} N_j = 1 \tag{10.16}$$

and

$$N_j(\xi_k, \eta_k, \zeta_k) = \begin{cases} 1 & j = k \\ 0 & j \neq k \end{cases} \tag{10.17}$$

For the following derivations it is convenient to write the coordinate mapping of the whole element assembly as

$$x = \mathbf{Nx}, \quad y = \mathbf{Ny}, \quad z = \mathbf{Nz} \tag{10.18}$$

where $\mathbf{x}, \mathbf{y}, \mathbf{z}$ are the vectors containing the coordinates of all nodes. It should be kept in mind, however, that this is only a notational convenience. The actual calculations will be done on an element by element basis.

The field variables and the weighting functions within an element are interpolated the same way as the coordinates. For a scalar field we have

$$\varphi = \mathbf{N}\boldsymbol{\varphi} \tag{10.19}$$

where $\boldsymbol{\varphi}$ is the vector of nodal values. Besides the field variable itself, also the first partial derivatives are needed. They are

$$\frac{\partial \varphi}{\partial x} = \mathbf{N}_x \boldsymbol{\varphi} \tag{10.20}$$

where

$$\mathbf{N}_x = \begin{bmatrix} \frac{\partial N_1}{\partial x} & \frac{\partial N_2}{\partial x} & \cdots \end{bmatrix} \tag{10.21}$$

contains the partial derivatives of the shape functions with respect to x. These are by the chain rule

$$\frac{\partial N_j}{\partial x} = \frac{\partial N_j}{\partial \xi}\frac{\partial \xi}{\partial x} + \frac{\partial N_j}{\partial \eta}\frac{\partial \eta}{\partial x} + \frac{\partial N_j}{\partial \zeta}\frac{\partial \zeta}{\partial x} \tag{10.22}$$

Similar expressions are valid for the other coordinates. The derivatives $\partial\xi/\partial x$, $\partial\eta/\partial x, \ldots$ are not known but can be calculated from the inverse Jacobian matrix.

$$\begin{bmatrix} \partial\xi/\partial x & \partial\xi/\partial y & \partial\xi\partial z \\ \partial\eta/\partial x & \partial\eta/\partial y & \partial\eta\partial z \\ \partial\zeta/\partial x & \partial\zeta/\partial y & \partial\zeta\partial z \end{bmatrix} = \mathbf{J}^{-1} = \begin{bmatrix} \partial x/\partial\xi & \partial x/\partial\eta & \partial x/\partial\zeta \\ \partial y/\partial\xi & \partial y/\partial\eta & \partial y/\partial\zeta \\ \partial z/\partial\xi & \partial z/\partial\eta & \partial z/\partial\zeta \end{bmatrix}^{-1} \tag{10.23}$$

The derivatives $\partial x/\partial\xi, \ldots$ are calculated from Equation (10.18) as

$$\partial x/\partial \xi = \mathbf{N}_\xi \mathbf{x} \tag{10.24}$$

with

$$\mathbf{N}_\xi = \begin{bmatrix} \partial N_1/\partial\xi & \partial N_2/\partial\xi & \ldots \end{bmatrix} \tag{10.25}$$

10.2 Fluid element matrices

The interpolated field variables and weighting functions are put into the weak form Equation (10.10). Writing $\psi = \boldsymbol{\psi}^{\mathrm{T}}\mathbf{N}^{\mathrm{T}}$ we have

$$\boldsymbol{\psi}^{\mathrm{T}} \Big(\rho \int_\Omega \left(\mathbf{N}_x^{\mathrm{T}}\mathbf{N}_x + \mathbf{N}_y^{\mathrm{T}}\mathbf{N}_y + \cdots \right) d\Omega\, \boldsymbol{\varphi} + \frac{\rho}{c^2}\int_\Omega \mathbf{N}^{\mathrm{T}}\mathbf{N}\, d\Omega \ddot{\boldsymbol{\varphi}}$$
$$+ q\rho \int_{\Gamma_v} \mathbf{N}^{\mathrm{T}}\mathbf{N}\, d\Gamma \dot{\boldsymbol{\varphi}} - \rho \int_{\Gamma_v} \mathbf{N}^{\mathrm{T}} v_n\, d\Gamma \Big) = 0 \tag{10.26}$$

The vector of the nodal weighting function ψ is arbitrary. Therefore, the equation has to be satisfied independently of it and ψ can be dropped. Introducing the matrix

$$\mathbf{B} = \begin{bmatrix} \mathbf{N}_x \\ \mathbf{N}_y \\ \mathbf{N}_z \end{bmatrix} \tag{10.27}$$

Equation (10.26) becomes

$$\rho \int_\Omega \mathbf{B}^\mathrm{T}\mathbf{B}\, d\Omega\, \varphi + \frac{\rho}{c^2}\int_\Omega \mathbf{N}^\mathrm{T}\mathbf{N}\, d\Omega\, \ddot{\varphi} + q\rho \int_{\Gamma_v} \mathbf{N}^\mathrm{T}\mathbf{N}\, d\Gamma\, \dot{\varphi} - \rho \int_{\Gamma_v} \mathbf{N}^\mathrm{T} v_n\, d\Gamma = 0 \tag{10.28}$$

This is an ordinary differential equation in time for the nodal variables φ. It can be written as

$$\mathbf{M}_{ff}\ddot{\varphi} + \mathbf{C}_{ff}\dot{\varphi} + \mathbf{K}_{ff}\varphi = \mathbf{V} \tag{10.29}$$

with $\mathbf{M}_{ff}, \mathbf{C}_{ff}, \mathbf{K}_{ff}$ the fluid mass, damping and stiffness matrices, respectively, and $\mathbf{V}$ the fluid right-hand vector. These matrices are

$$\mathbf{M}_{ff} = \frac{\rho}{c^2}\int_\Omega \mathbf{N}^\mathrm{T}\mathbf{N}\, d\Omega \tag{10.30}$$

$$\mathbf{C}_{ff} = q\rho \int_{\Gamma_v} \mathbf{N}^\mathrm{T}\mathbf{N}\, d\Gamma \tag{10.31}$$

$$\mathbf{K}_{ff} = \rho \int_\Omega \mathbf{B}^\mathrm{T}\mathbf{B}\, d\Omega \tag{10.32}$$

$$\mathbf{V} = \rho \int_{\Gamma_v} \mathbf{N}^\mathrm{T} v_n\, d\Gamma \tag{10.33}$$

The last equation is not used explicitly. Instead the velocity input $\mathbf{V}$ is calculated employing the interface element introduced in Section 10.4 below. The evaluation of the integrals is done element by element. For each element the integration is performed in the domain of the parent element.

$$\int_{\Omega^e} \mathbf{N}^{e\mathrm{T}}\mathbf{N}^e\, d\Omega = \int\int\int \mathbf{N}^{e\mathrm{T}}\mathbf{N}^e\, dx\, dy\, dz = \int_{-1}^{1}\int_{-1}^{1}\int_{-1}^{1} \mathbf{N}^{e\mathrm{T}}\mathbf{N}^e \det(\mathbf{J})\, d\xi\, d\eta\, d\zeta \tag{10.34}$$

Numerical integration methods are employed, usually the Gauss quadrature rule.

$$\int_{-1}^{1}\int_{-1}^{1}\int_{-1}^{1} \mathbf{N}^{e\mathrm{T}}\mathbf{N}^e \det(\mathbf{J})\, d\xi\, d\eta\, d\zeta \approx \sum_{j=1}^{K} \mathbf{N}^{e\mathrm{T}}(\xi_j,\eta_j,\zeta_j)\, \mathbf{N}^e(\xi_j,\eta_j,\zeta_j)\, \det(\mathbf{J}(\xi_j,\eta_j,\zeta_j))\, w_j \tag{10.35}$$

where (ξ_j, η_j, ζ_j) are the K integration points in the parent element, and w_j is the is the corresponding weight. The integral over the curved surface Γ_v is slightly different.

$$\int_{\Gamma_v^e} \mathbf{N}^{e\mathrm{T}}\mathbf{N}^e\, d\Gamma = \int_{-1}^{1}\int_{-1}^{1} \mathbf{N}^{e\mathrm{T}}\mathbf{N}^e\, \|\mathbf{t}_\xi \times \mathbf{t}_\eta\|\, d\xi\, d\eta \tag{10.36}$$

where $\mathbf{t}_\xi$ and $\mathbf{t}_\eta$ are vectors tangent to the surface, defined as

$$\mathbf{t}_\xi = \begin{bmatrix} \partial x/\partial\xi \\ \partial y/\partial\xi \\ \partial z/\partial\xi \end{bmatrix} \qquad \mathbf{t}_\eta = \begin{bmatrix} \partial x/\partial\eta \\ \partial y/\partial\eta \\ \partial z/\partial\eta \end{bmatrix} \tag{10.37}$$

These are the first two columns of the Jacobian matrix. The Euclidean length of the cross product $\|\mathbf{t}_\xi \times \mathbf{t}_\eta\|$ is analogous to the determinant of the Jacobian matrix $\det(\mathbf{J})$ in the case of a volume integral. The global matrices, corresponding to the sum of the element integrals, are found by appropriately assembling the element matrices.

Implemented are two-dimensional 4- and 8-node quadrilaterals and three-dimensional 8- and 20- nodes bricks for the mass matrix $\mathbf{M}_{ff}$ and the stiffness matrix $\mathbf{K}_{ff}$. The damping matrices $\mathbf{C}_{ff}$ are combined with the interface elements described later (see Section 10.4).

10.3 Solid element matrices

Linear elastic isotropic solid elements are considered. The field variables are the displacements. The formulation is somewhat more involved than for the fluid elements, because the field variable is a vector quantity. Since these elements are standard, only the main results are repeated here. For more details see Hughes [Hug87].

The geometric mapping is the same as for the fluid elements. The field variables, that is, the displacements u, v and w are interpolated individually. In matrix form

$$\begin{bmatrix} u \\ v \\ w \end{bmatrix} = \mathbf{N}_s^e \mathbf{u}^e = \begin{bmatrix} N_1 & & & \cdots & & \\ & N_1 & & & \cdots & \\ & & N_1 & & & \cdots \end{bmatrix} \begin{bmatrix} u_1 \\ v_1 \\ w_1 \\ \vdots \end{bmatrix} \tag{10.38}$$

where $\mathbf{N}_s^e$ denotes the matrix of the element shape functions for a solid element and $\mathbf{u}^e$ are the element nodal displacements using the local element numbering.

As for the fluid element, it is notationally convenient to use $\mathbf{N}_s$ for the assembled shape functions and $\mathbf{u}$ for the displacements of all nodes. Then the equation of motion is

$$\mathbf{M}_{ss}\ddot{\mathbf{u}} + \mathbf{C}_{ss}\dot{\mathbf{u}} + \mathbf{K}_{ss} = \mathbf{F}_s \tag{10.39}$$

with the element matrices/indexMass matrix!for solid

$$\mathbf{M}_{ss} = \int_\Omega \rho_s \mathbf{N}_s^{\mathrm{T}} \mathbf{N}_s \, d\Omega \tag{10.40}$$

$$\mathbf{K}_{ss} = \int_\Omega \mathbf{B}_s^{\mathrm{T}} \mathbf{D} \mathbf{B}_s \, d\Omega \tag{10.41}$$

$$\mathbf{F}_s = \int_{\Gamma_\sigma} \mathbf{N}_s^{\mathrm{T}} \mathbf{f} \, d\Gamma \tag{10.42}$$

In these matrices ρ_s is the mass density, Γ_σ the boundary with prescribed stresses, and $\mathbf{f}^{\mathrm{T}} = \begin{bmatrix} f_x & f_y & f_z \end{bmatrix}$ is the vector of applied distributed forces. The damping matrix is not formed explicitly but taken to be proportional to the stiffness matrix and the mass matrix as $\mathbf{C}_{ss} = \alpha_R \mathbf{M}_{ss} + \beta_R \mathbf{K}_{ss}$ (Rayleigh damping). The matrix $\mathbf{B}_s$ is the usual strain-displacement matrix for the solid and $\mathbf{D}$ the elasticity matrix containing the elastic material properties. See Hughes [Hug87] for further details.

Implemented are two-dimensional 4- and 8-node quadrilaterals and three-dimensional 8- and 20- nodes bricks for the mass matrix $\mathbf{M}_{ss}$ and the stiffness matrix $\mathbf{K}_{ss}$.

10.4 Interface element matrices

To couple the fluid elements to the solid elements, special interface elements are defined. At the fluid-solid interface the normal velocity is

$$v_n = \mathbf{n}^{\mathrm{T}} \mathbf{N}_s \dot{\mathbf{u}} \tag{10.43}$$

where $\dot{\mathbf{u}}$ are the nodal velocities (temporal derivative of nodal displacements) at the interface, $\mathbf{N}_s$ are the shape functions for the solid and $\mathbf{n}$ is the unit normal pointing outward from the fluid domain. Therefore, the last integral of Equation (10.28) can be written as

$$-\rho\int_{\Gamma_v}\mathbf{N}^{\mathrm{T}}v_n\,d\Gamma = -\rho\int_{\Gamma_v}\mathbf{N}^{\mathrm{T}}\mathbf{n}^{\mathrm{T}}\mathbf{N}_s\,d\Gamma\,\dot{\mathbf{u}} \tag{10.44}$$

The right-hand side integral of this equation can be interpreted as the coupling matrix of the displacement degree of freedoms with the right-hand side of the fluid equations.

$$\mathbf{C}_{fs} = -\rho\int_{\Gamma_v}\mathbf{N}^{\mathrm{T}}\mathbf{n}^{\mathrm{T}}\mathbf{N}_s\,d\Gamma \tag{10.45}$$

On the other hand, the fluid pressure is $p = -\rho\dot{\varphi} = -\rho\mathbf{N}\dot{\varphi}$ and the force applied to the the solid at the nodes is

$$-\int_{\Gamma_v}\mathbf{N}_s^{\mathrm{T}}\mathbf{n}p\,d\Gamma = \rho\int_{\Gamma_v}\mathbf{N}_s^{\mathrm{T}}\mathbf{n}\mathbf{N}\,d\Gamma\,\dot{\varphi} \tag{10.46}$$

The right-hand side integral can be interpreted as the matrix coupling the velocity-potential degrees of freedom with the right-hand side of the solid equations

$$\mathbf{C}_{sf} = \rho\int_{\Gamma_v}\mathbf{N}_s^{\mathrm{T}}\mathbf{n}\mathbf{N}\,d\Gamma \tag{10.47}$$

Note that $\mathbf{C}_{fs}^{\mathrm{T}} = -\mathbf{C}_{sf}$. Therefore, except for the sign, the coupling is symmetrical.

As for the fluid element matrix $\mathbf{C}_{ff}$, the integral over the curved surface Γ_v is formulated in the parent element as

$$\int_{\Gamma_v}\mathbf{N}_s^{\mathrm{T}}\mathbf{n}\mathbf{N}\,d\Gamma = \int_{-1}^{1}\int_{-1}^{1}\mathbf{N}_s^{\mathrm{T}}\mathbf{n}\mathbf{N}\|\mathbf{t}_\xi\times\mathbf{t}_\eta\|\,d\xi\,d\eta \tag{10.48}$$

The vectors $\mathbf{t}_\xi$ and $\mathbf{t}_\eta$ are also used to construct the unit normal. Since they are tangential to the surface, their cross product is, after proper scaling, the unit normal

$$\mathbf{n} = \frac{\mathbf{t}_\xi\times\mathbf{t}_\eta}{\|\mathbf{t}_\xi\times\mathbf{t}_\eta\|} \tag{10.49}$$

In the integral Equation (10.48) the Euclidean length of the cross product cancels and we are left with

$$\int_{\Gamma_v}\mathbf{N}_s^{\mathrm{T}}\mathbf{n}\mathbf{N}\,d\Gamma = \int_{-1}^{1}\int_{-1}^{1}\mathbf{N}_s^{\mathrm{T}}(\mathbf{t}_\xi\times\mathbf{t}_\eta)\mathbf{N}\,d\xi\,d\eta \tag{10.50}$$

The integration is again performed numerically by the Gauss quadrature rule.

For the implementation it is convenient to incorporate the damping matrix Equation (10.31) and the coupling matrix Equation (10.47) in the same element. This is possible because the domains for the corresponding integrals are the same and also the degrees of freedom involved are the same. The matrix $\mathbf{C}_{sf}$ couples the fluid degrees of freedom with the solid degrees of freedom and contains only off-diagonal terms. The matrix $\mathbf{C}_{ff}$ connects the fluid degrees of freedom to itself resulting in diagonal terms. The same element can also be used for the velocity input $\mathbf{V}$ (Equation 10.33). Rather than specifying the velocity input itself, a prescribed velocity can conveniently be specified at the nodes of the interface.

$$\mathbf{V} = \rho\int_{\Gamma_v}\mathbf{N}^{\mathrm{T}}\mathbf{n}^{\mathrm{T}}\mathbf{N}_s\,d\Gamma\,\dot{\mathbf{u}} = -\mathbf{C}_{fs}\dot{\mathbf{u}} \tag{10.51}$$

The element takes care of the integration in Equation (10.33).

The combined interface-damping elements are implemented as two-dimensional curved line elements with 2 or 3 nodes and as three-dimensional 4- or 8-node quadrilaterals curved in space.

10.5 Infinite fluid domain

The differential equation for the infinite channel is the same as for the finite fluid domain. However, only the cross-section is discretized by finite elements, whereas the upstream direction is formulated analytically. Because the solution has to satisfy the radiation condition, only a frequency-domain solution can be found. The concept has been applied to soil layers by Lysmer and Waas [LW72], to reservoirs by Hall and Chopra [HC82, HC83] and to a combination of soil and fluid layers by [LRT87].

10.5.1 Fluid element matrices for channel cross-section

The steady-state solution is written as

$$\varphi = \Phi(y,z)\, e^{i(\omega t - kx)} \tag{10.52}$$

Substituting Equation (10.52) into the wave equation (10.1) we obtain the differential equation for the cross-section Ω^I

$$\Delta\Phi + \left(\frac{\omega^2}{c^2} - k^2\right)\Phi = 0 \qquad \text{in } \Omega^I \tag{10.53}$$

the essential boundary condition at the free surface Γ_p^I

$$\Phi = 0 \qquad \text{on } \Gamma_p^I \tag{10.54}$$

and the natural boundary condition at the reservoir-foundation interface of the cross-section Γ_v^I

$$\frac{\partial\Phi}{\partial n} + i\omega q\Phi - \hat{v}_n = 0 \qquad \text{on } \Gamma_v^I \tag{10.55}$$

where $\hat{v}_n$ is the complex amplitude of the normal velocity acting on the boundary of the channel cross-section.

The weak form for the cross-section of the channel is, as for the finite fluid element, found by multiplying the differential equation and the natural boundary conditions by a weighting function ψ and integrating over the corresponding domains.

$$-\int_{\Omega^I} \psi\left(\Delta\Phi + \frac{\omega^2}{c^2}\Phi - k^2\Phi\right) d\Omega + \int_{\Gamma_v^I} \psi\left(\frac{\partial\Phi}{\partial n} + i\omega q\Phi - \hat{v}_n\right) d\Gamma \tag{10.56}$$

Using Green's theorem (Equation 10.7) this integral equation becomes

$$\int_{\Omega^I}\left(\frac{\partial\psi}{\partial x}\frac{\partial\Phi}{\partial x} + \frac{\partial\psi}{\partial y}\frac{\partial\Phi}{\partial y}\right) d\Omega + \int_{\Omega^I}\left(k^2 - \frac{\omega^2}{c^2}\right)\psi\Phi\, d\Omega + \int_{\Gamma_v^I}\psi(i\omega q\Phi - \hat{v}_n) d\Gamma - \int_{\Gamma_p^I}\psi\frac{\partial\Phi}{\partial n}\, d\Gamma = 0 \tag{10.57}$$

Again, ψ satisfies the essential boundary conditions and the last term vanishes. After multiplying by ρ the final form is

$$\rho\int_{\Omega^I}\left(\frac{\partial\psi}{\partial x}\frac{\partial\Phi}{\partial x} + \frac{\partial\psi}{\partial y}\frac{\partial\Phi}{\partial y}\right) d\Omega + \rho\left(k^2 - \frac{\omega^2}{c^2}\right)\int_{\Omega^I}\psi\Phi\, d\Omega + i\omega q\rho\int_{\Gamma_v^I}\psi\Phi\, d\Gamma - \rho\int_{\Gamma_v^I}\psi\hat{v}_n\, d\Gamma = 0 \tag{10.58}$$

The variables ψ and Φ are interpolated from nodal values the same way as for the finite fluid elements except that the dimension of the problem is reduced by one. We proceed as before and find the matrices

$$\begin{aligned}
\mathbf{A}_I &= \rho\int_{\Omega^I} \mathbf{N}^{\mathrm{T}}\mathbf{N}\,d\Omega && (10.59)\\
\mathbf{M}_I &= \frac{\rho}{c^2}\int_{\Omega^I} \mathbf{N}^{\mathrm{T}}\mathbf{N}\,d\Omega && (10.60)\\
\mathbf{C}_I &= q\rho\int_{\Gamma_v^I} \mathbf{N}^{\mathrm{T}}\mathbf{N}\,d\Gamma && (10.61)\\
\mathbf{K}_I &= \rho\int_{\Omega^I} \mathbf{B}^{\mathrm{T}}\mathbf{B}\,d\Omega && (10.62)\\
\hat{\mathbf{V}}_I &= \rho\int_{\Gamma_v^I} \mathbf{N}^{\mathrm{T}}\hat{v}_n\,d\Gamma && (10.63)
\end{aligned}$$

The matrices $\mathbf{M}_I$, $\mathbf{C}_I$ and $\mathbf{K}_I$ are direct counterparts of the fluid element matrices $\mathbf{M}_{ff}$, $\mathbf{C}_{ff}$, and $\mathbf{K}_{ff}$ with the space dimension reduced by one. The matrix $\mathbf{A}_I$ is a weighting matrix and is simply $\mathbf{A}_I = c^2\mathbf{M}_I$ at the element level. The damping matrix $\mathbf{C}_I$ includes the reservoir bottom absorption.

Expressing the velocity potential in terms of the nodal velocity potential as $\Phi = \mathbf{N}\hat{\mathbf{\Phi}}$, Equation (10.58) can be formulated with these matrices as

$$(\mathbf{K}_I + i\omega\mathbf{C}_I - \omega^2\mathbf{M}_I)\hat{\mathbf{\Phi}} = -k^2\mathbf{A}_I\hat{\mathbf{\Phi}} + \hat{\mathbf{V}}_I \tag{10.64}$$

As for the finite fluid, the velocity input (Equation 10.63) can be formulated using interface elements placed at the fluid-reservoir interface of the channel section. With

$$\hat{v}_n = i\omega\mathbf{n}^{\mathrm{T}}\mathbf{N}_s\hat{\mathbf{u}} \tag{10.65}$$

we have

$$\hat{\mathbf{V}}_I = i\omega\rho\int_{\Gamma_v^I} \mathbf{N}^{\mathrm{T}}\mathbf{n}^{\mathrm{T}}\mathbf{N}_s\,d\Gamma\;\hat{\mathbf{u}} = -i\omega\mathbf{C}_{fs}^I\hat{\mathbf{u}} \tag{10.66}$$

10.5.2 Semi-analytical solution for the whole channel

In the upstream direction (x-direction), the radiation condition has to be satisfied by choosing the proper sign of k. At the artificial boundary between the nearfield and the farfield, the velocity potential and its normal derivative have to match. For the discretized model this means that the nodal velocity potential $\hat{\mathbf{\Phi}}$ has to match the nodal velocity potential of the neafield, and the nodal velocity input from the farfield $\hat{\mathbf{V}}$ and the one form the nearfield have to cancel ("equilibrium"). It is convenient to choose for the farfield a coordinate system with $x = 0$ at the artificial boundary. Then the nodal values $\hat{\mathbf{\Phi}}$ and $\hat{\mathbf{V}}$ are given by

$$\varphi = \Phi e^{i\omega t} = \mathbf{N}\hat{\mathbf{\Phi}}e^{i\omega t} \qquad \text{at } x = 0 \tag{10.67}$$

and, observing that the outward normal points into the negative x-direction, by

$$\hat{\mathbf{V}}e^{i\omega t} = \mathbf{V} = -\rho\int_{\Gamma_v} \mathbf{N}^{\mathrm{T}}\frac{\partial\varphi}{\partial x}\,d\Gamma \qquad \text{at } x = 0 \tag{10.68}$$

The total solution $\hat{\mathbf{\Phi}}$ is obtained by superposition of the response $\hat{\mathbf{\Phi}}^h$ due to an excitation normal to the cross-section of the channel and the response $\hat{\mathbf{\Phi}}^p$ due to an excitation in the plane of the cross-section.

$$\hat{\mathbf{\Phi}} = \hat{\mathbf{\Phi}}^h + \hat{\mathbf{\Phi}}^p \tag{10.69}$$

Upstream Excitation. For an upstream excitation there is no velocity input into the cross-section, that is $\hat{\mathbf{V}}_I = 0$, and Equation (10.64) is a generalized eigenvalue problem for the nodal velocity potential on the cross-section of the infinite channel.

$$\boxed{(\mathbf{K}_I + i\omega\mathbf{C}_I - \omega^2\mathbf{M}_I)\mathbf{\Psi} = \mathbf{A}_I\mathbf{\Psi}\mathbf{k}^2} \tag{10.70}$$

The matrix $\mathbf{k}$ is diagonal and contains the values ik_n, $\mathbf{k} = \mathbf{diag}(ik_1, \ldots, ik_N)$. The eigenvalue problem has to be solved for several values of ω, yielding the wave numbers $k_n(\omega)$ as a function of ω. The root of $\mathbf{k}^2$ has to be taken such that the radiation condition is satisfied. As pointed out in Equation (9.43), the condition is $\Re(ik) > 0$ and $\Im(ik) > 0$. The matrix $\mathbf{\Psi}$ has been used to denote the eigenvectors. They are orthogonal and normalized with respect to $\mathbf{A}_I$

$$\mathbf{\Psi}^{\mathrm{T}}\mathbf{A}_I\mathbf{\Psi} = \mathbf{I} \tag{10.71}$$

Note, that for constant c and ρ, $\mathbf{M}_I = \mathbf{A}_I/c^2$ and the eigenvalue problem can also be written as

$$(\mathbf{K}_I + i\omega\mathbf{C}_I)\mathbf{\Psi} = \mathbf{A}_I\mathbf{\Psi}\mathbf{\Lambda}^2 \tag{10.72}$$

with $\mathbf{\Lambda}^2 = \mathbf{k}^2 - (\omega/c)^2\mathbf{I}$. Denoting the eigenvalues $\mathbf{\Lambda}$ as $\lambda_n^2 = (ik_n)^2 - (\omega/c)^2$, the previous notation (Equation 9.42) is recovered. We prefer, however, the more general form (Equation 10.70).

Expressing the nodal velocity potential as

$$\hat{\mathbf{\Phi}}^h = \mathbf{\Psi}\hat{\boldsymbol{\eta}}^h \tag{10.73}$$

the steady-state velocity potential (Equation 10.52) becomes

$$\varphi = \mathbf{N}\mathbf{\Psi}e^{-\mathbf{k}x}\hat{\boldsymbol{\eta}}^h e^{i\omega t} \tag{10.74}$$

where $e^{-\mathbf{k}x} = \mathbf{diag}(e^{-ik_1x}, \ldots, e^{-ik_Nx})$. The corresponding velocity in the upstream direction is

$$\frac{\partial\varphi}{\partial x} = -\mathbf{N}\mathbf{\Psi}\mathbf{k}e^{-\mathbf{k}x}\hat{\boldsymbol{\eta}}^h e^{i\omega t} \tag{10.75}$$

The velocity input from the farfield to the nearfield at $x = 0$ is (Equation 10.33)

$$\mathbf{V} = -\rho\int_{\Gamma_v} \mathbf{N}^{\mathrm{T}}\frac{\partial\varphi}{\partial x}\,d\Gamma = \rho\int_{\Gamma_v} \mathbf{N}^{\mathrm{T}}\mathbf{N}\,d\Gamma\;\mathbf{\Psi}\mathbf{k}\hat{\boldsymbol{\eta}}^h\,e^{i\omega t} = \hat{\mathbf{V}}e^{i\omega t} \tag{10.76}$$

Observing that the boundary of the finite fluid domain is the cross-section of the infinite channel, we see that $\Gamma_v = \Omega^I$ and

$$\hat{\mathbf{V}} = \rho\int_{\Omega^I} \mathbf{N}^{\mathrm{T}}\mathbf{N}\,d\Omega\;\mathbf{\Psi}\mathbf{k}\hat{\boldsymbol{\eta}}^h = \mathbf{A}_I\mathbf{\Psi}\mathbf{k}\hat{\boldsymbol{\eta}}^h \tag{10.77}$$

Premultiplying Equation (10.73) by $\mathbf{\Psi}^{\mathrm{T}}\mathbf{A}_I$, we find

$$\hat{\boldsymbol{\eta}}^h = \mathbf{\Psi}^{\mathrm{T}}\mathbf{A}_I\hat{\mathbf{\Phi}}^h \tag{10.78}$$

With this equation the velocity input to the nearfield is

$$\boxed{\hat{\mathbf{V}} = \mathbf{A}_I\mathbf{\Psi}\mathbf{k}\mathbf{\Psi}^{\mathrm{T}}\mathbf{A}_I\hat{\mathbf{\Phi}}^h} \tag{10.79}$$

The impedance matrix $\mathbf{A}_I\mathbf{\Psi}\mathbf{k}\mathbf{\Psi}^{\mathrm{T}}\mathbf{A}_I$ relates the velocity input $\hat{\mathbf{V}}$ to the velocity potential $\hat{\mathbf{\Phi}}^h$ at the transmitting boundary.

Vertical and Cross-stream Excitation. For an excitation in the plane of the channel cross-section, that is, vertical or cross-stream, the nodal velocity potential $\hat{\mathbf{\Phi}}^p$ can be calculated from the prescribed velocity on the fluid-foundation interface by Equation (10.64). If the excitation is independent of the upstream coordinate x, then $k = 0$ and the equation becomes

$$(\mathbf{K}_I + i\omega\mathbf{C}_I - \omega^2\mathbf{M}_I)\hat{\mathbf{\Phi}}^p = \hat{\mathbf{V}}_I \tag{10.80}$$

This equation can evaluated directly in the time domain. Setting $\hat{\mathbf{\Phi}}^h = \hat{\mathbf{\Phi}} - \hat{\mathbf{\Phi}}^p$, Equation (10.79) yields

$$\boxed{\hat{\mathbf{V}} = \mathbf{A}_I\mathbf{\Psi}\mathbf{k}\mathbf{\Psi}^{\mathrm{T}}\mathbf{A}_I\,(\hat{\mathbf{\Phi}} - \hat{\mathbf{\Phi}}^p)} \tag{10.81}$$

This is the formulation we will use in the further development. Alternatively, the solution can be determined using the eigenvector expansion

$$\hat{\mathbf{\Phi}}^p = \mathbf{\Psi}\hat{\eta}^p \tag{10.82}$$

Substituting this expansion into Equation (10.80) and using Equation (10.70) yields

$$\mathbf{\Psi}^{\mathrm{T}}(\mathbf{K}_I + i\omega\mathbf{C}_I - \omega^2\mathbf{M}_I)\mathbf{\Psi}\hat{\eta}^p = \mathbf{\Psi}^{\mathrm{T}}\mathbf{A}_I\mathbf{\Psi}\mathbf{k}^2\hat{\eta}^p = \mathbf{k}^2\hat{\eta}^p = \mathbf{\Psi}^{\mathrm{T}}\hat{\mathbf{V}}_I \tag{10.83}$$

Solving for $\hat{\eta}^p$ and substituting into Equation (10.82) leads to the alternative formulation for $\hat{\mathbf{\Phi}}^p$

$$\hat{\mathbf{\Phi}}^p = \mathbf{\Psi}\mathbf{k}^{-2}\mathbf{\Psi}^{\mathrm{T}}\hat{\mathbf{V}}_I \tag{10.84}$$

and the velocity input into the nearfield is

$$\hat{\mathbf{V}} = \mathbf{A}_I\mathbf{\Psi}\mathbf{k}\mathbf{\Psi}^{\mathrm{T}}\mathbf{A}_I\hat{\mathbf{\Phi}} - \mathbf{A}_I\mathbf{\Psi}\mathbf{k}^{-1}\mathbf{\Psi}^{\mathrm{T}}\hat{\mathbf{V}}_I \tag{10.85}$$

10.6 Coupled system

10.6.1 Nearfield

The equations of motion for the solid and for the fluid can easily be combined to a coupled system. Since the coupling matrices are symmetric except for the sign, the fluid equation is multiplied by -1 in order to get a symmetric coupled system.

$$\begin{bmatrix} \mathbf{M}_{ss} & \\ & -\mathbf{M}_{ff} \end{bmatrix}\begin{bmatrix} \ddot{\mathbf{u}} \\ \ddot{\varphi} \end{bmatrix} + \begin{bmatrix} \mathbf{C}_{ss} & \mathbf{C}_{sf} \\ \mathbf{C}_{sf}^{\mathrm{T}} & -\mathbf{C}_{ff} \end{bmatrix}\begin{bmatrix} \dot{\mathbf{u}} \\ \dot{\varphi} \end{bmatrix} + \begin{bmatrix} \mathbf{K}_{ss} & \\ & -\mathbf{K}_{ff} \end{bmatrix}\begin{bmatrix} \mathbf{u} \\ \varphi \end{bmatrix} = \begin{bmatrix} \mathbf{F}_s \\ -\mathbf{V} \end{bmatrix} \tag{10.86}$$

Although the matrices are now symmetric, they are no longer positive definite. This is not really a problem as long as the coupled system is well-behaved. Well-behaved means, firstly, that the system of equations to be solved in the time stepping algorithm is non-singular and, secondly, that the eigenvalues of the dynamic system have the same properties as a single-degree-of-freedom system.

For the first proof, consider the time stepping algorithm as given in Section 11.3. The matrix eventually used in the $\mathbf{LDL}^{\mathrm{T}}$ decomposition is the effective mass matrix $\mathbf{M}^* = \mathbf{M} + c_c\mathbf{C} + c_k\mathbf{K}$, where c_c and c_k are positive, non-zero constants (Equation 11.40). Without the fluid, the matrices $\mathbf{M}$ and $\mathbf{K}$ are positive definite, the matrix is $\mathbf{C}$ positive semi-definite (zero for the undamped case). The resulting effective mass matrix $\mathbf{M}^*$ is positive definite and hence non-singular. One way to show positive definiteness is to show that all eigenvalues of $\mathbf{M}^*$ are strictly positive. To see the difference between the regular positive definite case and the case of a coupled solid-fluid system, consider first the simple example

$$\mathbf{M}^* = \begin{bmatrix} a & c \\ c & b \end{bmatrix} \qquad a > 0, \quad b > 0 \tag{10.87}$$

For the positive definite case, a and b are considered strictly positive. The condition for c follows from the condition of strictly positive eigenvalues. The eigenvalues μ are given by the determinant

$$\begin{vmatrix} a-\mu & c \\ c & b-\mu \end{vmatrix} = \mu^2 - \mu\,(a+b) + (ab - c^2) = 0 \tag{10.88}$$

They are

$$\mu_{1,2} = \frac{a+b}{2} \pm \sqrt{\frac{(a+b)^2 - 4\,(ab-c^2)}{4}} \tag{10.89}$$

Because

$$(a+b)^2 - 4\,(ab-c^2) = (a-b)^2 + 4c^2 \; > 0 \tag{10.90}$$

the eigenvalues are real, which is also clear from symmetry. The eigenvalues are strictly positive if

$$\frac{a+b}{2} > \sqrt{\frac{(a+b)^2 - 4\,(ab-c^2)}{4}} \tag{10.91}$$

which is true if $ab > c^2$. This requirement is typically satisfied for dominant diagonals.

For the coupled solid-fluid system the example is modified as

$$\mathbf{M}^* = \begin{bmatrix} a & c \\ c & -b \end{bmatrix} \qquad a > 0, \quad b > 0 \tag{10.92}$$

The eigenvalues μ are given by

$$\begin{vmatrix} a-\mu & c \\ c & -b-\mu \end{vmatrix} = \mu^2 - \mu\,(a-b) - (ab + c^2) = 0 \tag{10.93}$$

They are

$$\mu_{1,2} = \frac{a-b}{2} \pm \sqrt{\frac{(a-b)^2 + 4\,(ab+c^2)}{4}} \tag{10.94}$$

Clearly, the eigenvalues are real because the expression under the square root is positive. Because

$$\sqrt{\frac{(a-b)^2 + 4(ab+c^2}{4}} > \left|\frac{a-b}{2}\right| \tag{10.95}$$

the eigenvalues are either strictly positive

$$\mu_1 = \frac{a-b}{2} + \sqrt{\frac{(a-b)^2 + 4(ab+c^2}{4}} > 0 \tag{10.96}$$

or strictly negative

$$\mu_2 = \frac{a-b}{2} - \sqrt{\frac{(a-b)^2 + 4(ab+c^2}{4}} < 0 \tag{10.97}$$

but non-zero and hence the matrix is non-singular, independently of the value of the off-diagonal term c.

For the general case of a coupled solid-fluid system, consider the eigenvalue problem

$$\begin{bmatrix} \mathbf{A} & \mathbf{C} \\ \mathbf{C}^{\mathrm{T}} & -\mathbf{B} \end{bmatrix} \begin{bmatrix} \mathbf{x} \\ \mathbf{y} \end{bmatrix} = \mu \begin{bmatrix} \mathbf{x} \\ \mathbf{y} \end{bmatrix} \tag{10.98}$$

The eigenvalue problem is premultiplied by $[\mathbf{x}^{\mathrm{T}} \ -\mathbf{y}^{\mathrm{T}}]$, which leads to

$$\mathbf{x}^{\mathrm{T}}\mathbf{A}\mathbf{x} + \mathbf{y}^{\mathrm{T}}\mathbf{B}\mathbf{y} = \mu\,(\mathbf{x}^{\mathrm{T}}\mathbf{x} - \mathbf{y}^{\mathrm{T}}\mathbf{y}) \tag{10.99}$$

The terms involving $\mathbf{C}$ cancel. For positive definite matrices $\mathbf{A}$ and $\mathbf{B}$, the left-hand side of Equation (10.99) is strictly positive. The expression $\mathbf{x}^{\mathrm{T}}\mathbf{x} - \mathbf{y}^{\mathrm{T}}\mathbf{y}$ is therefore non-zero and the eigenvalue μ is strictly positive or strictly negative but non-zero. This proves that the effective mass matrix of a coupled solid-fluid system is non-singular. It also follows, by Sylvester's inertia theorem [GvL83], that in an $\mathbf{LDL}^{\mathrm{T}}$ decomposition, some of the entries in $\mathbf{D}$ are positive, some negative, but none will be zero.

The second property of the coupled solid-fluid system we want to show, is the behavior of the eigenvalues of the dynamic system. Consider the quadratic eigenvalue problem

$$\left(\lambda^2 \begin{bmatrix} \mathbf{M}_{ss} & \\ & -\mathbf{M}_{ff} \end{bmatrix} + \lambda \begin{bmatrix} \mathbf{C}_{ss} & \mathbf{C}_{sf} \\ \mathbf{C}_{sf}^{\mathrm{T}} & -\mathbf{C}_{ff} \end{bmatrix} + \begin{bmatrix} \mathbf{K}_{ss} & \\ & -\mathbf{K}_{ff} \end{bmatrix}\right) \begin{bmatrix} \boldsymbol{\psi}_s \\ \boldsymbol{\psi}_f \end{bmatrix} = \mathbf{0} \tag{10.100}$$

It has been shown by Olson and Bathe [OB85a] that the eigenvalues λ^2 of the undamped system are real and positive. Here we show a similar proof, also valid for damped systems. Premultiplying Equation (10.100) by $[\bar{\boldsymbol{\psi}}_s^{\mathrm{T}},\ -\bar{\boldsymbol{\psi}}_f^{\mathrm{T}}]$ (bar denotes complex conjugate), we obtain

$$\lambda^2\,(\bar{\boldsymbol{\psi}}_s^{\mathrm{T}}\mathbf{M}_{ss}\boldsymbol{\psi}_s + \bar{\boldsymbol{\psi}}_f^{\mathrm{T}}\mathbf{M}_{ff}\boldsymbol{\psi}_f) + \lambda\,(\bar{\boldsymbol{\psi}}_s^{\mathrm{T}}\mathbf{C}_{ss}\boldsymbol{\psi}_s + \bar{\boldsymbol{\psi}}_f^{\mathrm{T}}\mathbf{C}_{ff}\boldsymbol{\psi}_f) + (\bar{\boldsymbol{\psi}}_s^{\mathrm{T}}\mathbf{K}_{ss}\boldsymbol{\psi}_s + \bar{\boldsymbol{\psi}}_f^{\mathrm{T}}\mathbf{K}_{ff}\boldsymbol{\psi}_f) = 0 \tag{10.101}$$

The terms $\bar{\boldsymbol{\psi}}_f^{\mathrm{T}}\mathbf{C}_{fs}\boldsymbol{\psi}_s$ and $\bar{\boldsymbol{\psi}}_s^{\mathrm{T}}\mathbf{C}_{fs}^{\mathrm{T}}\boldsymbol{\psi}_f$ cancel. The coefficients pertaining to λ^2 and the constant term are strictly positive because the matrices involved are positive definite. The coefficient pertaining to λ is positive because the damping matrices are positive semi-definite (zero for the undamped case). Equation (10.101) is thus the eigenvalue problem of a single-degree-of-freedom system with strictly positive mass and spring, and positive (possibly zero) damping. Because of the positive damping, the coupled system is stable.

10.6.2 Appending the farfield

Because the farfield can only be calculated in the frequency domain, the nearfield is also considered in the frequency domain. The fluid nodes are separated into nodes in the interior of the nearfield (superscript (1)) and nodes on the interface between the nearfield and the infinite channel (superscript (2)).

$$\begin{bmatrix} \mathbf{S}_{ss} & \mathbf{S}_{sf} & \\ \mathbf{S}_{fs} & -\mathbf{S}_{ff}^{(11)} & -\mathbf{S}_{ff}^{(12)} \\ & -\mathbf{S}_{ff}^{(21)} & -\mathbf{S}_{ff}^{(22)} \end{bmatrix} \begin{bmatrix} \hat{\mathbf{u}} \\ \hat{\varphi}^{(1)} \\ \hat{\varphi}^{(2)} \end{bmatrix} = \begin{bmatrix} \hat{\mathbf{F}}_s \\ -\hat{\mathbf{V}}^{(1)} \\ -\hat{\mathbf{V}}^{(2)} \end{bmatrix} \tag{10.102}$$

where

$$\mathbf{S}_{ss} = -\omega^2\mathbf{M}_{ss} + i\omega\mathbf{C}_{ss} + \mathbf{K}_{ss} \tag{10.103}$$

$$\mathbf{S}_{sf} = \mathbf{S}_{fs}^{\mathrm{T}} = i\omega\mathbf{C}_{sf} \tag{10.104}$$

$$\mathbf{S}_{ff} = \begin{bmatrix} \mathbf{S}_{ff}^{(11)} & \mathbf{S}_{ff}^{(12)} \\ \mathbf{S}_{ff}^{(21)} & \mathbf{S}_{ff}^{(22)} \end{bmatrix} = -\omega^2\mathbf{M}_{ff} + i\omega\mathbf{C}_{ff} + \mathbf{K}_{ff} \tag{10.105}$$

$\hat{\mathbf{F}}_s$ and $\hat{\mathbf{V}}$ are the complex amplitudes corresponding to $\mathbf{F}_s$ and $\mathbf{V}$, respectively.

To include the influence of the infinite channel, $\hat{\mathbf{V}}^{(2)}$ is replaced by the relation Equation (10.81) with the substitutions $\hat{\mathbf{V}}^{(2)} = -\hat{\mathbf{V}}$ ("equilibrium") and $\hat{\varphi}^{(2)} = \hat{\mathbf{\Phi}}$.

$$\begin{bmatrix} \mathbf{S}_{ss} & \mathbf{S}_{sf} & \\ \mathbf{S}_{fs} & -\mathbf{S}_{ff}^{(11)} & -\mathbf{S}_{ff}^{(12)} \\ & -\mathbf{S}_{ff}^{(21)} & -(\mathbf{S}_{ff}^{(22)} + \mathbf{A}_I \mathbf{\Psi} \mathbf{k} \mathbf{\Psi}^{\mathrm{T}} \mathbf{A}_I) \end{bmatrix} \begin{bmatrix} \hat{\mathbf{u}} \\ \hat{\varphi}^{(1)} \\ \hat{\varphi}^{(2)} \end{bmatrix} = \begin{bmatrix} \hat{\mathbf{F}}_s \\ -\hat{\mathbf{V}}^{(1)} \\ -\mathbf{A}_I \mathbf{\Psi} \mathbf{k} \mathbf{\Psi}^{\mathrm{T}} \mathbf{A}_I \hat{\mathbf{\Phi}}^p \end{bmatrix} \tag{10.106}$$

The $p \times p$ impedance matrix $\mathbf{A}_I \mathbf{\Psi} \mathbf{k} \mathbf{\Psi}^{\mathrm{T}} \mathbf{A}_I$ has only full rank if all p eigenvectors are included. The usually employed strategy of taking only $N_\Psi < p$ eigenvectors does not reduce the size of the matrix but only its rank because the eigenvectors appear as an outer product. The commonly used form for frequency-domain analyses [HC83] is obtained by substituting the eigenvector expansion $\hat{\varphi}^{(2)} = \mathbf{\Psi} \hat{\boldsymbol{\eta}}$ and by premultiplying the last block row by $\mathbf{\Psi}^{\mathrm{T}}$. The bottom term of the right-hand-side vector of Equation (10.106) can be written differently, by substituting Equation (10.84) for $\hat{\mathbf{\Phi}}^p$. This formulation leads to

$$\begin{bmatrix} \mathbf{S}_{ss} & \mathbf{S}_{sf} & \\ \mathbf{S}_{fs} & -\mathbf{S}_{ff}^{(11)} & -\mathbf{S}_{ff}^{(12)} \mathbf{\Psi} \\ & -\mathbf{\Psi}^{\mathrm{T}} \mathbf{S}_{ff}^{(21)} & -(\mathbf{\Psi}^{\mathrm{T}} \mathbf{S}_{ff}^{(22)} \mathbf{\Psi} + \mathbf{k}) \end{bmatrix} \begin{bmatrix} \hat{\mathbf{u}} \\ \hat{\varphi}^{(1)} \\ \hat{\boldsymbol{\eta}} \end{bmatrix} = \begin{bmatrix} \hat{\mathbf{F}}_s \\ -\hat{\mathbf{V}}^{(1)} \\ -\mathbf{k}^{-1} \mathbf{\Psi}^{\mathrm{T}} \hat{\mathbf{V}}_I \end{bmatrix} \tag{10.107}$$

If only N_ψ out of all p eigenvectors are considered in $\mathbf{\Psi}$, this form has fewer degrees of freedom because the vector $\hat{\boldsymbol{\eta}}$ is only of size N_ψ compared to the vector $\varphi^{(2)}$ which is of size p. For a frequency-domain analysis this is the form usually used. For a time-domain analysis however, if $\mathbf{\Psi}$ is frequency-dependent, this form is not useful because not only the matrix $\mathbf{k}$ has to be approximated by a linear time-invariant system but also all submatrices involving $\mathbf{\Psi}$.

10.6.3 Ritz vectors

A form also useful for a time-domain analysis can be derived using the Rayleigh-Ritz approach [Hug87, p. 574]. The introduction of the Ritz vectors is not essential for the frequency-domain analysis but has certain advantages: The size of the eigenvalue problem is reduced and the same form can later be used for the time-domain analysis.

The eigenvectors $\mathbf{\Psi}$ are expressed as a linear combination of the Ritz vectors $\mathbf{R}$ as

$$\mathbf{\Psi} = \mathbf{R} \mathbf{\Psi}^* \tag{10.108}$$

where the matrix $\mathbf{R}$ is $p \times N_\Psi$ and the matrix $\mathbf{\Psi}^*$ is $N_\Psi \times N_\Psi$. Substituting into Equation (10.70) and premultiplying by $\mathbf{R}^{\mathrm{T}}$, yields the reduced eigenvalue problem for the cross-section of the channel

$$\mathbf{R}^{\mathrm{T}} (\mathbf{K}_I + i\omega \mathbf{C}_I - \omega^2 \mathbf{M}_I) \mathbf{R} \mathbf{\Psi}^* = \mathbf{R}^{\mathrm{T}} \mathbf{A}_I \mathbf{R} \mathbf{\Psi}^* \mathbf{k}^2 \tag{10.109}$$

This equation can be written in the form

$$(\mathbf{K}_I^* + i\omega \mathbf{C}_I^* - \omega^2 \mathbf{M}_I^*) \mathbf{\Psi}^* = \mathbf{A}_I^* \mathbf{\Psi}^* \mathbf{k}^2 \tag{10.110}$$

where now the matrices $\mathbf{A}_I^*$, $\mathbf{M}_I^*$, $\mathbf{C}_I^*$, $\mathbf{K}_I^*$ are smaller ($N_\Psi \times N_\Psi$) than the original matrices $\mathbf{A}_I$, $\mathbf{M}_I$, $\mathbf{C}_I$, $\mathbf{K}_I$. In the following we take for the Ritz vectors $\mathbf{R} = \mathbf{\Psi}_0$, the N_Ψ eigenvectors of the full problem at $\omega = 0$. With this strategy the full-size eigenvalue problem Equation (10.70) has to be solved only once for $\omega = 0$. It takes the form

$$\mathbf{K}_I \mathbf{\Psi}_0 = \mathbf{A}_I \mathbf{\Psi}_0 \mathbf{k}_0^2 \tag{10.111}$$

For other frequencies the reduced problem can be solved. The reduced eigenvalue problem is then

$$\boxed{(\mathbf{K}_I^* + i\omega\mathbf{C}_I^* - \omega^2\mathbf{M}_I^*)\mathbf{\Psi}^* = \mathbf{\Psi}^*\mathbf{k}^2} \tag{10.112}$$

with the reduced matrices

$$\mathbf{M}_I^* = \mathbf{\Psi}_0^{\mathrm{T}}\mathbf{M}_I\mathbf{\Psi}_0 \tag{10.113}$$
$$\mathbf{C}_I^* = \mathbf{\Psi}_0^{\mathrm{T}}\mathbf{C}_I\mathbf{\Psi}_0 \tag{10.114}$$
$$\mathbf{K}_I^* = \mathbf{\Psi}_0^{\mathrm{T}}\mathbf{K}_I\mathbf{\Psi}_0 \tag{10.115}$$

The matrix $\mathbf{A}_I^* = \mathbf{\Psi}_0^{\mathrm{T}}\mathbf{A}_I\mathbf{\Psi}_0 = \mathbf{I}$ disappears. To avoid overflow for high frequencies, a frequency shift σ can be included in the eigenvalue problem to compensate for large values of $\omega^2\mathbf{M}_I^*$. The modified eigenvalue problem is

$$(\mathbf{K}_I^* + i\omega\mathbf{C}_I^* - \omega^2\mathbf{M}_I^* + \sigma^2\mathbf{I})\mathbf{\Psi}^* = \mathbf{\Psi}^*\mathbf{\Lambda}^2 \tag{10.116}$$

with $\mathbf{\Lambda}^2 = \sigma^2\mathbf{I} + \mathbf{k}^2$. The eigenvectors are normalized consistent with the full-size problem as

$$\mathbf{\Psi}^{\mathrm{T}}\mathbf{A}_I\mathbf{\Psi} = \mathbf{\Psi}^{*\mathrm{T}}\mathbf{\Psi}_0^{\mathrm{T}}\mathbf{A}_I\mathbf{\Psi}_0\mathbf{\Psi}^* = \mathbf{\Psi}^{*\mathrm{T}}\mathbf{\Psi}^* = \mathbf{I} \tag{10.117}$$

The eigenvectors of the original problem are approximated by

$$\mathbf{\Psi} = \mathbf{\Psi}_0\mathbf{\Psi}^* \tag{10.118}$$

With this expansion the impedance matrix becomes

$$\mathbf{A}_I\mathbf{\Psi}\mathbf{k}\mathbf{\Psi}^{\mathrm{T}}\mathbf{A}_I = \mathbf{A}_I\mathbf{\Psi}_0\mathbf{\Psi}^*\mathbf{k}\mathbf{\Psi}^{*\mathrm{T}}\mathbf{\Psi}_0^{\mathrm{T}}\mathbf{A}_I = \mathbf{A}_I\mathbf{\Psi}_0\mathbf{k}^*\mathbf{\Psi}_0^{\mathrm{T}}\mathbf{A}_I \tag{10.119}$$

with

$$\mathbf{k}^* = \mathbf{\Psi}^*\mathbf{k}\mathbf{\Psi}^{*\mathrm{T}} \tag{10.120}$$

Note that $\mathbf{k}^*$ is now a full matrix. The general equation for the coupled system (Equation 10.106) becomes

$$\begin{bmatrix} \mathbf{S}_{ss} & \mathbf{S}_{sf} & \\ \mathbf{S}_{fs} & -\mathbf{S}_{ff}^{(11)} & -\mathbf{S}_{ff}^{(12)} \\ & -\mathbf{S}_{ff}^{(21)} & -(\mathbf{S}_{ff}^{(22)} + \mathbf{A}_I\mathbf{\Psi}_0\mathbf{k}^*\mathbf{\Psi}_0^{\mathrm{T}}\mathbf{A}_I) \end{bmatrix} \begin{bmatrix} \hat{\mathbf{u}} \\ \hat{\varphi}^{(1)} \\ \hat{\varphi}^{(2)} \end{bmatrix} = \begin{bmatrix} \hat{\mathbf{F}}_s \\ -\hat{\mathbf{V}}^{(1)} \\ -\mathbf{A}_I\mathbf{\Psi}_0\mathbf{k}^*\mathbf{\Psi}_0^{\mathrm{T}}\mathbf{A}_I\hat{\mathbf{\Phi}}^p \end{bmatrix} \tag{10.121}$$

The advantage of this form is that now the frequency-independent matrix $\mathbf{\Psi}_0$ appears instead of the frequency-dependent matrix $\mathbf{\Psi}$ and only the $N_\Psi \times N_\Psi$ square matrix $\mathbf{k}^*$ is frequency-dependent and has to approximated. However, this matrix is no longer diagonal as was the matrix $\mathbf{k}$ in Equation (10.106).

10.6.4 Right-hand-side input

For an earthquake analysis with uniform input ground motion the different components of the right-hand-side input vector can be expanded further. If we consider the solid degrees of freedom $\hat{\mathbf{u}}$ to be relative to the ground motion, the earthquake input can be formulated as a force applied to the dam as

$$\hat{\mathbf{F}}_s = -\mathbf{M}_{ss}\,\hat{\mathbf{a}}_g \tag{10.122}$$

As usual, results are shown as relative displacement, relative velocity or absolute acceleration. The input vector to the reservoir $\hat{\mathbf{V}}^{(1)}$ can be derived from the velocity $\hat{\mathbf{v}}_g$ at the dam-foundation interface by Equation (10.51) as

$$\hat{\mathbf{V}}^{(1)} = -\mathbf{C}_{fs}\,\hat{\mathbf{v}}_g = \mathbf{C}_{sf}^{\mathrm{T}}\,\hat{\mathbf{v}}_g \tag{10.123}$$

The velocity potential $\hat{\mathbf{\Phi}}^p$ can be calculated by the differential equation for the cross-section Equation (10.80)

$$(\mathbf{K}_I + i\omega\mathbf{C}_I - \omega^2\mathbf{M}_I)\hat{\mathbf{\Phi}}^p = \hat{\mathbf{V}}_I \tag{10.124}$$

As for the nearfield the velocity potential $\hat{\mathbf{V}}_I$ can be evaluated from the velocity $\hat{\mathbf{v}}_g$ at the dam-foundation interface of the cross-section by Equation (10.66)

$$\hat{\mathbf{V}}_I = -\mathbf{C}_{fs}^I\,\hat{\mathbf{v}}_g \tag{10.125}$$

10.6.5 Special cases

The incompressible fluid and the viscous boundary can be implemented as special cases of the general frequency-domain equations. Here we give the form derived from the system reduced by the Ritz vector modification, Equation (10.121).

Incompressible fluid. As pointed out earlier (Chapter 9) the special case of a reservoir with incompressible fluid and rigid bottom (no absorption) can be obtained by taking the zero-frequency solution of the compressible case for all frequencies. With $\mathbf{k}_0^* = \mathbf{k}_0$ (Equation 10.111) the corresponding impedance matrix is written as

$$\mathbf{A}_I\mathbf{\Psi}_0\mathbf{k}_0\mathbf{\Psi}_0^{\mathrm{T}}\mathbf{A}_I = \mathbf{G}_2 \tag{10.126}$$

For an incompressible fluid the nearfield matrices $\mathbf{M}_{ff}$ and $\mathbf{C}_{ff}$ disappear. This can be seen by setting $c \to \infty$ and $q = 0$ in Equations (10.30–10.33). The equation for the coupled system is then

$$\begin{bmatrix} \mathbf{S}_{ss} & i\omega\mathbf{C}_{sf} & \\ i\omega\mathbf{C}_{sf}^{\mathrm{T}} & -\mathbf{K}_{ff}^{(11)} & -\mathbf{K}_{ff}^{(12)} \\ & -\mathbf{K}_{ff}^{(21)} & -(\mathbf{K}_{ff}^{(22)} + \mathbf{G}_2) \end{bmatrix} \begin{bmatrix} \hat{\mathbf{u}} \\ \hat{\varphi}^{(1)} \\ \hat{\varphi}^{(2)} \end{bmatrix} = \begin{bmatrix} \hat{\mathbf{F}}_s \\ -\hat{\mathbf{V}}^{(1)} \\ -\mathbf{G}_2\hat{\mathbf{\Phi}}^p \end{bmatrix} \tag{10.127}$$

To show that these equations are equivalent to an added mass, they are written out assuming no right-hand-side input.

$$(-\omega^2\mathbf{M}_{ss} + i\omega\mathbf{C}_{ss} + \mathbf{K}_{ss})\,\hat{\mathbf{u}} + i\omega\mathbf{C}_{sf}\,\hat{\varphi}^{(1)} = \mathbf{0} \tag{10.128}$$

$$\mathbf{K}_{ff}^{(11)}\,\varphi^{(1)} + \mathbf{K}_{ff}^{(12)}\,\varphi^{(2)} = i\omega\mathbf{C}_{sf}^{\mathrm{T}}\,\hat{\mathbf{u}} \tag{10.129}$$

$$\mathbf{K}_{ff}^{(21)}\,\varphi^{(1)} + \left(\mathbf{K}_{ff}^{(22)} + \mathbf{G}_2\right)\varphi^{(2)} = \mathbf{0} \tag{10.130}$$

We first solve Equation (10.130) for $\varphi^{(2)}$.

$$\varphi^{(2)} = -\left(\mathbf{K}_{ff}^{(22)} + \mathbf{G}_2\right)^{-1}\mathbf{K}_{ff}^{(21)}\,\varphi^{(1)} \tag{10.131}$$

This equation is substituted into Equation (10.129).

$$\left[\mathbf{K}_{ff}^{(11)} - \mathbf{K}_{ff}^{(12)}\left(\mathbf{K}_{ff}^{(22)} + \mathbf{G}_2\right)^{-1}\mathbf{K}_{ff}^{(21)}\right]\varphi^{(1)} = i\omega\mathbf{C}_{sf}^{\mathrm{T}}\,\hat{\mathbf{u}} \tag{10.132}$$

Solved for $\varphi^{(1)}$, this yields for the term $i\omega\mathbf{C}_{sf}$ in Equation (10.128)

$$i\omega\mathbf{C}_{sf}\,\hat{\varphi}^{(1)} = -\omega^2\mathbf{C}_{sf}\left[\mathbf{K}_{ff}^{(11)} - \mathbf{K}_{ff}^{(12)}\left(\mathbf{K}_{ff}^{(22)} + \mathbf{G}_2\right)^{-1}\mathbf{K}_{ff}^{(21)}\right]^{-1}\mathbf{C}_{sf}^{\mathrm{T}}\,\hat{\mathbf{u}} \tag{10.133}$$

This expression depends only on $\omega^2\hat{\mathbf{u}}$ and can therefore be interpreted as an added mass.

It should be noted that the formulation presented here is not the usual way of analyzing the incompressible model. The conventional method is to take a long reservoir with a fixed (or any other) boundary condition at the artificial boundary. Because no radiation occurs, the pressure field is confined to a region close to the dam and the type of boundary condition at a sufficiently large distance has no effect. The practical importance of the implementation presented here lies in the fact that only a small nearfield is needed because the transmitting boundary rigorously accounts for the farfield.

Viscous boundary. Another special case is the viscous boundary condition. One way to describe the viscous boundary is to consider it as the high-frequency behavior. For $\omega \to \infty$ we divide the eigenvalue problem Equation (10.112) by ω^2 and drop terms with $1/\omega$ and $1/\omega^2$ on the left-hand side to get the eigenvalue problem

$$\mathbf{M}_I^*\mathbf{\Psi}_\infty^* = \mathbf{\Psi}_\infty^*\mathbf{\Lambda}_\infty^2 \tag{10.134}$$

with $\mathbf{\Lambda}_\infty = \mathbf{k}/(i\omega)$. The matrix $\mathbf{k}_\infty$ is therefore

$$\mathbf{k}_\infty = i\omega\mathbf{\Lambda}_\infty \tag{10.135}$$

The matrix $\mathbf{k}_\infty^*$ is

$$\mathbf{k}_\infty^* = \mathbf{\Psi}_\infty^*\mathbf{k}_\infty\mathbf{\Psi}_\infty^{*\mathrm{T}} = i\omega\mathbf{\Psi}_\infty^*\mathbf{\Lambda}_\infty\mathbf{\Psi}_\infty^{*\mathrm{T}} \tag{10.136}$$

and the corresponding impedance matrix is written as

$$\mathbf{A}_I\mathbf{\Psi}_0\mathbf{k}_\infty^*\mathbf{\Psi}_0^{\mathrm{T}}\mathbf{A}_I = i\omega\mathbf{A}_I\mathbf{\Psi}_0\mathbf{\Psi}_\infty^*\mathbf{\Lambda}_\infty\mathbf{\Psi}_\infty^{*\mathrm{T}}\mathbf{\Psi}_0^{\mathrm{T}}\mathbf{A}_I = i\omega\mathbf{G}_1 \tag{10.137}$$

The coupled system takes the form

$$\begin{bmatrix} \mathbf{S}_{ss} & \mathbf{S}_{sf} & \\ \mathbf{S}_{fs} & -\mathbf{S}_{ff}^{(11)} & -\mathbf{S}_{ff}^{(12)} \\ & -\mathbf{S}_{ff}^{(21)} & -(\mathbf{S}_{ff}^{(22)} + i\omega\mathbf{G}_1) \end{bmatrix}\begin{bmatrix} \hat{\mathbf{u}} \\ \hat{\varphi}^{(1)} \\ \hat{\varphi}^{(2)} \end{bmatrix} = \begin{bmatrix} \hat{\mathbf{F}}_s \\ -\hat{\mathbf{V}}^{(1)} \\ \mathbf{0} \end{bmatrix} \tag{10.138}$$

The velocity input from the farfield has been discarded to be consistent with the conventional model of a finite channel.

The other, the conventional interpretation of a viscous boundary is the local boundary condition $\partial\varphi/\partial x = \partial\varphi/c\partial t$. The corresponding velocity input is (Equation 10.33)

$$\mathbf{V}^{(2)} = \rho\int_{\Gamma_v}\mathbf{N}^{\mathrm{T}}\frac{\partial\varphi}{\partial x}\,d\Gamma = \rho\int_{\Gamma_v}\mathbf{N}^{\mathrm{T}}\frac{\partial\varphi}{c\partial t}\,d\Gamma = \rho\frac{i\omega}{c}\int_{\Gamma_v}\mathbf{N}^{\mathrm{T}}\mathbf{N}\,d\Gamma\,\hat{\varphi}^{(2)}\,e^{i\omega t} = \frac{i\omega}{c}\mathbf{A}_I\hat{\varphi}^{(2)}\,e^{i\omega t} \tag{10.139}$$

On the other hand, we can expand $\hat{\varphi}^{(2)}$ as

$$\hat{\varphi}^{(2)} = \mathbf{\Psi}_\infty\hat{\boldsymbol{\eta}} = \mathbf{\Psi}_0\mathbf{\Psi}_\infty^*\hat{\boldsymbol{\eta}} \tag{10.140}$$

and use the high-frequency formulation Equation (10.137) to calculate $\hat{\mathbf{V}}^{(2)}$. Observing the normalizations $\mathbf{\Psi}_0^{\mathrm{T}}\mathbf{A}_I\mathbf{\Psi}_0^{\mathrm{T}} = \mathbf{I}$ and $\mathbf{\Psi}_\infty^{*\mathrm{T}}\mathbf{\Psi}_\infty^* = \mathbf{I}$ we obtain

$$\hat{\mathbf{V}}^{(2)} = i\omega\mathbf{G}_1\hat{\varphi}^{(2)} = \mathbf{A}_I\mathbf{\Psi}_0\mathbf{\Psi}_\infty^*\mathbf{\Lambda}\hat{\boldsymbol{\eta}} \tag{10.141}$$

The equivalence to the first interpretation is easily seen for the case of constant material properties c and ρ. In that case $\mathbf{M}_I = \mathbf{A}_I/c^2$ and

$$\mathbf{M}_I^* = \mathbf{\Psi}_0 \mathbf{M}_I \mathbf{\Psi}_0^{\mathrm{T}} = \frac{1}{c^2}\mathbf{\Psi}_0 \mathbf{A}_I \mathbf{\Psi}_0^{\mathrm{T}} = \frac{1}{c^2}\mathbf{I} \tag{10.142}$$

The eigenvalue problem Equation (10.134) yields

$$\mathbf{\Lambda}_\infty = \frac{1}{c}\mathbf{I} \tag{10.143}$$

and the velocity input Equation (10.141) is

$$\hat{\mathbf{v}}^{(2)} = \frac{i\omega}{c}\mathbf{A}_I \mathbf{\Psi}_0 \mathbf{\Psi}_\infty^* \hat{\eta} = \frac{i\omega}{c}\mathbf{A}_I \hat{\varphi}^{(2)} \tag{10.144}$$

as in Equation (10.139), which concludes the proof.

The comparison of the high-frequency solution with the conventional viscous boundary is only of theoretical interest. From a practical point of view, there is no advantage in the high-frequency implementation. This is in contrast to the incompressible case, where the zero-frequency solution substantially reduces the problem size.

10.7 Frequency-domain analysis, simplified modeling

The remaining part of this chapter is used to demonstrate the finite element analysis in the frequency-domain. Besides the rigorous solutions, simplified models are analyzed and compared with the former to get some idea of the influence of such simplifications. Simplified models, in particular the incompressible fluid model and the viscous boundary, have been investigated in Chapter 9 for the two-dimensional reservoir with a rigid dam. Here, we first analyze the two-dimensional problem using finite elements to verify the fluid equations for the coupled nearfield-farfield system. This verification also includes the special cases of the incompressible model and the viscous boundary. Other examples introduced in Chapter 9 (rectangular and semi-circular channel) are considered in the next chapter in the context of rational approximation.

In the second part of this section we analyze three-dimensional dam-reservoir systems. The conclusions drawn for the two-dimensional case are also valid for the three-dimensional case. However, including a flexible dam gives a different picture. Although an incompressible reservoir behaves frequency-independently, the combined system of a flexible dam with an incompressible reservoir leads to a frequency-dependent solution.

The analyses were performed using the computer program DANAID described in Appendix A.

10.7.1 Two-dimensional fluid problem

In this subsection the two-dimensional semi-infinite model of a reservoir with constant depth is examined. The mathematical model and the analytical solutions have beeen presented in Section 9.2. The finite element discretization is shown in Figure 10.2. The fluid consists of five quadrilaterals. Five line elements are employed for the dam-reservoir interface and one line element for the foundation-reservoir interface. The reflection coefficient for the dam-reservoir interface is $\alpha = 0.8$ and for the dam-reservoir interface $\alpha = 1$. For the farfield there are five line elements for the fluid and one point element for the foundation-reservoir interface. The excitation is formulated as a prescribed velocity applied to the displacement degrees of freedom of the interfaces. Results are evaluated as

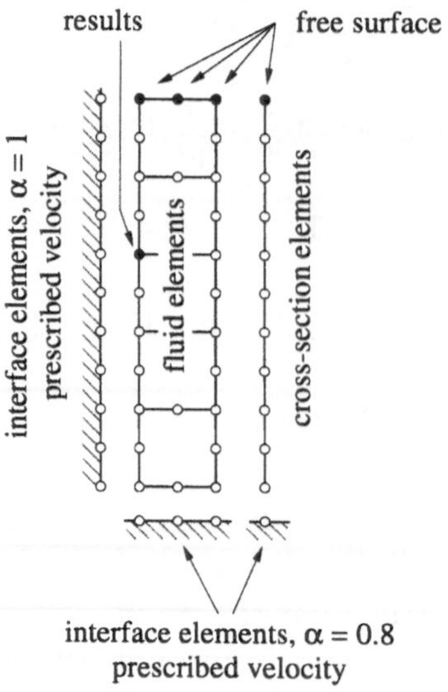

Figure 10.2: Finite element model for 2-D fluid domain

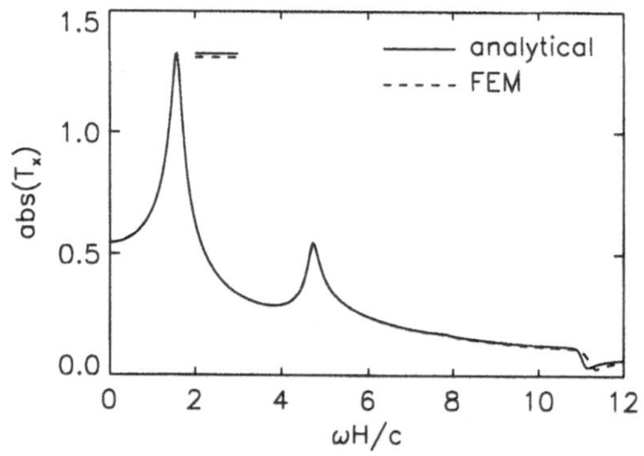

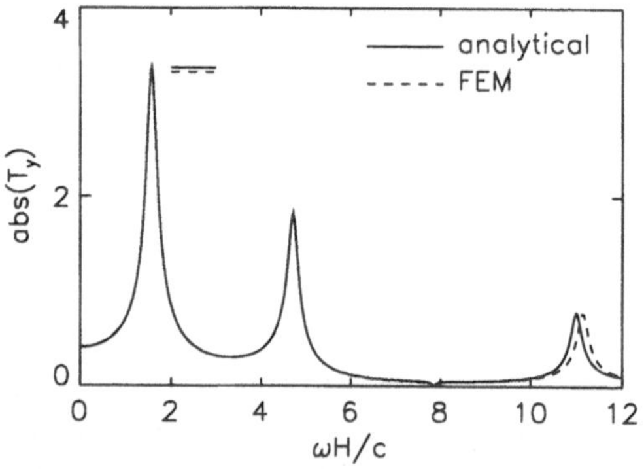

Figure 10.3: 2-D fluid problem: rigorous solution. Left: upstream excitation. Right: vertical excitation.

pressure transfer functions at the point indicated. For the calculations the reservoir depth is taken to be 100 m and the wave velocity $c = 1440$ m/s. Results are shown, however, in dimensionless form. As in Chapter 9, the transfer functions are defined as $T_x = \hat{p}/(H\rho\hat{a}_x)$ and analogously for the y-direction. Six Ritz vectors corresponding to six modes are included. Figure 10.3 shows the pressure transfer functions of the analytical and of the finite element solution. The figure shows the response due to an upstream excitation on the left, and due to a vertical excitation on the right. The finite element solution matches the analytical one very well except for a frequency shift in the higher frequency range, which is due to the discretization.

Figure 10.4 shows the pressure transfer function of the incompressible model for an upstream (left) and for a vertical (right) excitation. The farfield is modeled by the first six zero-frequency eigenvectors according to Equation (10.127). The transfer functions are frequency-independent and match the zero-frequency solution of the analytical compressible solution very well, which confirms the implementation.

For the viscous boundary the finite element model is extended to $L = H$ as shown in Figure 10.5. Two models for the viscous boundary are compared with the analytical solution derived in Section 9.3. One is the high-frequency solution given by Equation (10.138),

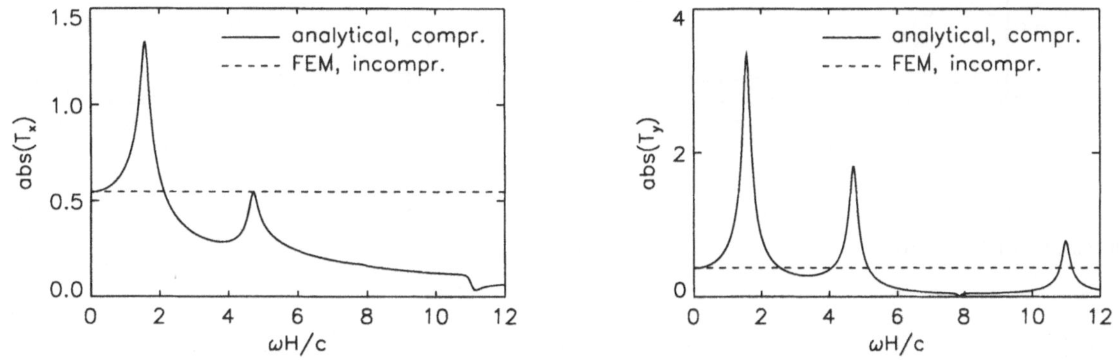

Figure 10.4: 2-D fluid problem: incompressible model. Left: upstream excitation. Right: vertical excitation.

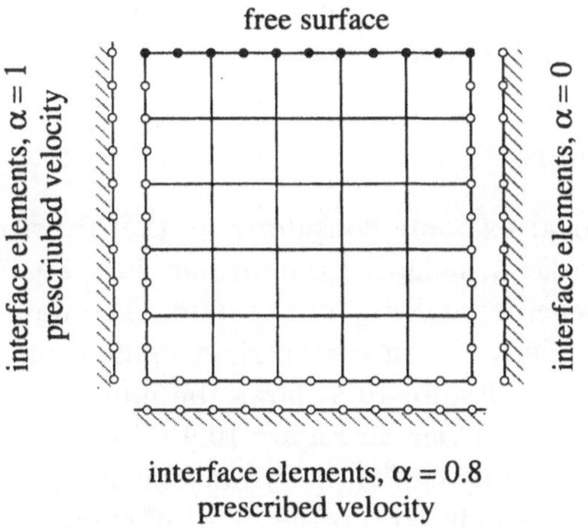

Figure 10.5: Finite element model for 2-D fluid domain with viscous boundary

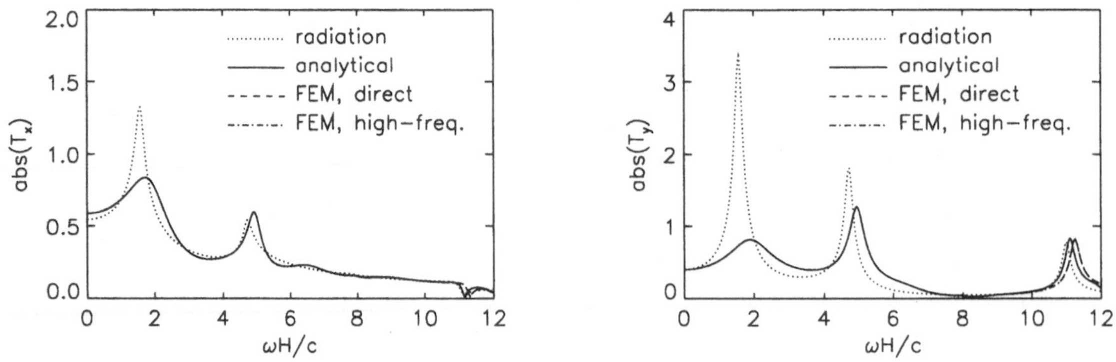

Figure 10.6: 2-D fluid problem: viscous boundary. Left: upstream excitation. Right: vertical excitation.

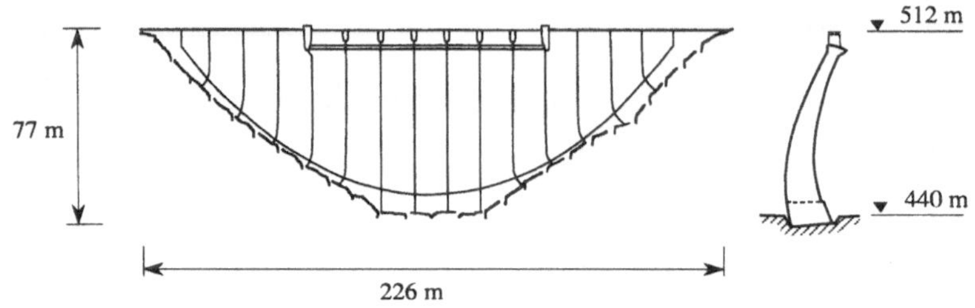

Figure 10.7: Talvacchia dam: geometry [Cas87]

the other is the conventional viscous boundary implemented by interface elements with a reflection coefficient $\alpha = 0$. For the conventional model the prescribed velocity is only applied to the interface elements belonging to the dam and to the foundation. The displacement degrees of freedom of the viscous boundary are fixed. The pressure transfer functions are shown in Figure 10.6. Both finite element solutions are very close to the analytcal solution, which verifies the implementation. The analytical solution for the radiation boundary is given for comparison.

10.7.2 Talvacchia dam

The first three-dimensional example considered is the Talvacchia dam which has been analyzed by different workers, including the author, in a benchmark workshop [ISM93]. Figure 10.7 shows the general geometry of the dam. The dam has a height of 77 m and a crest length of 226 m. Only the case with rigid foundation (fixed boundary condition for the dam) is treated here. Figure 10.8 shows the finite element mesh of the dam with a short nearfield. The exploded view in Figure 10.9 also shows the interface elements and the transmitting boundary. Only half of the symmetric model is used for the computations with appropriate boundary conditions on the plane of symmetry. The dam is modeled by 42 solid bricks, the fluid by 114 fluid bricks. For the reservoir foundation and for the dam-reservoir interface there are 65 fluid-solid interface elements. This results in 1439 degrees of freedom for symmetric loading. The cross-section of the channel includes 38 2D-fluid elements and 9 2D-fluid-solid interface elements. The cross-section model has 118 degrees of freedom for symmetric loading. Material properties are given in Table 10.1. The factors α_R for mass proportional and β_R for stiffness proportional damping are chosen such that the prescribed damping of $\xi = 5\%$ is obtained at 3 Hz and 5 Hz. The same damping value

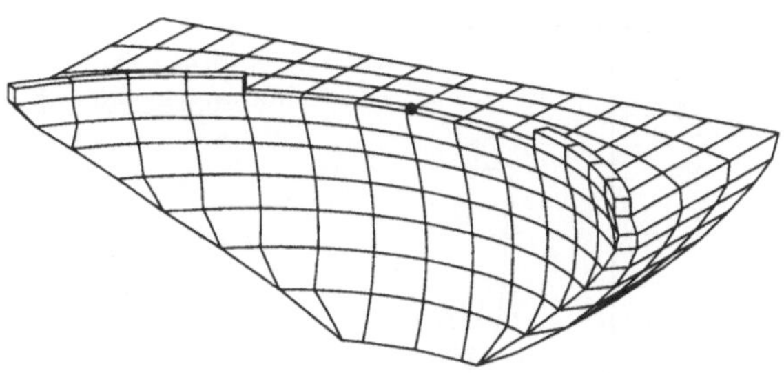

Figure 10.8: Talvacchia dam: short nearfield mesh

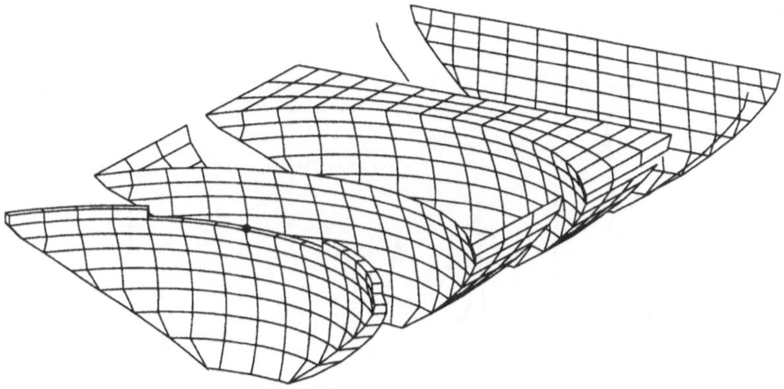

Figure 10.9: Talvacchia dam: exploded view of short nearfield mesh

results approximately for the frequencies in between. No bottom absorption is included (although it has been included in the workshop) to have comparable results between the compressible and the incompressible models.

water	wave velocity	$c = 1440$ m/s
	unit weight	$\rho = 1000$ kg/m^3
concrete	Young's modulus	$E = 3.6 \cdot 10^{10}$ N/m^2
	unit weight	$\rho_s = 2400$ kg/m^3
	Poisson's ratio	$\nu = 0.2$
	Rayleigh damping	$\alpha_R = 1.18$ 1/s, $\beta_R = 0.002$ s

Table 10.1: Material properties for Talvacchia dam

First, the compressible and the incompressible behavior are compared. Figure 10.10 shows the transfer functions for the radial absolute acceleration at the center of the dam crest (indicated by a dot in Figure 10.8) due to an upstream excitation. The fundamental frequency of the complete model is seen to be at about 3 Hz, much below the fundamental frequency of the reservoir farfield at 5.84 Hz, indicated by a short vertical line. For this case, the compressibility effects are not expected to be significant. This is, however, only true for the frequency range up to the fundamental frequency of the system, as can seen from the figure. At this main resonance, the compressible and the incompressible case compare reasonably. The maximum values are emphasized by small horizontal marks.

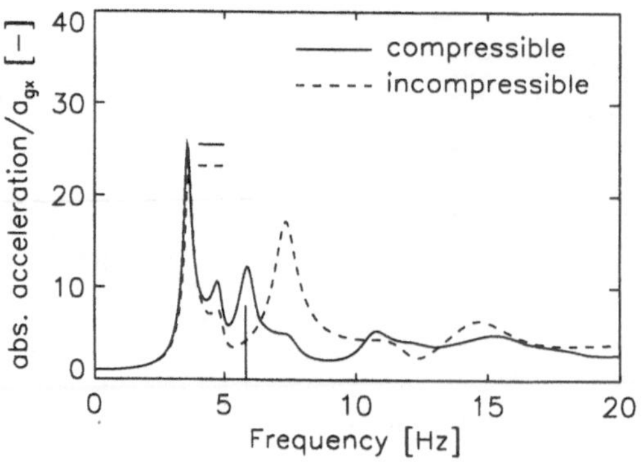

Figure 10.10: Talvacchia dam: compressible versus incompressible behavior

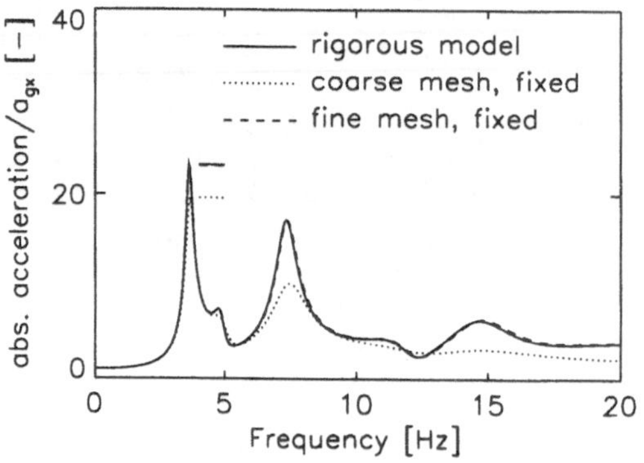

Figure 10.11: Talvacchia dam: three models for incompressible case

The main resonance is due to the dam flexibility and the mass effects of the reservoir. The fundamental frequency of the reservoir itself shows up only as a small bump in the compressible model. In the benchmark workshop an earthquake record with a main frequency content at 10 Hz has been specified (case with rigid foundation). Clearly, with this input, comparisons between the compressible and the incompressible model are not conclusive.

The next example, Figure 10.11, is a comparison of different models with incompressible fluid. The short nearfield mesh (Figure 10.8) combined with the semi-analytical farfield (Equation 10.127) is considered the rigorous solution. This model is compared to a long mesh with a fixed boundary condition. The long mesh is the one originally proposed for the workshop and used by most participants. It is shown in Figure 10.12. Although a large nearfield has been modeled, the results differ significantly from the rigorous solution. The reason is that the fluid elements are too coarse to appropriately represent the fluid motion especially in the vicinity of the dam. This argument is confirmed by the long model using a fine mesh shown in Figure 10.13. Although the mesh is shorter, much better results are obtained with the same number of elements. This comparison shows that, for the incompressible case, the short model with a semi-analytical farfield is equivalent to the conventional long model with a fixed boundary. However, the long model is less efficient because it has more degrees of freedom (2071 compared to 1439 for the short model) and also requires an appropriate meshing.

In next example, the long, fine mesh (Figure 10.13) has been used to model a compressible nearfield with a fixed boundary condition. For frequencies below the cut-off frequency, there is no radiation and compressibility effects take place mainly in the nearfield. Figure 10.14 shows the comparison with the rigorous solution. It is striking how well the

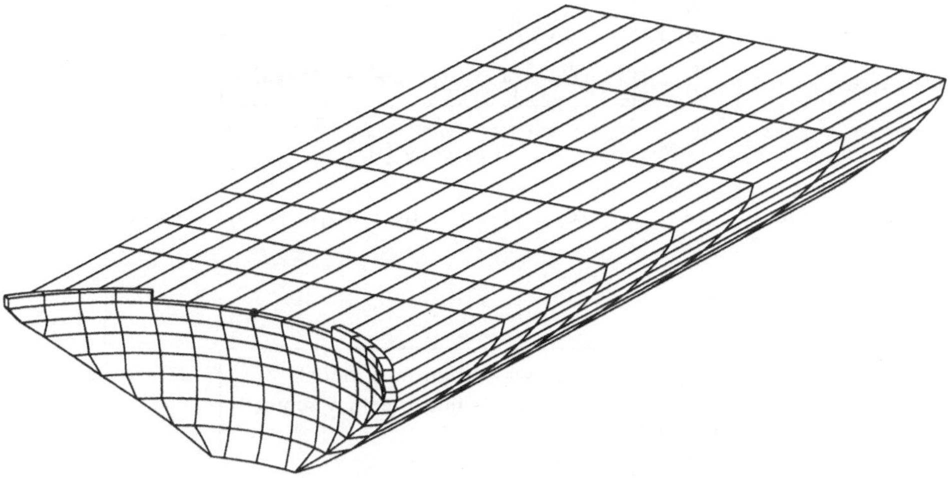

Figure 10.12: Talvacchia dam: long nearfield with coarse mesh

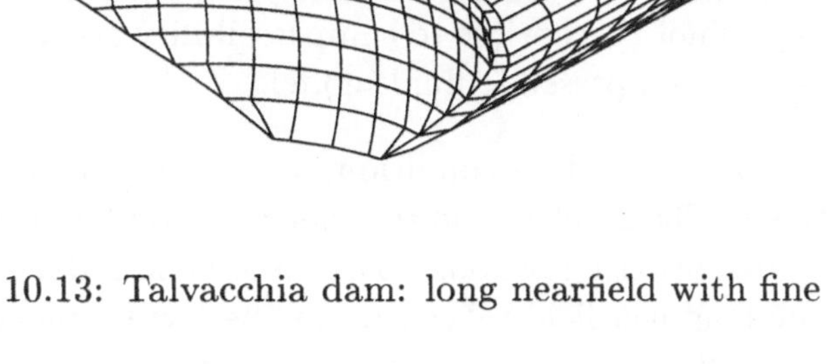

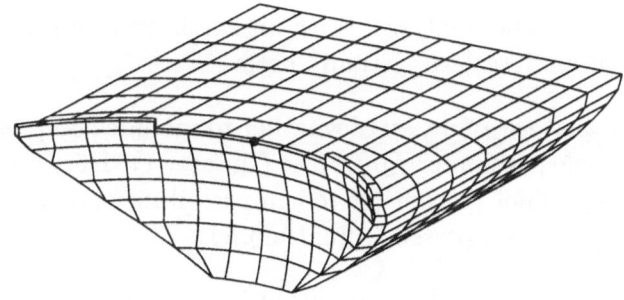

Figure 10.13: Talvacchia dam: long nearfield with fine mesh

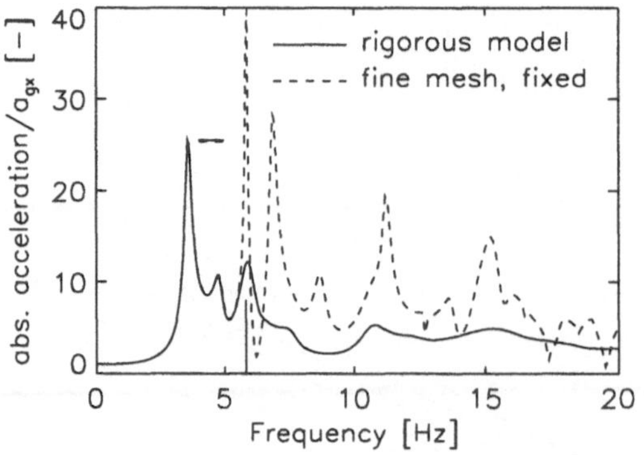

Figure 10.14: Talvacchia dam: rigorous versus fine compressible model with fixed end

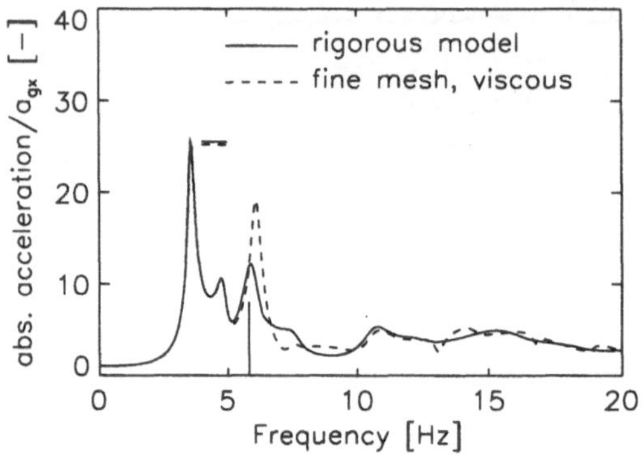

Figure 10.15: Talvacchia dam: rigorous versus fine compressible model with viscous boundary

simplified model compares to the rigorous solution exactly up to the cut-off frequency (indicated by the vertical line around 6 Hz) and how badly for higher frequencies. A similar model (in fact the coarse mesh) has been used by many participants of the workshop for the compressible fluid analysis. With an earthquake input at around 10 Hz, the comparison of different analyses are virtually useless.

The example show that the question of whether to include compressibility effects or not is not captured by a single parameter f_d/f_w (see Chapter 1). Depending on the earthquake input, even a dam which is not susceptible to compressibility according to Table 1.1 might need to be analyzed with a compressible fluid model.

Instead of having a fixed boundary condition at the far end we can also use a viscous boundary condition there. The acceleration transfer function for this case is shown in Figure 10.15. The results are surprisingly good. The viscous boundary takes care of the high frequencies whereas the long nearfield takes care of the low frequencies. The only critical region is around the cut-off frequency. A similar behavior has been observed earlier for two-dimensional problems (Figure 9.8 on the right). Whether the viscous boundary model leads to acceptable results depends on the frequency content of the actual earthquake input. It should also be noted, that the good results in the acceleration do not necessarily imply good results in pressure or other variables.

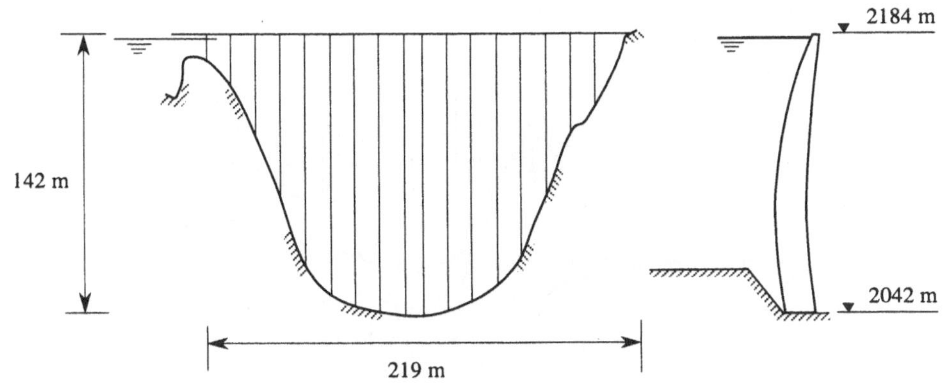

Figure 10.16: Morrow Point dam: geometry [DH88]

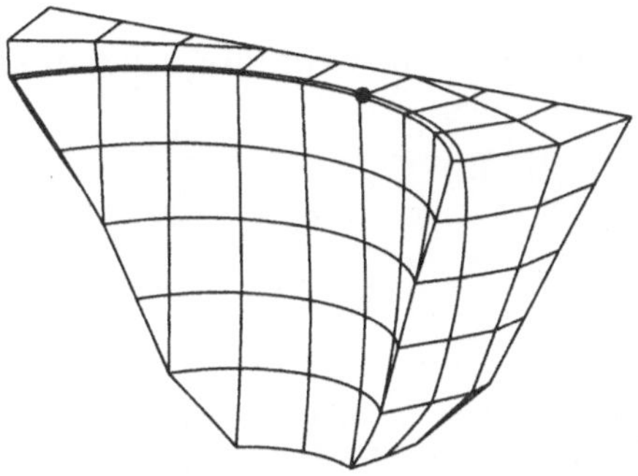

Figure 10.17: Morrow Point dam: short nearfield mesh

10.7.3 Morrow Point dam

The second dam investigated is the Morrow Point dam. This example is taken from [FC80]. The general geometry of this arch dam is shown in Figure 10.16. The dam has a height of 142 m and a crest length of 219 m. The finite element mesh is shown in Figure 10.17. The exploded view in Figure 10.18 also shows the interface elements and the transmitting boundary. Only half the model with symmetric boundary conditions is used for computation. The dam is modeled by 16 solid brick elements and has a fixed boundary condition at the foundation (rigid foundation). The nearfield of the reservoir consists of 27 fluid brick elements. The dam-reservoir and the foundation-reservoir interface consist of 27 quadrilateral elements. The complete dam-reservoir model for the nearfield has 468 degrees of freedom. The farfield cross-section is modeled by 16 quadrilaterals for the fluid and 6 line elements for the fluid-foundation interface. It has 52 degrees of freedom. The material properties are given in Table 10.2. The factors for the Rayleigh damping are adjusted for $\xi = 5\%$ at 2.5 Hz and 3.5 Hz. For the reflection coefficient, two values are considered, $\alpha = 0.5$ and $\alpha = 1.0$.

First, we compare the compressible case without bottom absorption to the incompressible case. Figure 10.19 shows the transfer functions for the radial absolute acceleration at the point indicated in Figure 10.17 due to horizontal earthquake input. The cut-off frequency is denoted by a short vertical line around 3 Hz. As seen from this figure, the fundamental frequency of the dam-reservoir system is only slightly below the cut-off frequency and the differences between the two cases are substantially as expected.

Next, we compare two different incompressible models. One model has a short incom-

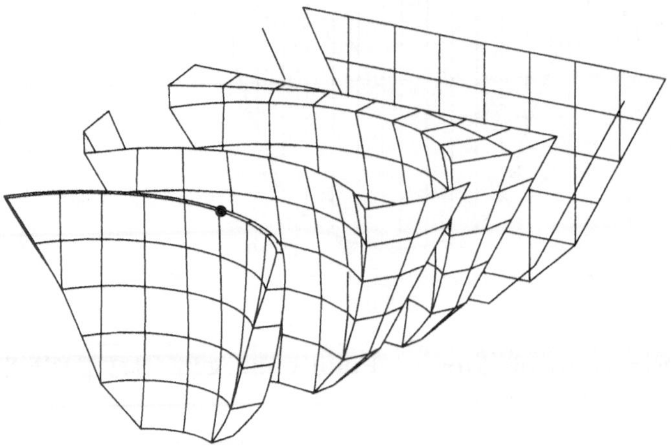

Figure 10.18: Morrow Point dam: exploded view of short nearfield mesh

water	wave velocity	$c = 1440$ m/s (4720 ft/sec)
	unit weight	$\rho = 1000$ kg/m^3 (62.4 pcf = 0.00193 ks^2/ft^4)
concrete	Young's modulus	$E = 2.76 \cdot 10^{10}$ N/m^2 ($4 \cdot 10^6$ psi = 576000 ksf)
	unit weight	$\rho_s = 2478$ kg/m^3 (155 pcf = 0.00482 ks^2/ft^4)
	Poisson's ratio	$\nu = 0.2$
	Rayleigh damping	$\alpha_R = 0.916$ 1/s, $\beta_R = 0.0027$ s
foundation	reflection	$\alpha = 0.5$ ($q = 0.0.0000706$ sec/ft)
		$\alpha = 1.0$ ($q = 0.0$ sec/ft)

Table 10.2: Material properties for Morrow Point dam

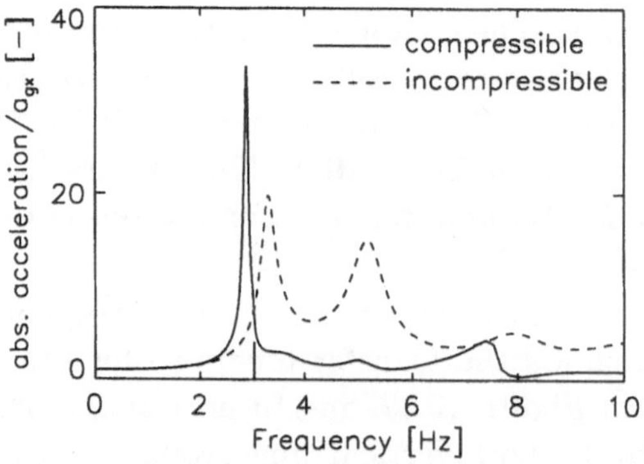

Figure 10.19: Morrow Point dam: compressible versus incompressible behavior

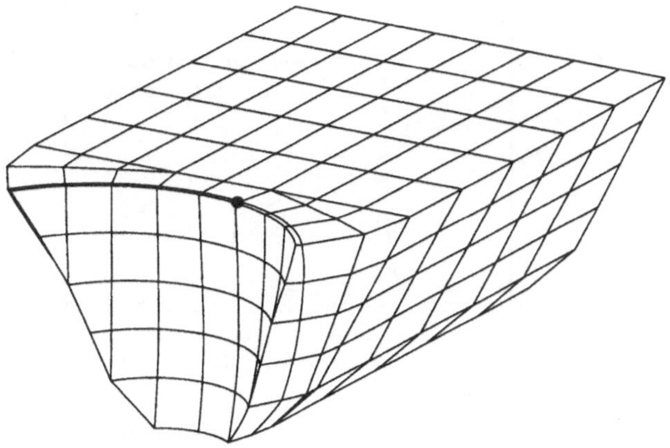

Figure 10.20: Morrow Point dam: long nearfield mesh

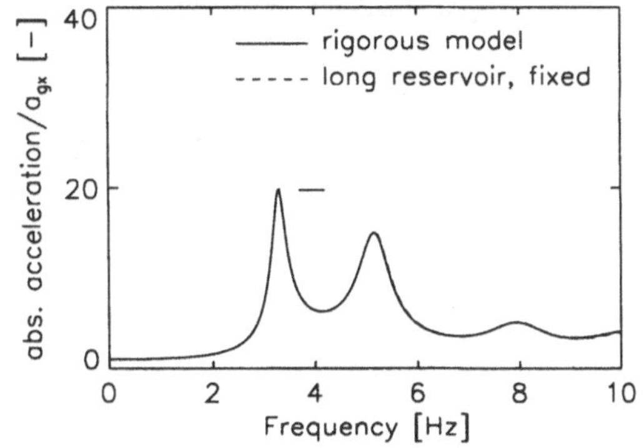

Figure 10.21: Morrow Point dam: incompressible reservoir, rigorous model versus long reservoir with fixed boundary

pressible nearfield as shown in Figure 10.17. The farfield is captured semi-analytically by the zero-frequency eigenvalues (Equation 10.127). The second model is a conventional incompressible long reservoir as shown in Figure 10.20 with a fixed boundary condition at the far end. The comparison between these two models is shown in Figure 10.21. The two models behave almost equally, which confirms the applicability of the semi-analytical solution. The model with the semi-amalytical solution is much smaller (468 degress of freedom) than the conventional long model (823 degress of freedom).

The next analysis investigates viscous boundaries. Two different nearfield models are used, the short nearfield (Figure 10.17) and the long one (Figure 10.20). The viscous boundary is modeled directly by interface elements with a reflection coefficient $\alpha = 0$. The two models are compared to the rigorous compressible model in Figure 10.22. As already observed for the Talvacchia dam, the viscous boundary combined with a long nearfield leads to reasonable results except near the cut-off frequency. This is because the low-frequency range is modeled by the long compressible nearfield while the high-frequency range is captured by the viscous boundary. Besides the fact that this model is not accurate at the cut-off frequency it is also not efficient compared to the short nearfield with a rigorous boundary condition. The short model with a viscous boundary does not exhibit any resonance at all and is completely useless.

As a concluding remark we can state that the rigorous compressible solution is only available in the frequency domain, whereas the simplified models can easily be implemented

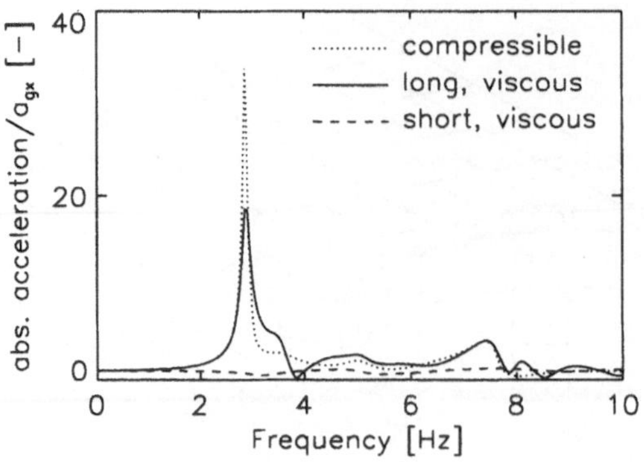

Figure 10.22: Morrow Point dam: viscous boundary, long versus short compressible nearfield

in a time-domain analysis. The error involved in these simplifications, however, appears to be substantial. In the next chapter we will show similar results obtained from a time-stepping analysis.

Chapter 11

Time-Domain Implementation

In this chapter the different parts developed in the previous chapters are combined to a complete dam-reservoir model for the time-domain analysis. The impedance matrix, which was reduced in the last chapter using Ritz vectors, is further modified to be compatible with the subsequent approximation procedure. Then the balanced realization of Chapter 7 is applied to the modified impedance matrix yielding a time-domain system which can be coupled to the total dam-reservoir model. The time integration scheme is described briefly. The second part of the chapter gives a review of the whole procedure repeating the essential equations from previous chapters. In the last part, several examples are given ranging from simple two-dimensional geometries to three-dimensional arch dam models. The results are compared to analytical results and to results from the literature to confirm the accuracy and efficiency of the procedure.

11.1 Further modifications for impedance matrix

After having reduced the impedance matrix of the farfield using a Rayleigh-Ritz formulation as described in the last chapter, two further modifications have to be applied before changing to the time domain. The first is to scale the reduced matrix for better numerical performance and the second is to extract the high-frequency behavior.

11.1.1 Scaling

The matrix $\mathbf{k}^*$ (Equation 10.120) contains the transfer functions of several modes. The transfer functions of the higher modes have typically higher values and therefore gain more importance in the rational approximation. This is not desirable because these transfer functions are in fact less important. The problem can be avoided by an appropriate scaling. The scaled form is found to lead to much better approximations than the original form.

The matrix $\mathbf{k}^*$ is scaled by the static value $\mathbf{k}_0$ (Equation 10.111) as

$$\bar{\mathbf{k}}^* = \mathbf{k}_0^{-1/2}\mathbf{k}^*\mathbf{k}_0^{-1/2} \tag{11.1}$$

The scaling matrix $\mathbf{k}_0 = \mathbf{k}_0^*$ is diagonal because the Ritz vectors $\mathbf{\Psi}_0$ are equal to the eigenvectors for the frequency $\omega = 0$. With this scaling the impedance matrix is written as

$$\mathbf{A}_I\mathbf{\Psi}_0\mathbf{k}^*\mathbf{\Psi}_0^{\mathrm{T}}\mathbf{A}_I = \mathbf{A}_I\mathbf{\Psi}_0\mathbf{k}_0^{1/2}\bar{\mathbf{k}}^*\mathbf{k}_0^{1/2}\mathbf{\Psi}_0^{\mathrm{T}}\mathbf{A}_I = \mathbf{T}\bar{\mathbf{k}}^*\mathbf{T}^{\mathrm{T}} \tag{11.2}$$

with the transformation matrix

$$\mathbf{T} = \mathbf{A}_I\mathbf{\Psi}_0\mathbf{k}_0^{1/2} \tag{11.3}$$

11.1.2 High-frequency behavior

The second adjustment is to extract the high-frequency behavior of $\bar{\mathbf{k}}^*$. This is necessary because only proper transfer functions can be approximated by the balanced realization. The high-frequency behavior is added back after the approximation and can immediately be implemented in the time domain by a damping matrix.

The high-frequency behavior was already investigated in the context of the viscous boundary in Chapter 10. As shown there (Equation 10.135)

$$\mathbf{k}_\infty = i\omega\mathbf{\Lambda}_\infty \tag{11.4}$$

where $\mathbf{\Lambda}_\infty$ is the solution of the high-frequency eigenvalue problem Equation (10.134). In scaled form this is

$$\bar{\mathbf{k}}^*_\infty = \mathbf{k}_0^{-1/2}\mathbf{\Psi}^*_\infty\,\mathbf{k}_\infty\,\mathbf{\Psi}^{*\mathrm{T}}_\infty\mathbf{k}_0^{-1/2} = i\omega\mathbf{k}_0^{-1/2}\mathbf{\Psi}^*_\infty\,\mathbf{\Lambda}_\infty\,\mathbf{\Psi}^{*\mathrm{T}}_\infty\mathbf{k}_0^{-1/2} = i\omega\mathbf{D}_1 \tag{11.5}$$

with the damping matrix

$$\mathbf{D}_1 = \mathbf{k}_0^{-1/2}\mathbf{\Psi}^*_\infty\,\mathbf{\Lambda}_\infty\,\mathbf{\Psi}^{*\mathrm{T}}_\infty\mathbf{k}_0^{-1/2} \tag{11.6}$$

The general form of the modified impedance matrix is then written as the sum of two parts, the first one decaying to zero for high frequencies and the second one considering the high-frequency behavior.

$$\bar{\mathbf{k}}^* = \boldsymbol{\kappa} + i\omega\mathbf{D}_1 \tag{11.7}$$

Care has to be taken when subtracting the value $i\omega\mathbf{D}_1$ from $\bar{\mathbf{k}}^*$ to evaluate $\boldsymbol{\kappa}$ for high frequencies. The difference of large numbers may not decay to zero if the number of significant digits is too low. On the side we remark that this definition of $\boldsymbol{\kappa}$ is consistent with the dimensionless parameter $\kappa = ik/\lambda - i\omega/(c\lambda)$ used in Chapter 5 because here we have $\lambda_0 = k_0$.

With this notation the equation for the complete coupled system is

$$\begin{bmatrix} \mathbf{S}_{ss} & \mathbf{S}_{sf} & \\ \mathbf{S}_{fs} & -\mathbf{S}_{ff}^{(11)} & -\mathbf{S}_{ff}^{(12)} \\ & -\mathbf{S}_{ff}^{(21)} & -(\mathbf{S}_{ff}^{(22)} + \mathbf{T}(\boldsymbol{\kappa} + s\mathbf{D}_1)\mathbf{T}^\mathrm{T}) \end{bmatrix} \begin{bmatrix} \hat{\mathbf{u}} \\ \hat{\boldsymbol{\varphi}}^{(1)} \\ \hat{\boldsymbol{\varphi}}^{(2)} \end{bmatrix} = \begin{bmatrix} \hat{\mathbf{F}}_s \\ -\hat{\mathbf{V}}^{(1)} \\ -\mathbf{T}(\boldsymbol{\kappa} + s\mathbf{D}_1)\mathbf{T}^\mathrm{T}\hat{\mathbf{\Phi}}^p \end{bmatrix} \tag{11.8}$$

where the variable $i\omega$ has been changed to s to be consistent with the system notation used in the next section.

11.2 Coupled system with approximated farfield

11.2.1 General case

To couple the farfield to the nearfield, the frequency-dependent matrix $\boldsymbol{\kappa}$ is first approximated by a first-order system as described in Chapter 7. Because the high-frequency behavior has been removed, the matrix $\boldsymbol{\kappa}$ is a proper transfer function (that is, it does not grow to infinity for high frequencies) and can be approximated by the standard form

$$\boldsymbol{\kappa} = \mathbf{C}(s\mathbf{I} - \mathbf{A})^{-1}\mathbf{B} + \mathbf{D} \tag{11.9}$$

The second-order system is calculated by the procedure described in Chapter 8. This transformation results in a matrix $\mathbf{D}_2$ but not in a matrix $\mathbf{D}_1$ (see Equation 8.79). For the matrix $\bar{\mathbf{k}}^*$ the high-frequency part $s\mathbf{D}_1$ is added back to get the second-order formulation

$$\bar{\mathbf{k}}^* = \boldsymbol{\kappa} + s\mathbf{D}_1 = (s\mathbf{B}_1 + \mathbf{B}_2)^\mathrm{T}(s^2\mathbf{A}_0 + s\mathbf{A}_1 + \mathbf{A}_2)^{-1}(s\mathbf{B}_1 + \mathbf{B}_2) + s\mathbf{D}_1 + \mathbf{D}_2 \tag{11.10}$$

For a more compact notation we can further write

$$\mathbf{T}(\boldsymbol{\kappa} + s\mathbf{D}_1)\mathbf{T}^{\mathrm{T}} = \mathbf{R}^{\mathrm{T}}\mathbf{S}^{-1}\mathbf{R} + \mathbf{Q} \tag{11.11}$$

with

$$\begin{aligned} \mathbf{R} &= (s\mathbf{B}_1 + \mathbf{B}_2)\mathbf{T}^{\mathrm{T}} && (11.12)\\ \mathbf{S} &= s^2\mathbf{A}_0 + s\mathbf{A}_1 + \mathbf{A}_2 && (11.13)\\ \mathbf{Q} &= \mathbf{T}(s\mathbf{D}_1 + \mathbf{D}_2)\mathbf{T}^{\mathrm{T}} && (11.14) \end{aligned}$$

To implement the impedance relation at the transmitting boundary,

$$\mathbf{T}(\boldsymbol{\kappa} + s\mathbf{D}_1)\mathbf{T}^{\mathrm{T}}\,(\hat{\varphi}^{(2)} - \hat{\mathbf{\Phi}}^p) = (\mathbf{R}^{\mathrm{T}}\mathbf{S}^{-1}\mathbf{R} + \mathbf{Q})\,(\hat{\varphi}^{(2)} - \hat{\mathbf{\Phi}}^p) \tag{11.15}$$

we introduce the variable $\hat{\boldsymbol{\zeta}}$ as

$$\hat{\boldsymbol{\zeta}} = \mathbf{S}^{-1}\mathbf{R}\,(\hat{\varphi}^{(2)} - \hat{\mathbf{\Phi}}^p) \tag{11.16}$$

This removes the matrix inversion of $\mathbf{S}$ and replaces Equation (11.15) by the two equations

$$\mathbf{T}(\boldsymbol{\kappa} + s\mathbf{D}_1)\mathbf{T}^{\mathrm{T}}(\hat{\varphi}^{(2)} - \hat{\mathbf{\Phi}}^p) = \mathbf{R}^{\mathrm{T}}\hat{\boldsymbol{\zeta}} + \mathbf{Q}\,(\hat{\varphi}^{(2)} - \hat{\mathbf{\Phi}}^p) \tag{11.17}$$

and

$$\mathbf{S}\hat{\boldsymbol{\zeta}} = \mathbf{R}\,(\hat{\varphi}^{(2)} - \hat{\mathbf{\Phi}}^p) \tag{11.18}$$

These two equations are substituted into the coupled system Equation (10.121) to give the final result

$$\begin{bmatrix} \mathbf{S}_{ss} & \mathbf{S}_{sf} & & \\ \mathbf{S}_{fs} & -\mathbf{S}_{ff}^{(11)} & -\mathbf{S}_{ff}^{(12)} & \\ & -\mathbf{S}_{ff}^{(21)} & -(\mathbf{S}_{ff}^{(22)}+\mathbf{Q}) & -\mathbf{R}^{\mathrm{T}} \\ & & -\mathbf{R} & \mathbf{S} \end{bmatrix} \begin{bmatrix} \hat{\mathbf{u}} \\ \hat{\varphi}^{(1)} \\ \hat{\varphi}^{(2)} \\ \hat{\boldsymbol{\zeta}} \end{bmatrix} = \begin{bmatrix} \hat{\mathbf{F}}_s \\ -\hat{\mathbf{V}}^{(1)} \\ -\mathbf{Q}\hat{\mathbf{\Phi}}^p \\ -\mathbf{R}\hat{\mathbf{\Phi}}^p \end{bmatrix} \tag{11.19}$$

For compact notation this equation is written in the frequency domain, but since all terms are matrix polynomials of degree at most two, it represents a system of second-order differential equations and has the same form as the standard equations of motion. The corresponding matrices are

$$\mathbf{M} = \begin{bmatrix} \mathbf{M}_{ss} & & & \\ & -\mathbf{M}_{ff}^{(11)} & -\mathbf{M}_{ff}^{(12)} & \\ & -\mathbf{M}_{ff}^{(21)} & -\mathbf{M}_{ff}^{(22)} & \\ & & & \mathbf{A}_0 \end{bmatrix} \tag{11.20}$$

$$\mathbf{C} = \begin{bmatrix} \mathbf{C}_{ss} & \mathbf{C}_{sf} & & \\ \mathbf{C}_{sf}^{\mathrm{T}} & -\mathbf{C}_{ff}^{(11)} & -\mathbf{C}_{ff}^{(12)} & \\ & -\mathbf{C}_{ff}^{(21)} & -(\mathbf{C}_{ff}^{(22)}+\mathbf{T}\mathbf{D}_1\mathbf{T}^{\mathrm{T}}) & -\mathbf{T}\mathbf{B}_1^{\mathrm{T}} \\ & & -\mathbf{B}_1\mathbf{T}^{\mathrm{T}} & \mathbf{A}_1 \end{bmatrix} \tag{11.21}$$

$$\mathbf{K} = \begin{bmatrix} \mathbf{K}_{ss} & & & \\ & -\mathbf{K}_{ff}^{(11)} & -\mathbf{K}_{ff}^{(12)} & \\ & -\mathbf{K}_{ff}^{(21)} & -(\mathbf{K}_{ff}^{(22)}+\mathbf{T}\mathbf{D}_2\mathbf{T}^{\mathrm{T}}) & -\mathbf{T}\mathbf{B}_2^{\mathrm{T}} \\ & & -\mathbf{B}_2\mathbf{T}^{\mathrm{T}} & \mathbf{A}_2 \end{bmatrix} \tag{11.22}$$

$$\mathbf{F} = \begin{bmatrix} \mathbf{F}_s \\ -\mathbf{V}^{(1)} \\ -\mathbf{T}\mathbf{D}_1\mathbf{T}^{\mathrm{T}}\dot{\mathbf{\Phi}}^p - \mathbf{T}\mathbf{D}_2\mathbf{T}^{\mathrm{T}}\mathbf{\Phi}^p \\ -\mathbf{B}_1\mathbf{T}^{\mathrm{T}}\dot{\mathbf{\Phi}}^p - \mathbf{B}_2\mathbf{T}^{\mathrm{T}}\mathbf{\Phi}^p \end{bmatrix} \tag{11.23}$$

Right-hand-side vector. As already described for the frequency domain in Section 10.6.4 the different components of the right-hand-side vector can be expanded further. In a earthquake analysis with uniform input ground motion we can consider the solid degrees of freedom $\mathbf{u}$, $\dot{\mathbf{u}}$ and $\ddot{\mathbf{u}}$ to be relative to the input motion. Then the input acceleration prescribed for the dam, $\mathbf{a}_g$, leads to

$$\mathbf{F}_s = -\mathbf{M}_{ss}\,\mathbf{a}_g \tag{11.24}$$

As usual, results are shown as relative displacement, relative velocity or absolute acceleration. The velocity input $\mathbf{V}^{(1)}$ can be written as (Equation 10.51)

$$\mathbf{V}^{(1)} = -\mathbf{C}_{fs}\,\mathbf{v}_g = \mathbf{C}_{sf}^{\mathrm{T}}\,\mathbf{v}_g \tag{11.25}$$

where $\mathbf{v}_g$ is the velocity prescribed at the interfaces between the the dam and the reservoir and between the reservoir and the foundation. The velocity potential $\mathbf{\Phi}^p$ used to determine the input from the farfield into the nearfield is found by solving the differential equation (10.80)

$$\mathbf{M}_I\,\ddot{\mathbf{\Phi}}^p + \mathbf{C}_I\,\dot{\mathbf{\Phi}}^p + \mathbf{K}_I\,\mathbf{\Phi}^p = \mathbf{V}_I \tag{11.26}$$

As for the nearfield, the velocity input of the cross-section can be formulated using interface elements as (Equation 10.66)

$$\mathbf{V}_I = -\mathbf{C}_{fs}^I\,\mathbf{v}_g \tag{11.27}$$

In the time-stepping procedure the right-hand-side vector is not $\mathbf{F}$ itself but the residual vector (see Section 11.3)

$$\mathbf{R} = \mathbf{F} - \mathbf{M}\mathbf{a} - \mathbf{C}\mathbf{v} - \mathbf{K}\mathbf{d} \tag{11.28}$$

Many contributions can be simplified using the absolute quantities $(\ddot{\mathbf{u}} + \mathbf{a}_g)$ and $(\dot{\mathbf{u}} + \mathbf{v}_g)$, and the relative quantities $(\boldsymbol{\varphi}^{(2)} - \mathbf{\Phi}^p)$ and $(\dot{\boldsymbol{\varphi}}^{(2)} - \dot{\mathbf{\Phi}}^p)$ as indicated in the following.

$$\mathbf{R} = \begin{bmatrix} -\mathbf{M}_{ss}\,(\ddot{\mathbf{u}} + \mathbf{a}_g) + \cdots \\ -\mathbf{C}_{sf}^{\mathrm{T}}\,(\dot{\mathbf{u}} + \mathbf{v}_g) + \cdots \\ \mathbf{T}\mathbf{D}_1\mathbf{T}^{\mathrm{T}}\,(\dot{\boldsymbol{\varphi}}^{(2)} - \dot{\mathbf{\Phi}}^p) + \mathbf{T}\mathbf{D}_2\mathbf{T}^{\mathrm{T}}\,(\boldsymbol{\varphi}^{(2)} - \mathbf{\Phi}^p) + \cdots \\ \mathbf{B}_1\mathbf{T}^{\mathrm{T}}\,(\dot{\boldsymbol{\varphi}}^{(2)} - \dot{\mathbf{\Phi}}^p) + \mathbf{B}_2\mathbf{T}^{\mathrm{T}}\,(\boldsymbol{\varphi}^{(2)} - \mathbf{\Phi}^p) + \cdots \end{bmatrix} \tag{11.29}$$

This also applies to the case with numerical α-damping. Because the problem is linear, the residual force (Equation 11.37) is

$$\begin{aligned} \mathbf{R} &= \mathbf{F}_{n+1+\alpha} - \mathbf{M}\mathbf{a}_{n+1} - (1+\alpha)\,(\mathbf{C}\mathbf{v}_{n+1} + \mathbf{K}\mathbf{d}_{n+1}) + \alpha\,(\mathbf{C}\mathbf{v}_n + \mathbf{K}\mathbf{d}_n) \\ &= (1+\alpha)\,\mathbf{F}_{n+1} - \alpha\,\mathbf{F}_n - \mathbf{M}\mathbf{a}_{n+1} - \mathbf{C}\Big((1+\alpha)\mathbf{v}_{n+1} - \alpha\mathbf{v}_n\Big) - \mathbf{K}\Big((1+\alpha)\mathbf{d}_{n+1} - \alpha\mathbf{d}_n\Big) \\ &= (1+\alpha)\,(\mathbf{F}_{n+1} - \mathbf{C}\mathbf{v}_{n+1} - \mathbf{K}\mathbf{d}_{n+1}) - \alpha\,(\mathbf{F}_n - \mathbf{C}\mathbf{v}_n - \mathbf{K}\mathbf{d}_n) \end{aligned} \tag{11.30}$$

11.2.2 Diagonal approximation

When using the Ritz vectors, the matrix $\mathbf{k}^* = \mathbf{\Psi}^* \mathbf{k} \mathbf{\Psi}^{*\mathrm{T}}$ is no longer diagonal. However, for small values of bottom absorption the off-diagonal terms are generally small and it is possible to neglect them. Then each mode (diagonal entry) can be approximated individually, which simplifies the procedure considerably, because the Hankel matrix is only small. Numerical examples are shown later in this chapter.

Taking the diagonal terms of $\mathbf{k}^*$ is not quite the same as using directly the diagonal matrix $\mathbf{k}$, not even for case without bottom absorption. The point is that in a numerical procedure the eigenvalues and eigenvectors may not be ordered properly for all frequencies. The ordering is, however, essential for the subsequent approximation. The problem occurs only for three-dimensional problems, if eigenvalues belonging to different modes cross each other. In the case without bottom absorption the eigenvectors $\mathbf{\Psi} = \mathbf{\Psi}_0 \mathbf{\Psi}^*$ are the same as the vectors $\mathbf{\Psi}_0$ except for the ordering. The matrix $\mathbf{\Psi}^*$ can therefore be considered as a permutation matrix enforcing that the matrix $\mathbf{\Psi}^* \mathbf{k} \mathbf{\Psi}^{*\mathrm{T}}$ is ordered properly.

11.2.3 Special cases

The special cases of incompressible fluid and viscous boundaries have been already discussed in Section 10.6.5. Here they are adjusted to take care of the modifications introduced in Section 11.1 for the time-domain analysis.

Incompressible fluid. With the transforamtion matrix $\mathbf{T}$ introduced in Equation (11.3), the zero-frequency impedance matrix is

$$\mathbf{A}_I \mathbf{\Psi}_0 \mathbf{k}_0 \mathbf{\Psi}_0^{\mathrm{T}} \mathbf{A}_I = \mathbf{T} \boldsymbol{\kappa}_0 \mathbf{T}^{\mathrm{T}} \tag{11.31}$$

Substituting this into Equation (10.106), the coupled system becomes

$$\begin{bmatrix} \mathbf{S}_{ss} & \mathbf{S}_{sf} & \\ \mathbf{S}_{fs} & -\mathbf{K}_{ff}^{(11)} & -\mathbf{K}_{ff}^{(12)} \\ & -\mathbf{K}_{ff}^{(21)} & -(\mathbf{K}_{ff}^{(22)} + \mathbf{T}\boldsymbol{\kappa}_0\mathbf{T}^{\mathrm{T}}) \end{bmatrix} \begin{bmatrix} \hat{\mathbf{u}} \\ \hat{\varphi}^{(1)} \\ \hat{\varphi}^{(2)} \end{bmatrix} = \begin{bmatrix} \hat{\mathbf{F}}_s \\ -\hat{\mathbf{V}}^{(1)} \\ -\mathbf{T}\boldsymbol{\kappa}_0\mathbf{T}^{\mathrm{T}}\hat{\mathbf{\Phi}}^p \end{bmatrix} \tag{11.32}$$

The matrix $\boldsymbol{\kappa}_0$ is the unit matrix for the chosen scaling. It is, however, still written in the formulas for clarity. The equivalence to an added mass has been shown in Section 10.6.5.

Viscous boundary. The viscous boundary corresponds to the high-frequency behavior. As seen from Equation (11.8) the high-frequency term is $s\mathbf{T}\mathbf{D}_1\mathbf{T}^{\mathrm{T}}$ and the coupled system is

$$\begin{bmatrix} \mathbf{S}_{ss} & \mathbf{S}_{sf} & \\ \mathbf{S}_{fs} & -\mathbf{S}_{ff}^{(11)} & -\mathbf{S}_{ff}^{(12)} \\ & -\mathbf{S}_{ff}^{(21)} & -(\mathbf{S}_{ff}^{(22)} + s\mathbf{T}\mathbf{D}_1\mathbf{T}^{\mathrm{T}}) \end{bmatrix} \begin{bmatrix} \hat{\mathbf{u}} \\ \hat{\varphi}^{(1)} \\ \hat{\varphi}^{(2)} \end{bmatrix} = \begin{bmatrix} \hat{\mathbf{F}}_s \\ -\hat{\mathbf{V}}^{(1)} \\ \mathbf{0} \end{bmatrix} \tag{11.33}$$

It has been shown in Section 10.6.5 that this formulation is equivalent to the viscous boundary condition $\partial\varphi/\partial x + \partial\varphi/c\partial t = 0$

11.3 Time Integration

For the integration of the equations of motion many different time-stepping schemes are presented in the literature. An extensive overview of the classical methods with their advantages and disadvantages is given in Hughes [Hug87]. Because the subject is not of primary interest in this work, no further investigations are made here. The Hilber-Hughes-Taylor method (α-method) [HHT77] has been chosen for its accuracy, its algorithmic damping of higher modes and its flexibility. It has been implemented in the predictor-corrector form [MFH89], which allows combining the implicit method with a consistent explicit method. The partition between implicit and explicit domains is done on an element basis. The parameter α can be chosen in the range $-1/3 \leq \alpha \leq 0$, with $\alpha = -1/3$ resulting in the highest algorithmic damping. If the parameters β and γ are chosen as $\gamma = (1-2\alpha)/2$ and $\beta = (1-\alpha)^2/4$, a second-order accurate, unconditionally stable scheme results.

The nodal variables at time step n are denoted by $\mathbf{d}_n$, the first derivatives by $\mathbf{v}_n$ and the second derivatives by $\mathbf{a}_n$. Stepping from the current time step to the next, we first calculate the predictors as

$$\mathbf{d}_{n+1}^{(i)} = \mathbf{d}_n + \Delta t\, \mathbf{v}_n + \frac{1}{2}(1-2\beta)\,\Delta t^2\, \mathbf{a}_n \tag{11.34}$$

$$\mathbf{v}_{n+1}^{(i)} = \mathbf{v}_n + (1-\gamma)\,\Delta t\, \mathbf{a}_n \tag{11.35}$$

$$\mathbf{a}_{n+1}^{(i)} = \mathbf{0} \tag{11.36}$$

For kinematic boundary conditions, the predictors have to be specified differently [MFH89].

The multi-corrector loop with superscript (i) starts with the calculation of the residual forces $\mathbf{R}$ from the predictors

$$\mathbf{R}^{(i)} = \mathbf{F}_{n+1+\alpha}^{ext} - \mathbf{M}\mathbf{a}_{n+1}^{(i)} - (1+\alpha)\,\mathbf{F}^{int}(\mathbf{d}_{n+1}^{(i)}, \mathbf{v}_{n+1}^{(i)}) + \alpha\,\mathbf{F}^{int}(\mathbf{d}_n, \mathbf{v}_n) \tag{11.37}$$

where $\mathbf{F}_{n+1+\alpha}^{ext}$ is the vector of externally applied loads, and $\mathbf{F}^{int}$ the vector of internal element forces. In the linear case we have

$$\mathbf{F}^{int}(\mathbf{d}, \mathbf{v}) = \mathbf{C}\mathbf{v} + \mathbf{K}\mathbf{d} \tag{11.38}$$

The matrices do not have to be formed explicitly but the internal element forces can be evaluated element by element.

The acceleration increment $\Delta\mathbf{a}^{(i)}$ is found by solving the equation

$$\mathbf{M}^*\Delta\mathbf{a}^{(i)} = \mathbf{R}^{(i)} \tag{11.39}$$

with the effective mass matrix

$$\mathbf{M}^* = \mathbf{M} + \gamma\,\Delta t\,(1+\alpha)\,\mathbf{C}^I\big(\mathbf{d}_{n+1}^{(i)}, \mathbf{v}_{n+1}^{(i)}\big) + \beta\,\Delta t^2\,(1+\alpha)\,\mathbf{K}^I\big(\mathbf{d}_{n+1}^{(i)}, \mathbf{v}_{n+1}^{(i)}\big) \tag{11.40}$$

$\mathbf{K}^I$ and $\mathbf{C}^I$ are the tangent stiffness and damping matrices, respectively, for the implicit elements. The explicit elements contribute only to the mass matrix but not to the damping and stiffness matrices. If the damping and stiffness matrices are calculated for the current configuration, the iterative solution technique is a Newton-Raphson method. However, modified Newton or quasi-Newton methods are also possible. For a linear analysis the matrix $\mathbf{M}^*$ is factored only once at the beginning and only back-substitution is necessary for each time step.

In the corrector phase, the variables $\mathbf{d}_{n+1}^{(i)}$ and $\mathbf{v}_{n+1}^{(i)}$ are updated to $\mathbf{d}_{n+1}^{(i+1)}$ and $\mathbf{v}_{n+1}^{(i+1)}$.

$$\mathbf{d}_{n+1}^{(i+1)} = \mathbf{d}_{n+1}^{(i)} + \beta\,\Delta t^2\,\Delta\mathbf{a}^{(i)} \tag{11.41}$$

$$\mathbf{v}_{n+1}^{(i+1)} = \mathbf{v}_{n+1}^{(i)} + \gamma\,\Delta t\,\Delta\mathbf{a}^{(i)} \tag{11.42}$$

$$\mathbf{a}_{n+1}^{(i+1)} = \mathbf{a}_{n+1}^{(i)} + \Delta\mathbf{a}^{(i)} \tag{11.43}$$

The multi-corrector loop is repeated until the residual forces are smaller than a prescribed tolerance or until a prescribed maximum number of iterations is reached. For a detailed description of the algorithm the reader is referred to the literature [MFH89].

Matrix Profile. Because additional degrees of freedom are introduced through the internal variables, optimizing the ordering of the equations for profile reduction of $\mathbf{M}^*$ is essential. The algorithm proposed by Gibbs, Poole and Stockmeyer [GPS76] has been used.

11.4 Review of procedure

The following section gives a review of the implemented procedure. The important equations are repeated from previous chapters. Some parameters have been given more suggestive names, which are also used in the computer program. The procedure is implemented in the computer program DANAID, which has been developed specifically for that purpose. The program is described in Appendix A.

The procedure comprises the following steps:

1) Find frequency-domain solution for infinite channel (Chapter 10)

a) Compute finite element matrices for channel cross-section (Equations 10.59–10.62)

$$\mathbf{A}_I = \rho\int_{\Omega^I} \mathbf{N}^{\mathrm{T}}\mathbf{N}\,d\Omega \tag{11.44}$$

$$\mathbf{M}_I = \frac{\rho}{c^2}\int_{\Omega^I} \mathbf{N}^{\mathrm{T}}\mathbf{N}\,d\Omega \tag{11.45}$$

$$\mathbf{C}_I = q\rho\int_{\Gamma_v^I} \mathbf{N}^{\mathrm{T}}\mathbf{N}\,d\Gamma \tag{11.46}$$

$$\mathbf{K}_I = \rho\int_{\Omega^I} \mathbf{B}^{\mathrm{T}}\mathbf{B}\,d\Omega \tag{11.47}$$

b) With the finite element matrices for the cross-section, solve the eigenvalue problem Equation (10.70) for $\omega = 0$

$$\mathbf{K}_I\boldsymbol{\Psi}_0 = \mathbf{A}_I\boldsymbol{\Psi}_0\mathbf{k}_0^2 \tag{11.48}$$

c) For frequencies $0 < \omega < \infty$, solve the reduced eigenvalue problem Equation (10.116) including a frequency shift σ for improved numerical performance

$$(\mathbf{K}_I^* + i\omega\mathbf{C}_I^* - \omega^2\mathbf{M}_I^* + \sigma^2\mathbf{I})\boldsymbol{\Psi}^* = \boldsymbol{\Psi}^*\boldsymbol{\Lambda}^2 \tag{11.49}$$

Calculate $\mathbf{k}$ from

$$\boldsymbol{\Lambda}^2 = \mathbf{k}^2 + \sigma\mathbf{I}^2 \tag{11.50}$$

taking into account the radiation condition, and form the transfer matrix

$$\mathbf{k}^* = \boldsymbol{\Psi}^*\mathbf{k}\boldsymbol{\Psi}^{*\mathrm{T}} \tag{11.51}$$

The frequency points ω_k have to chosen such that the matrix $\mathbf{k}^*$ is accurately represented. In view of the subsequent Fast Fourier Transform, the points are currently selected as $\omega_k = \alpha_{bil}\tan(k(\pi/2)/N_{freq})$. Another possibility would be an automatic adaption of the frequency step depending on the variation of the transfer function [FC80]. For efficiency, the number of frequencies N_{freq} for which the transfer function is evaluated is typically less than the number of frequencies considered in the Fourier transform ($N_{FFT}/2$). The values used in the FFT can be found by interpolation.

d) Solve the reduced eigenvalue problem for $\omega \to \infty$ (Equation 10.134)

$$\mathbf{M}_I^* \mathbf{\Psi}_\infty^* = \mathbf{\Psi}_\infty^* \mathbf{\Lambda}_\infty^2 \tag{11.52}$$

e) Scale $\mathbf{k}^*$ for all frequencies (Equation 11.1)

$$\bar{\mathbf{k}}^* = \mathbf{k}_0^{-1/2}\mathbf{k}^*\mathbf{k}_0^{-1/2} \tag{11.53}$$

and split into regular and high-frequency behavior (avoiding round-off errors at high frequencies)

$$\bar{\mathbf{k}}^* = \boldsymbol{\kappa}(i\omega) + i\omega\mathbf{D}_1 \tag{11.54}$$

where (Equation 11.6)

$$\mathbf{D}_1 = \mathbf{k}_0^{-1/2}\mathbf{\Psi}_\infty^* \mathbf{\Lambda}_\infty \mathbf{\Psi}_\infty^{*\mathrm{T}}\mathbf{k}_0^{-1/2} \tag{11.55}$$

2) Approximate the transfer matrix $\boldsymbol{\kappa}$ by a linear system (Chapter 7)

a) Find discrete-time impulse response matrices $\mathbf{f}_k$ of the transfer matrix $\boldsymbol{\kappa}$ by the Fast Fourier Transform (Equation 4.117) treating each entry individually:

$$f_n = \frac{1}{N_{FFT}} \sum_{k=0}^{N_{FFT}-1} \kappa(i\alpha_{bil}\tan k\pi/N_{FFT})\, e^{2\pi ikn/N_{FFT}} \tag{11.56}$$

Each entry of the transfer function has to be evaluated (by interpolation) for the frequencies (Equation 4.116)

$$\omega_k = \alpha_{bil}\tan(k\pi/N_{FFT}) \qquad k = 0, \ldots, N_{FFT}/2 \tag{11.57}$$

The frequency point for $k = N_{FFT}/2$ is at $\omega \to \infty$. The transfer function for the other frequencies can be found by the complex conjugate as $F(k) = \bar{F}(N_{FFT} - k)$.

b) Form discrete-time Hankel matrix (Equation 7.1) with N_{Hank} blocks and of total size $N_\sigma \times N_\sigma$.

$$\mathbf{\Gamma} = \begin{bmatrix} \mathbf{f}_1 & \ldots & \mathbf{f}_{N_{Hank}} \\ \vdots & & \vdots \\ \mathbf{f}_{N_{Hank}} & \ldots & \mathbf{f}_{2N_{Hank}-1} \end{bmatrix} \tag{11.58}$$

and shifted Hankel matrix (Equation 7.2)

$$\check{\mathbf{\Gamma}} = \begin{bmatrix} \mathbf{f}_2 & \ldots & \mathbf{f}_{N_{Hank}+1} \\ \vdots & & \vdots \\ \mathbf{f}_{N_{Hank}+1} & \ldots & \mathbf{f}_{2N_{Hank}} \end{bmatrix} \tag{11.59}$$

c) Perform singular value decomposition of Hankel matrix (Equation 7.4)

$$\mathbf{\Gamma} = \begin{bmatrix} \mathbf{U}_1 & \mathbf{U}_2 \end{bmatrix} \begin{bmatrix} \mathbf{\Sigma}_1 & \\ & \mathbf{\Sigma}_2 \end{bmatrix} \begin{bmatrix} \mathbf{V}_1^{\mathrm{T}} \\ \mathbf{V}_2^{\mathrm{T}} \end{bmatrix} \tag{11.60}$$

d) Determine the degree N of the approximating system by the condition on the "sum of the tail"

$$\sum_{i=N+1}^{N_\sigma} \sigma_i \leq tol\, \sigma_1 \tag{11.61}$$

A suggested value for the tolerance is $tol = 0.005$.

e) Find system matrices by the balanced realization method Equations (7.10–7.13)

$$\tilde{\mathbf{A}} = \mathbf{\Sigma}_1^{-1/2} \mathbf{U}_1^{\mathrm{T}} \check{\mathbf{\Gamma}} \mathbf{V}_1 \mathbf{\Sigma}_1^{-1/2} \tag{11.62}$$

$$\tilde{\mathbf{B}} = \text{first block column of } \mathbf{\Sigma}_1^{1/2} \mathbf{V}_1^{\mathrm{T}} \tag{11.63}$$

$$\tilde{\mathbf{C}} = \text{first block row of } \mathbf{U}_1 \mathbf{\Sigma}_1^{1/2} \tag{11.64}$$

$$\tilde{\mathbf{D}} = \mathbf{h}_0 \tag{11.65}$$

f) Transform discrete-time system matrices into continuous-time matrices (Equations 6.163–6.166)

$$\mathbf{A} = \alpha_{bil} (\mathbf{I} + \tilde{\mathbf{A}})^{-1} (\tilde{\mathbf{A}} - \mathbf{I}) \tag{11.66}$$

$$\mathbf{B} = \sqrt{2\alpha_{bil}}\, (\mathbf{I} + \tilde{\mathbf{A}})^{-1} \tilde{\mathbf{B}} \tag{11.67}$$

$$\mathbf{C} = \sqrt{2\alpha_{bil}}\, \tilde{\mathbf{C}} (\mathbf{I} + \tilde{\mathbf{A}})^{-1} \tag{11.68}$$

$$\mathbf{D} = \tilde{\mathbf{D}} - \tilde{\mathbf{C}} (\mathbf{I} + \tilde{\mathbf{A}})^{-1} \tilde{\mathbf{B}} \tag{11.69}$$

3) Change linear system to symmetric second-order form (Chapter 8)

a) Diagonalize the approximated system (Equation 8.53)

$$\hat{\mathbf{C}}(s\mathbf{I} - \mathbf{\Lambda})^{-1} \hat{\mathbf{B}} = \begin{bmatrix} \gamma_1 & \gamma_2 & \cdots \end{bmatrix} \begin{bmatrix} s - \lambda_1 & & \\ & s - \lambda_2 & \\ & & \ddots \end{bmatrix} \begin{bmatrix} \boldsymbol{\beta}_1^{\mathrm{T}} \\ \boldsymbol{\beta}_2^{\mathrm{T}} \\ \vdots \end{bmatrix} \tag{11.70}$$

b) For each real pole λ_k compute (Equation 8.42)

$$a_0^{(k)} = \pm 1 \qquad a_1^{(k)} = -a_0^{(k)} \lambda_k \qquad a_2^{(k)} = 0 \tag{11.71}$$

With the temporary variable (Equation 8.37)

$$t_k = \frac{\boldsymbol{\gamma}_k^{\mathrm{T}} \boldsymbol{\beta}_k}{\boldsymbol{\beta}_k^{\mathrm{T}} \boldsymbol{\beta}_k} \tag{11.72}$$

compute (Equation 8.41)

$$\mathbf{b}_1^{(k)} = \boldsymbol{\beta}_k \sqrt{\frac{a_0^{(k)}\, t_k}{\lambda_k}} \qquad \mathbf{b}_2^{(k)} = \mathbf{0} \tag{11.73}$$

The sign of $a_0^{(k)}$ has to chosen such that $\mathbf{b}_1^{(k)}$ is real.

c) For each pair of complex conjugate poles λ_k and $\bar{\lambda}_k$ compute (Equation 8.47)

$$a_0^{(k)} = 1 \qquad a_1^{(k)} = -(\lambda_k + \bar{\lambda}_k) \qquad a_2^{(k)} = \lambda_k \bar{\lambda}_k \tag{11.74}$$

With the temporary variables (Equation 8.37)

$$t_k = \frac{\boldsymbol{\gamma}_k^{\mathrm{T}} \boldsymbol{\beta}_k}{\boldsymbol{\beta}_k^{\mathrm{T}} \boldsymbol{\beta}_k} \tag{11.75}$$

and (Equation 8.45)

$$\mathbf{e}_k = \boldsymbol{\beta}_k \sqrt{\frac{t_k}{\lambda_k - \bar{\lambda}_k}} \tag{11.76}$$

compute (Equation 8.46)

$$\mathbf{b}_1^{(k)} = \mathbf{e}_k + \bar{\mathbf{e}}_k \qquad \mathbf{b}_2^{(k)} = -(\bar{\lambda}_k \, \mathbf{e}_k + \lambda \bar{\mathbf{e}}_k) \tag{11.77}$$

d) Assemble matrices for approximated farfield, neglecting unstable poles

$$\mathbf{A}_0 = \mathbf{diag}(a_0^{(1)}, a_0^{(2)}, \ldots) \tag{11.78}$$

$$\mathbf{A}_1 = \mathbf{diag}(a_1^{(1)}, a_1^{(2)}, \ldots) \tag{11.79}$$

$$\mathbf{A}_2 = \mathbf{diag}(a_2^{(1)}, a_2^{(2)}, \ldots) \tag{11.80}$$

$$\mathbf{B}_1 = \begin{bmatrix} \mathbf{b}_1^{(1)\mathrm{T}} \\ \mathbf{b}_1^{(2)\mathrm{T}} \\ \vdots \end{bmatrix} \qquad \mathbf{B}_2 = \begin{bmatrix} \mathbf{b}_2^{(1)\mathrm{T}} \\ \mathbf{b}_2^{(2)\mathrm{T}} \\ \vdots \end{bmatrix} \tag{11.81}$$

$$\mathbf{D}_2 = \mathbf{D} - \mathbf{B}_1^{\mathrm{T}} \mathbf{B}_1 \tag{11.82}$$

4) Perform analysis of whole fluid-solid system (Chapter 11)

a) Compute finite element matrices for the nearfield fluid (Equations 10.30–10.33)

$$\mathbf{M}_{ff} = \frac{\rho}{c^2} \int_\Omega \mathbf{N}^{\mathrm{T}} \mathbf{N} \, d\Omega \tag{11.83}$$

$$\mathbf{C}_{ff} = q\rho \int_{\Gamma_v} \mathbf{N}^{\mathrm{T}} \mathbf{N} \, d\Gamma \tag{11.84}$$

$$\mathbf{K}_{ff} = \rho \int_\Omega \mathbf{B}^{\mathrm{T}} \mathbf{B} \, d\Omega \tag{11.85}$$

$$\mathbf{V} = \rho \int_{\Gamma_v} \mathbf{N}^{\mathrm{T}} v_n \, d\Gamma \tag{11.86}$$

for the solid (Equations 10.40–10.42)

$$\mathbf{M}_{ss} = \int_\Omega \mathbf{N}_s^{\mathrm{T}} \mathbf{N}_s \, d\Omega \tag{11.87}$$

$$\mathbf{K}_{ss} = \int_\Omega \mathbf{B}_s^{\mathrm{T}} \mathbf{D} \mathbf{B}_s \, d\Omega \tag{11.88}$$

$$\mathbf{F}_s = \int_{\Gamma_\sigma} \mathbf{N}_s^{\mathrm{T}} \mathbf{f} \, d\Gamma \tag{11.89}$$

and for the interface (Equations 10.45 and 10.47)

$$\mathbf{C}_{sf} = -\mathbf{C}_{fs}^{\mathrm{T}} = \rho \int_{\Gamma_v} \mathbf{N}_s^{\mathrm{T}} \mathbf{n} \mathbf{N} \, d\Gamma \tag{11.90}$$

b) Assemble matrices for coupled fluid-solid system (Equations 11.20–11.23)

$$\mathbf{M} = \begin{bmatrix} \mathbf{M}_{ss} & & & \\ & -\mathbf{M}_{ff}^{(11)} & -\mathbf{M}_{ff}^{(12)} & \\ & -\mathbf{M}_{ff}^{(21)} & -\mathbf{M}_{ff}^{(22)} & \\ & & & \mathbf{A}_0 \end{bmatrix} \tag{11.91}$$

$$\mathbf{C} = \begin{bmatrix} \mathbf{C}_{ss} & \mathbf{C}_{sf} & & \\ \mathbf{C}_{sf}^{\mathrm{T}} & -\mathbf{C}_{ff}^{(11)} & -\mathbf{C}_{ff}^{(12)} & \\ & -\mathbf{C}_{ff}^{(21)} & -(\mathbf{C}_{ff}^{(22)} + \mathbf{T}\mathbf{D}_1\mathbf{T}^{\mathrm{T}}) & -\mathbf{T}\mathbf{B}_1^{\mathrm{T}} \\ & & -\mathbf{B}_1\mathbf{T}^{\mathrm{T}} & \mathbf{A}_1 \end{bmatrix} \tag{11.92}$$

$$\mathbf{K} = \begin{bmatrix} \mathbf{K}_{ss} & & & \\ & -\mathbf{K}_{ff}^{(11)} & -\mathbf{K}_{ff}^{(12)} & \\ & -\mathbf{K}_{ff}^{(21)} & -(\mathbf{K}_{ff}^{(22)} + \mathbf{T}\mathbf{D}_2\mathbf{T}^{\mathrm{T}}) & -\mathbf{T}\mathbf{B}_2^{\mathrm{T}} \\ & & -\mathbf{B}_2\mathbf{T}^{\mathrm{T}} & \mathbf{A}_2 \end{bmatrix} \tag{11.93}$$

$$\mathbf{F} = \begin{bmatrix} \mathbf{F}_s \\ -\mathbf{V}^{(1)} \\ -\mathbf{T}\mathbf{D}_1\mathbf{T}^{\mathrm{T}}\dot{\mathbf{\Phi}}^p - \mathbf{T}\mathbf{D}_2\mathbf{T}^{\mathrm{T}}\mathbf{\Phi}^p \\ -\mathbf{B}_1\mathbf{T}^{\mathrm{T}}\dot{\mathbf{\Phi}}^p - \mathbf{B}_2\mathbf{T}^{\mathrm{T}}\mathbf{\Phi}^p \end{bmatrix} \tag{11.94}$$

c) Perform time-stepping procedure (Section 11.3) for the two systems, that is, the cross-section of the channel and the complete dam-reservoir system, in parallel. The differential equation for the cross-section is (Equations 11.26 and 11.27)

$$\mathbf{M}_I \, \ddot{\mathbf{\Phi}}^p + \mathbf{C}_I \, \dot{\mathbf{\Phi}}^p + \mathbf{K}_I \, \mathbf{\Phi}^p = -\mathbf{C}_{fs}^I \, \mathbf{v}_g \tag{11.95}$$

The results $\mathbf{\Phi}^p$ are used in the right-hand-side vector of the dam-reservoir system. The differential equation for the dam-reservoir system is

$$\mathbf{M}\,\ddot{\mathbf{d}} + \mathbf{C}\,\dot{\mathbf{d}} + \mathbf{K}\,\mathbf{d} = \mathbf{F} \tag{11.96}$$

with the matrices $\mathbf{M}, \mathbf{C}, \mathbf{K}$ and the vector $\mathbf{F}$ given above, and the vector of nodal variables $\mathbf{d} = [\mathbf{u} \;\; \varphi^{(1)} \;\; \hat{\varphi}^{(2)} \;\; \hat{\zeta}]^{\mathrm{T}}$.

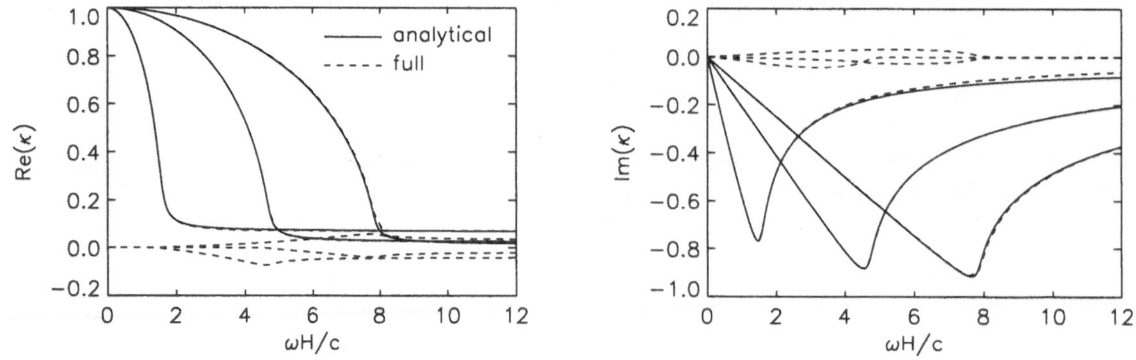

Figure 11.1: Transfer matrix $\boldsymbol{\kappa}$

11.5 Examples with rigid dam

In this section numerical results are compared to analytical results derived in Chapter 9. These examples assume a rigid dam. Examples with a flexible dam are given in the next section. The influence of the different approximations introduced in different stages of the analysis are investigated. These approximations involve: finite element discretization, diagonalization of the transfer matrix $\boldsymbol{\kappa}$ and rational approximation of $\boldsymbol{\kappa}$.

Identical parameters are used for the system approximation of all cases: $N_{freq} = 100$, $N_{FFT} = 512$, $N_{Hank} = 10$, $\alpha_{bil}H/c = 10 \cdot 2\pi \cdot 100/1440 = 4.36$. However, the number of modes and the degree of the system approximation depend on the individual examples. The degree N of the (first-order) system approximation is determined by the ("sum of the tail") criterion $\sum_{i=N+1}^{N_\sigma} \sigma_i \leq 0.005\,\sigma_1$.

11.5.1 Two-dimensional fluid domain with constant depth

In this subsection the two-dimensional fluid problem introduced in Chapter 10 is analyzed further. The mathematical model and the analytical solutions have beeen presented in Section 9.2 and the finite element discretization has been shown in Figure 10.2. The reflection coefficient at the foundation-reservoir interface is $\alpha = 0.8$. For all calculations with this model, six Ritz vectors, corresponding to six modes, are considered for the farfield.

Diagonalization. Before analyzing the influence of the rational system approximation we first address the question concerning the effects of diagonalizing the transfer matrix $\boldsymbol{\kappa}$ in the frequency domain. Figure 11.1 shows the full matrix $\boldsymbol{\kappa}$. The diagonal terms of the first three modes and their coupling terms are plotted as dashed lines. The diagonal terms are almost identical to the analytical curves and the coupling terms are very small indicating that the eigenvectors are almost frequency-independent for the small bottom absorption considered here ($\alpha = 0.8$). Therefore, it may be concluded that the diagonal part of the matrix could be used neglecting the coupling terms.

To investigate the question further, transfer functions of dynamic pressure are compared in Figure 11.2. The figure shows the pressure amplitudes at the point indicated in the finite element mesh (Figure 10.2). As in Chapter 9, the transfer functions are defined as $T_x = \hat{p}/(H\rho\hat{a}_x)$ and analogously for the y-direction. The left figure is for a horizontal, the right for a vertical excitation. Both curves, the one using the full matrix and the one using the diagonal matrix, compare well with the analytical solution, the full matrix showing only slightly better results. The diagonal form is much simpler for the approximation procedure because each mode can be treated individually and only scalar systems

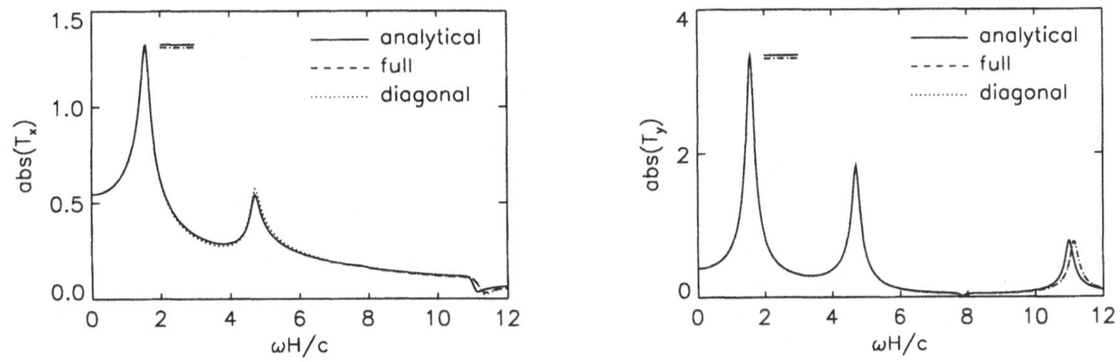

Figure 11.2: Full versus diagonal transfer matrix $\boldsymbol{\kappa}$. Left: horizontal excitation. Right: vertical excitation.

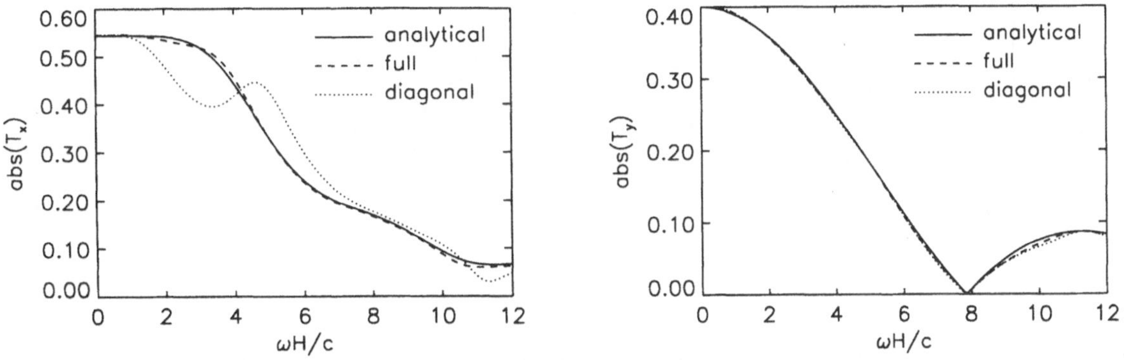

Figure 11.3: Full versus diagonal transfer matrix $\boldsymbol{\kappa}$ for full bottom absorption. Left: upstream excitation. Right: vertical excitation.

are involved whereas the full matrix requires the approximation of a multivariable system. Looking more closely, one can see a frequency shift between the finite element curves ('full' and 'diagonal') and the analytical curves due to the finite element discretization, which has already been observed earlier (Figure 10.3). The diagonalization itself introduces almost no additional error so that the dashed and the dotted lines appear as one dash-dotted line.

An example where the diagonalization is no longer appropriate is the model with full bottom absorption (no reflection, $\alpha = 0$, $cq = 1$). Figure 11.3 shows the pressure transfer functions due to an upstream (left) and due to a vertical (right) excitation. The transfer functions obtained by the finite element analysis using a full or diagonal transfer matrix $\boldsymbol{\kappa}$ are compared to the analytical results. Clearly, using the full transfer matrix $\boldsymbol{\kappa}$ yields a result close to the analytical one, whereas a diagonal matrix leads to different values, at least for the horizontal excitation, and should not be used here. The difference occurs because the eigenvectors are strongly frequency-dependent due to the frequency-dependent boundary condition at the bottom.

Rational approximation. Sofar we have not employed the rational system approximation yet. This is done in the following investigation. Figure 11.4 shows the approximation of the first three of the total six diagonal terms. The curves compare very well with the analytical solution. The 6 diagonal terms, approximated individually, lead to a total of 14 internal variables (2 real and 12 pairs of complex conjugate poles).

Although the rational approximation is performed for the transfer matrix $\boldsymbol{\kappa}$, the important curves to compare are the transfer functions for the physical variables. Figure 11.5

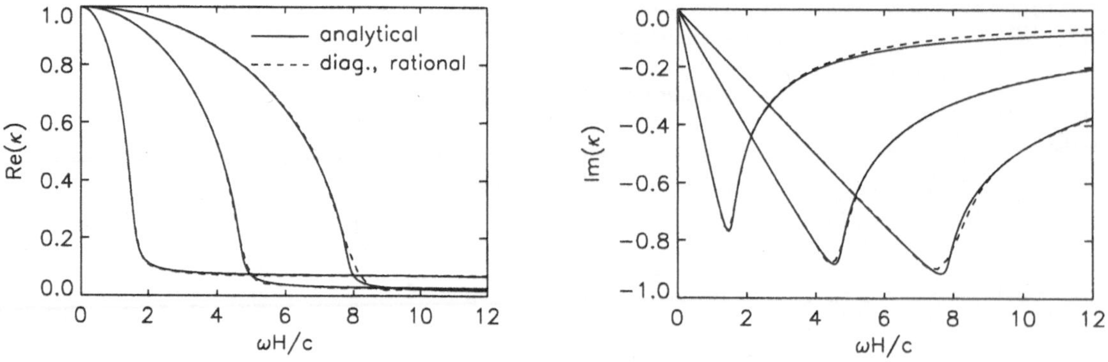

Figure 11.4: System approximation of transfer matrix $\boldsymbol{\kappa}$

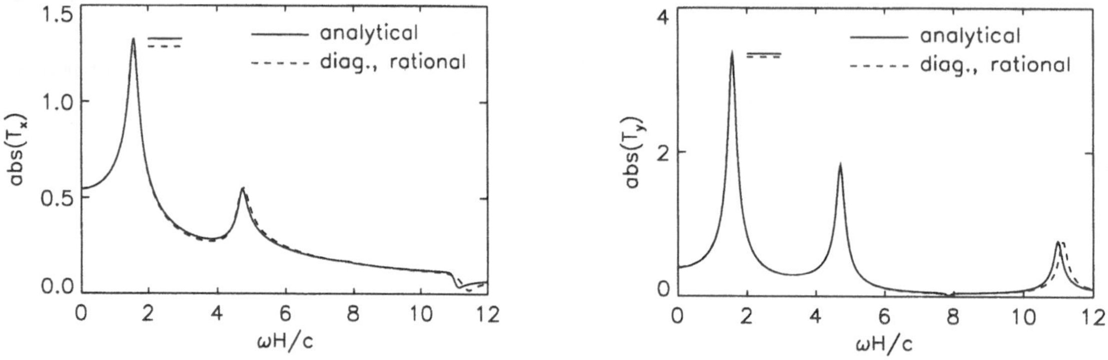

Figure 11.5: Effect of rational approximation. Left: upstream excitation. Right: vertical excitation

shows the pressure amplitudes for a horizontal (left) and for a vertical (right) excitation. The values for an approximated diagonal system are compared to the analytical solution. The approximated solution matches the analytical one very well. As observed earlier, the finite element discretization introduces a frequency shift at higher frequencies. The approximation by the linear system, however, is practically identical to the finite element solution. Obviously, the error introduced by system approximation is much smaller than the one introduced by the finite element discretization. This tendency appears throughout the examples.

As we have seen earlier, the diagonal approximation is no longer appropriate for the semi-infinite reservoir with full bottom absorption (no reflection, $\alpha = 0$). First, we look at the rational approximation of the transfer matrix $\boldsymbol{\kappa}$. Figure 11.6 shows the first 3 diagonal terms and the corresponding coupling terms (symmetry of the matrix) of the finite element solution and their system approximations. For the calculated 6 modes there would be another 15 curves which are not shown. All these 21 curves are approximated simultaneously by a system with 15 internal variables (5 real and 10 pairs of comples conjugate poles). Figure 11.7 shows the transfer functions for an upstream (left) and for a vertical (right) excitation. The approximated solution compares well with the analytical one.

11.5.2 Three-dimensional examples

Rectangular channel. The simplest three-dimensional case is the rectangular channel. Figure 11.8 shows the finite element mesh. The reservoir depth is taken as $H = 100$ m

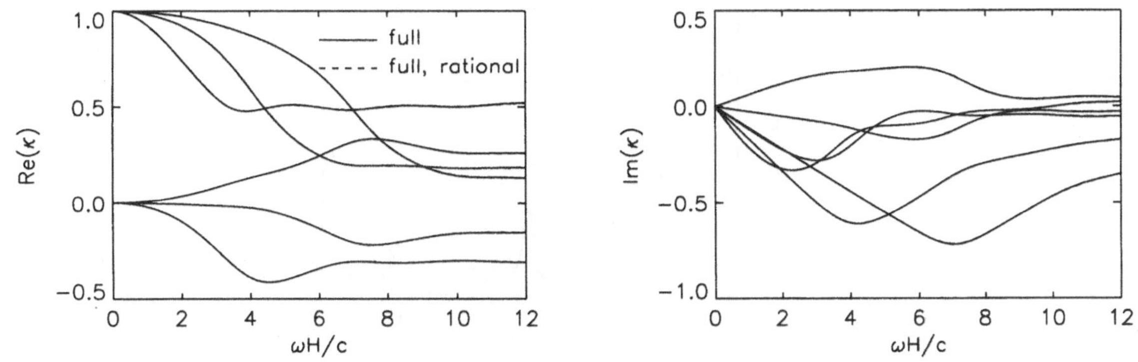

Figure 11.6: Full absorption: rational approximation of transfer matrix $\boldsymbol{\kappa}$

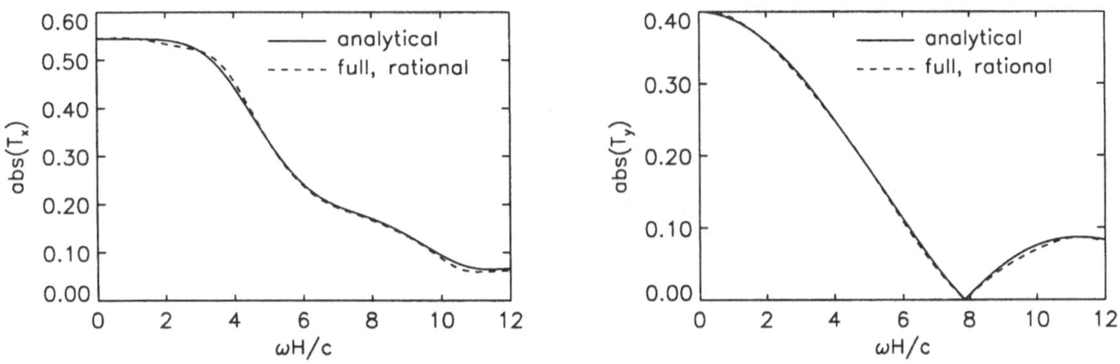

Figure 11.7: Full absorption. Left: upstream excitation. Right: vertical excitation.

and the width as $2W = 160$ m. Results are, however, shown in dimensionless form. They are given at the point indicated in the figure ($y = 60$ m, $z = W = 80$ m). For the analytical solution it has been shown, that for rigid side walls, the solutions for upstream and vertical excitation are the same as in the two dimensional case. These solutions can therefore also be used to verify the three-dimensional model. Because the results are similar to the ones obtained for the two-dimensional case, these comparisons are not shown here. What is shown, however, is the transfer function for a cross-stream excitation. Figure 11.9 compares the approximation with the analytical solution. The 6 diagonal terms are approximated by a total of 17 degrees of freedom (5 real and 12 pairs of complex conjugate poles). The agreement is very good except at high frequencies where the shift due to the finite element discretization shows up.

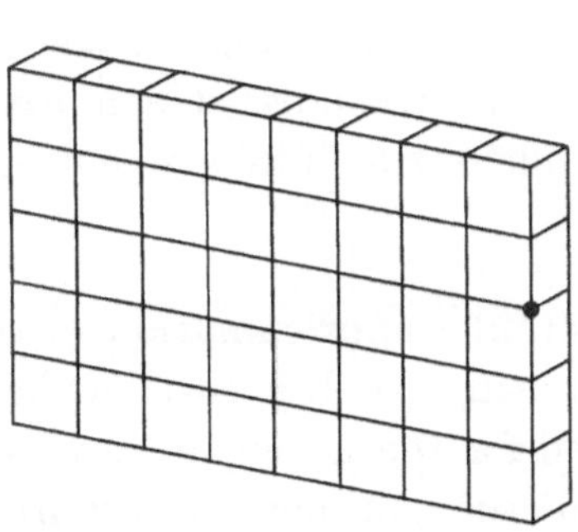

Figure 11.8: Rectangular channel: finite element mesh

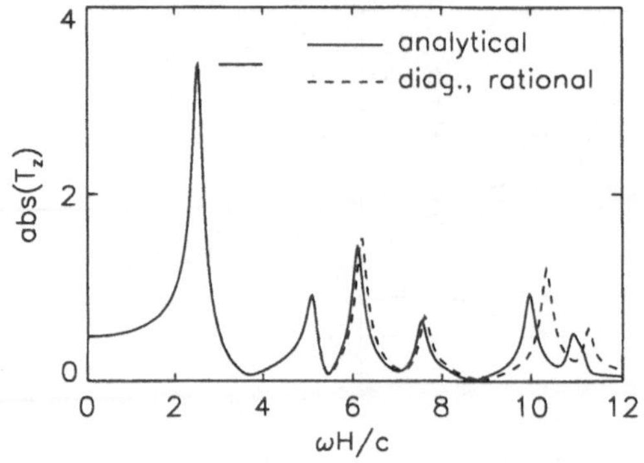

Figure 11.9: Rectangular channel: cross-stream excitation

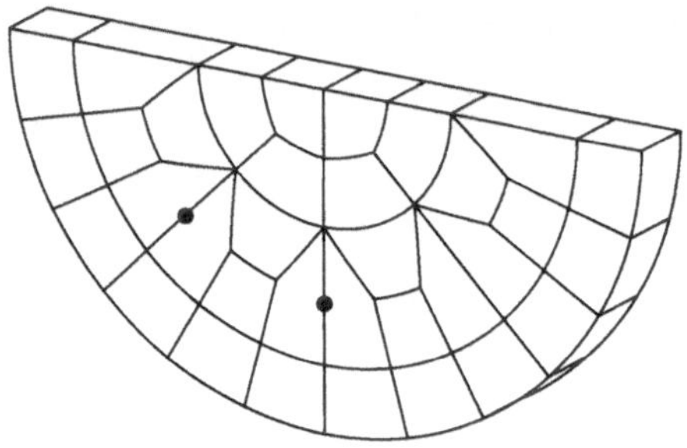

Figure 11.10: Semi-circular channel: finite element mesh

Semi-circular channel. The next verification example is the semi-circular channel. The finite element mesh is shown in Figure 11.10. Although the whole model is shown, only half of it is used for the computations, taking account of symmetry and anti-symmetry by corresponding boundary conditions. The nearfield is modeled by 17 fluid brick elements and by 23 interface elements. For the symmetric case (upstream and vertical) this results in 134, for the anti-symmetric case (cross-stream) in 144 degrees of freedom. The cross-section of the farfield consists of 17 fluid quadrilaterals and 6 interface line elements, resulting in 57 or 49 degrees of freedom depending on the boundary conditions. The diagonal transfer matrix $\boldsymbol{\kappa}$ with 9 modes is used. For the rational approximation, 24 variables are used for the symmetric model and 23 for the anti-symmetric model, respectively. These numbers have to be considered in relation to the number of degrees of freedom of the finite element model. The transfer functions for the upstream and vertical excitation are compared in Figure 11.11 and for the cross-stream excitation in Figure 11.12. Besides the shift for higher frequencies due to the finite element discretization, the curves are very close.

After having verified the rational approximation in the frequency domain, the time-domain analysis is investigated using the Taft earthquake (Figures 11.14 and 11.16). The horizontal component is applied for the upstream direction (first 10 seconds). The time step used for integration is 0.005 sec. As shown in Figure 11.13, the time-domain computation using the system approximation for the farfield virtually coincides with the solution obtained by the analytical solution followed by a Fast Fourier Transform [SW92].

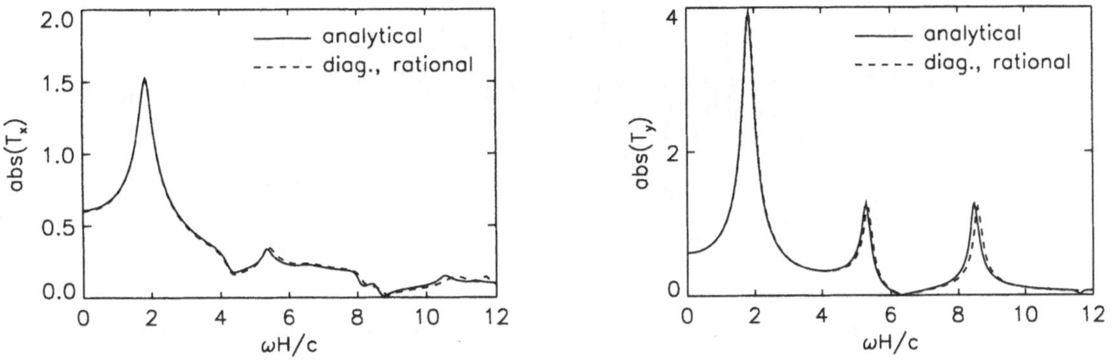

Figure 11.11: Semi-circular channel. Left: upstream excitation. Right: vertical excitation

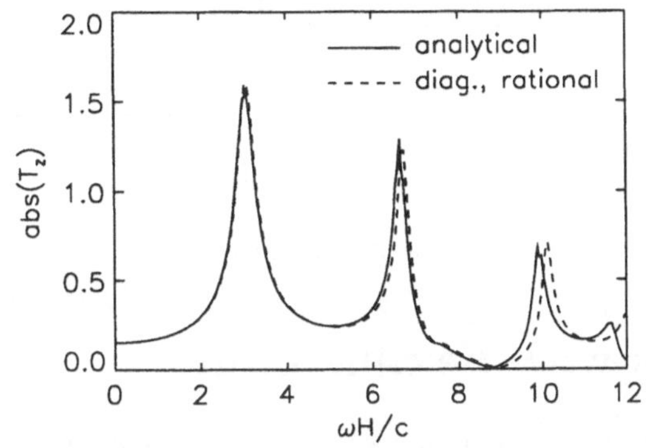

Figure 11.12: Semi-circular channel: cross-stream excitation

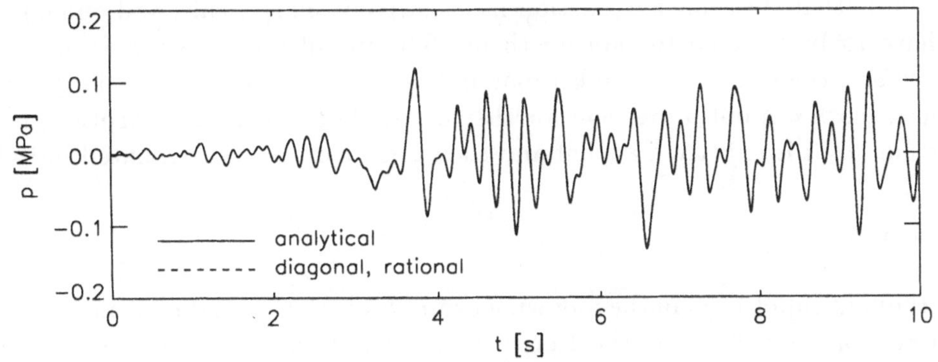

Figure 11.13: Semi-circular channel: upstream excitation

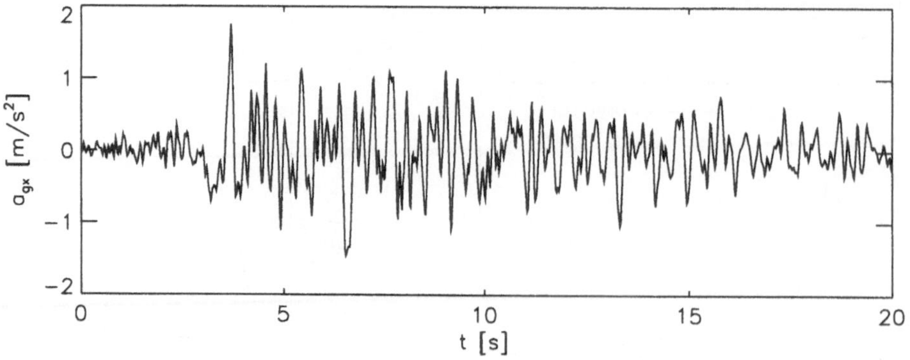

Figure 11.14: Horizontal (S69E component) acceleration of Taft earthquake

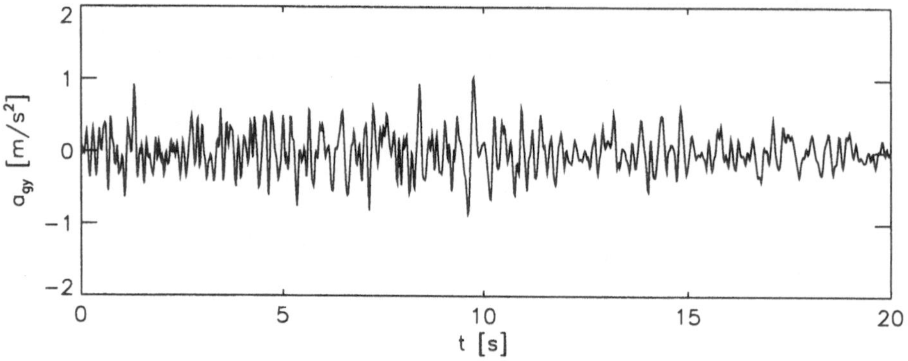

Figure 11.15: Vertical acceleration of Taft earthquake

11.6 Examples with flexible dam

The following examples give a verification of the complete dam-reservoir model. Different analyses in the time domain are compared to each other and to reference solutions from the literature.

For the farfield approximation of models with bottom absorption, parameters are taken the same as in the examples with a rigid dam (Section 11.5), that is: $N_{freq} = 100$, $N_{FFT} = 512$, $N_{Hank} = 10$, $\alpha_{bil} = 10 \cdot 2\pi = 62.8$ and $tol = 0.005$. For examples without bottom absorption, the parameters that are taken differently are $N_{freq} = 250$ and $N_{Hankel} = 80$. The reason is that the transfer function has relatively sharp peaks and enough frequency points have to be selected to capture them. The impulse response dies out very slowly and, therefore, the size of the Hankel matrix has to be taken relatively large. The size of the Hankel matrix is not a problem for the case without bottom absorption, because the transfer matrix $\boldsymbol{\kappa}$ is exactly diagonal and each term can be approximated individually.

11.6.1 Taft earthquake records

The earthquake input used in the following examples is the ground motion recorded at Taft Lincoln School Tunnel during the Kern County, California, earthquake of 21 July 1952. This record has primarily been chosen, to be able to make a comparison with the examples shown in the literature. The horizontal (S69E component) and vertical acceleration of the Taft earthquake is plotted in Figures 11.14 and 11.15, respectively. The corresponding response spectra are shown in Figure 11.16.

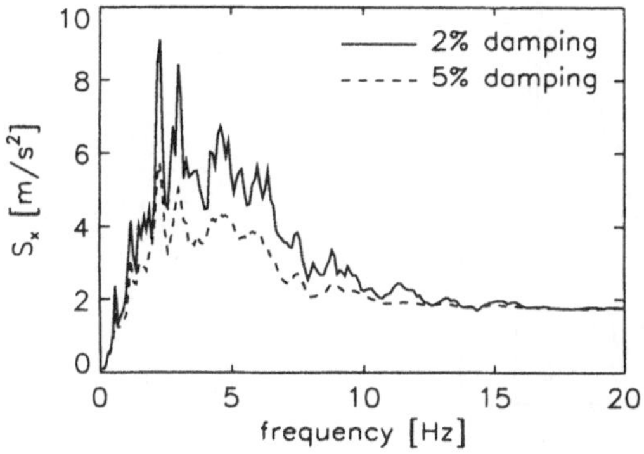

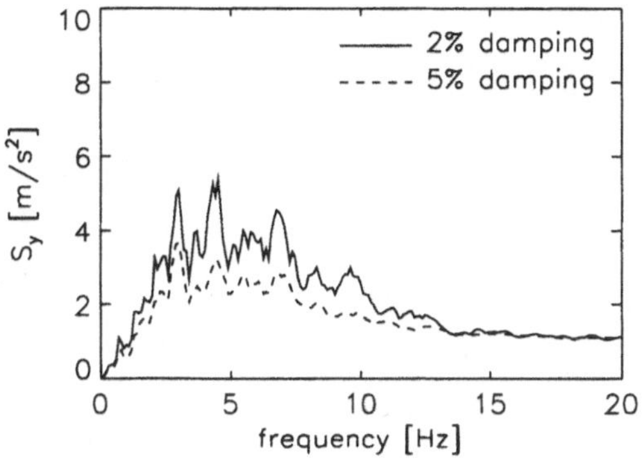

Figure 11.16: Response spectra of horizontal (right) and vertical (left) Taft earthquake

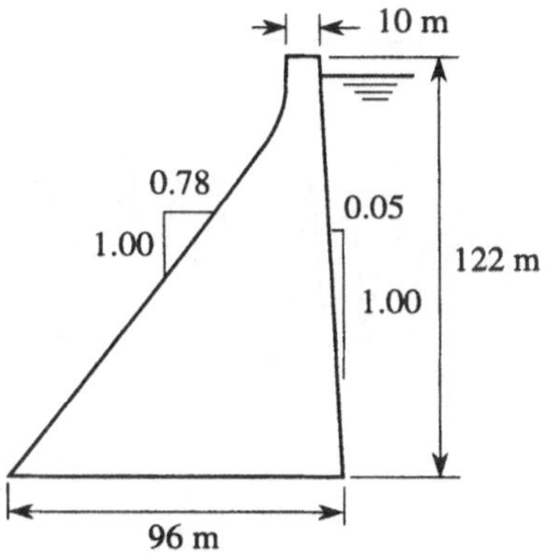

Figure 11.17: Pine Flat dam: geometry of tallest non-overflow monolith [FC84]

11.6.2 Two-dimensional example: Pine Flat dam

The 122 m (400 ft) high Pine Flat dam is described in [FC84]. The geometry of the tallest non-overflow monolith is shown in Figure 11.17. The finite element mesh consisting of 136 quadrilateral solid elements, 16 fluid elements and 17 interface elements is shown in Figure 11.18. A fixed boundary condition is applied to the bottom of the dam (rigid foundation). The number of degrees of freedom is 338 for the nearfield and 16 for the cross-section of the farfield. The material properties are shown in Table 11.1. The factors α_R for mass proportional and β_R for stiffness proportional damping are chosen such that the prescribed damping of $\xi = 5\%$ is obtained at 2 Hz and 3 Hz. The same damping value results approximately for intermediate frequencies. For the approximation of the farfield

water	wave velocity	$c = 1440$ m/s (4720 ft/sec)
	unit weight	$\rho = 1000$ kg/m^3 (62.4 pcf $= 0.00193$ ks^2/ft^4)
concrete	Young's modulus	$E = 2.24 \cdot 10^{10}$ N/m^2 ($3.25 \cdot 10^6$ psi $= 468000$ ksf)
	unit weight	$\rho_s = 2478$ kg/m^3 (155 pcf $= 0.00482$ ks^2/ft^4)
	Poisson's ratio	$\nu = 0.2$
	Rayleigh damping	$\alpha_R = 0.751$ 1/s, $\beta_R = 0.0032$ s
foundation	reflection	$\alpha = 1.0$

Table 11.1: Material properties for Pine Flat dam

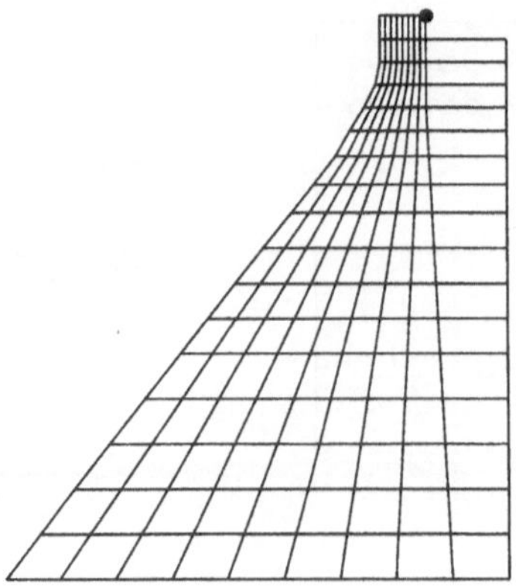

Figure 11.18: Pine Flat dam: finite element mesh

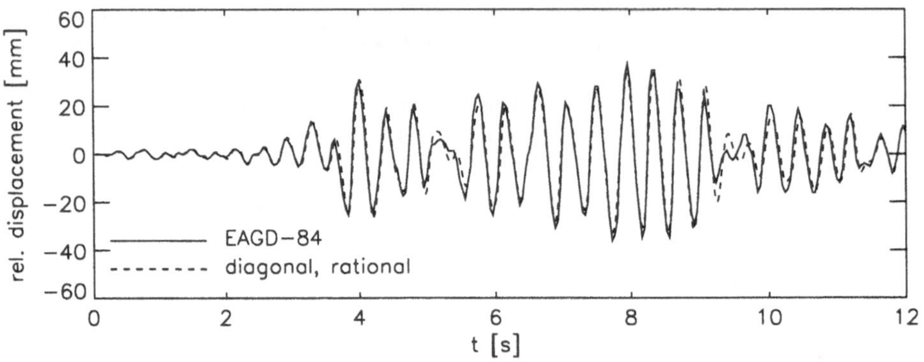

Figure 11.19: Pine Flat dam: upstream excitation

the parameters given at the beginning of this section (p. 216) are taken ($N_{freq} = 250$ and $N_{Hankel} = 80$). The 6 diagonal terms of the transfer matrix $\boldsymbol{\kappa}$ are approximated by 24 internal degrees of freedom (6 real and 18 pairs of complex conjugate poles).

Figure 11.19 shows the horizontal relative displacement at the top of the dam due to the horizontal Taft earthquake record. The time history is compared to the results taken from [FC84, p. 112] (labeled 'EAGD-84'). Considering that a different damping model was used and that the results are discretized from a tiny figure in the report, the agreement is quite satisfactory.

11.6.3 Three-dimensional example: Morrow Point dam

This arch dam example is taken from [FC80] and has already been introduced in Section 10.7.3, where also the general geometry and two finite element meshes, that is, the short and the long reservoir, are shown.

Incompressible reservoir. First, we consider the incompressible case. To verify the implementation we compare two different analyses. For the first analysis a short nearfield (Figure 10.17) is modeled by incompressible fluid elements and the farfield is captured by 9 zero-frequency eigenvalues of the channel cross-section (Equation 11.32). Because, for the incompressible case, the farfield does not introduce any additional degrees of freedom, the total system has 468 degrees of freedom, that is, those of the nearfield. The second analysis is performed by the conventional method using a long reservoir (Figure 10.20) with a fixed boundary condition at the far end. Any boundary condition is valid there because no waves occur and the pressure variation is restricted to the region close to the dam.

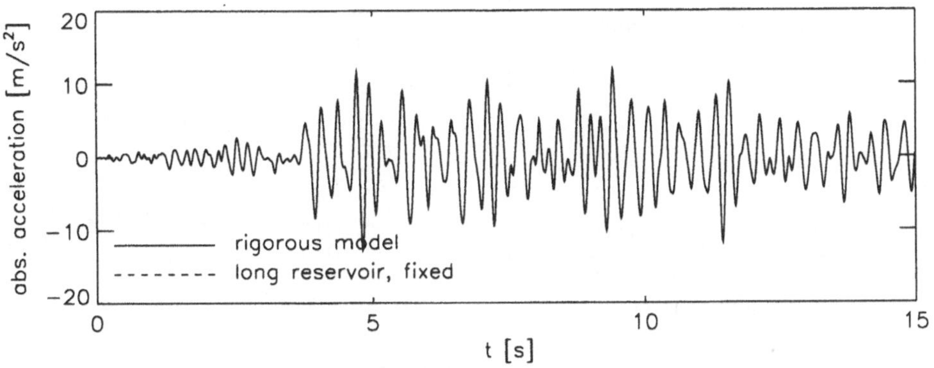

Figure 11.20: Morrow Point dam: incompressible reservoir, rigorous model versus long model with fixed boundary

This model has 823 degrees of freedom. The same comparison has been shown earlier for the frequency domain (Figure 10.21). The time histories of absolute accelerations for the two incompressible analyses are plotted in Figure 11.20. The identical results show that the two formulations are indeed equivalent. The formulation capturing the farfield by the eigenmodes of the channel cross-section is convenient because the same short mesh can be used for both the compressible and the incompressible cases. Also, the short model is much smaller and uses therefore less computer time. In this example the factor in computer time was 1.9, which is approximately proportional to the numbers of degrees of freedom. From an algorithmic view, there is no advantage of the eigenmode solution over the conventional method, because no rational approximation is involved for the incompressible case.

Compressible reservoir. We now turn to the more interesting analysis of compressible water. The short nearfield reservoir model (Figure 10.17) is used. We first investigate the case without bottom absorption ($\alpha = 1$). Then the eigenvectors are frequency-independent and the transfer matrix $\boldsymbol{\kappa}$ is diagonal. We can approximate each diagonal term individually, which simplifies the approximation because only scalar systems are involved. On the other hand, the approximation is more demanding because the transfer functions are not smooth and the impulse response dies out only slowly. Therefore we have to take many frequency points ($N_{freq} = 250$) and many entries in the Hankel matrices ($N_{Hankel} = 80$). Because the entries are only scalars the size of the Hankel matrix is no problem. The other parameters for the farfield approximation are those given at the beginning of this section (p. 216). Taking 9 modes, the number of internal degrees of freedom for the approximation of the farfield is 36 (9 real and 27 pairs of comples conjugate poles). The approximation in the frequency domain is shown in Figure 11.21 on the left, where transfer functions for the radial total acceleration at the dam crest are plotted. The curve labeled 'diagonal' represents the finite element solution without approximation. That curve is compared to the one obtained using the rational approximation (labeled 'rational'). The two transfer functions are almost identical except for a small difference in the peak values as indicated by small horizontal lines.

The second case considered is the one including bottom absorption ($\alpha = 0.5$). The transfer matrix $\boldsymbol{\kappa}$ is no longer diagonal but the impulse response dies out more quickly because of the loss of energy. For the rational approximation the same parameters are used as in the other examples with bottom absorption ($N_{freq} = 100$ and $N_{Hank} = 10$). Again, 9 Ritz vectors are considered. Two analyses are performed, one by approximating

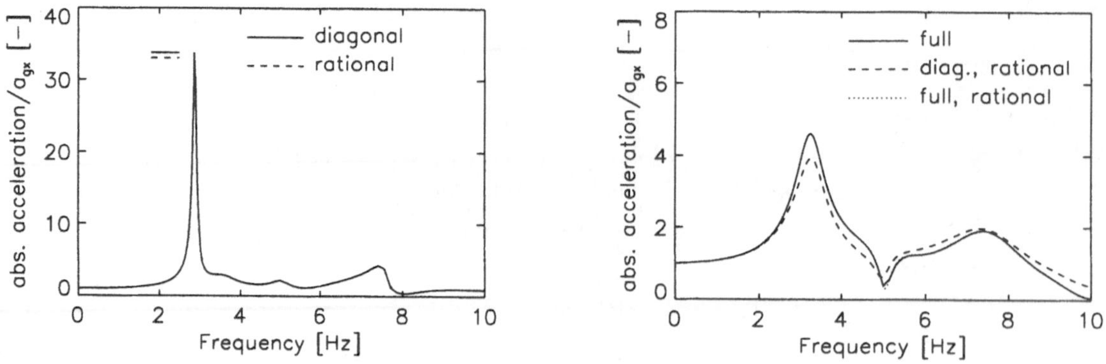

Figure 11.21: Morrow Point dam: transfer functions for radial crest acceleration due to upstream excitation. Left: $\alpha = 1$. Right: $\alpha = 0.5$.

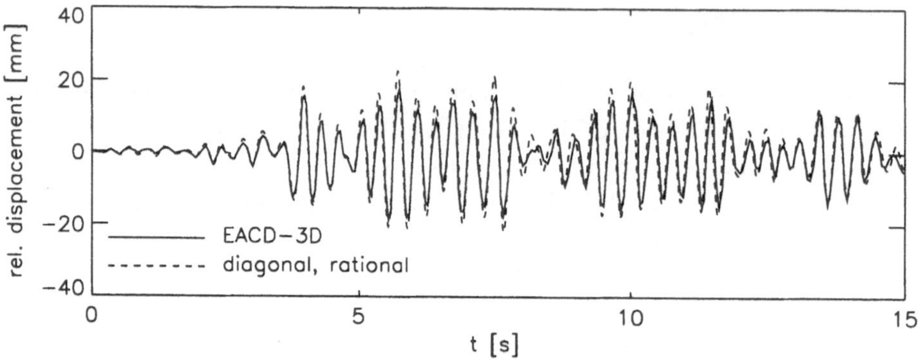

Figure 11.22: Morrow Point dam: $\alpha = 1.0$, upstream excitation

the full matrix by a multi-variable system and the other by considering only the diagonal terms, approximating each of them individually. The two approximations are compared to the exact frequency-domain solution using the full transfer matrix $\boldsymbol{\kappa}$. The comparison is shown as a transfer function for the horizontal acceleration in Figure 11.21 on the right. The approximation of the full transfer matrix $\boldsymbol{\kappa}$ yields results almost identical to the exact solution, whereas the diagonal approximation is somewhat different but still satisfactory. Approximating the full matrix involves a 90×90 Hankel matrix and leads to 26 internal degrees of freedom (6 real 20 paires of complex conjugate poles). It takes considerably more computer time (about 50 times) than the diagonal approximation, which involves 9 Hankel matrices of size 10×10 each and results in 24 (5 first-order and 19 second-order) internal degrees of freedom. The computational effort for the subsequent time-stepping procedure is, however, practically identical for both approximations because they use approximately the same number of internal degrees of freedom. Also, because the rational approximation takes up only a negligible part of the whole computer time, the greater effort for the approximation of the full transfer matrix $\boldsymbol{\kappa}$ is not really a drawback.

Instead of showing the same comparison in the time domain, the results of the time-stepping analysis are compared to results from the literature [FC80, p. 124]. We first consider the case without bottom absorption. Figure 11.22 shows the radial crest displacement due to the horizontal component of the Taft earthquake record. Figure 11.23 shows the same result for the vertical excitation. The curves compare quite well with the results taken from the literature (labeled 'EACD-3D').

The next two figures pertain to the case with bottom absorption ($\alpha = 0.5$) and diagonal

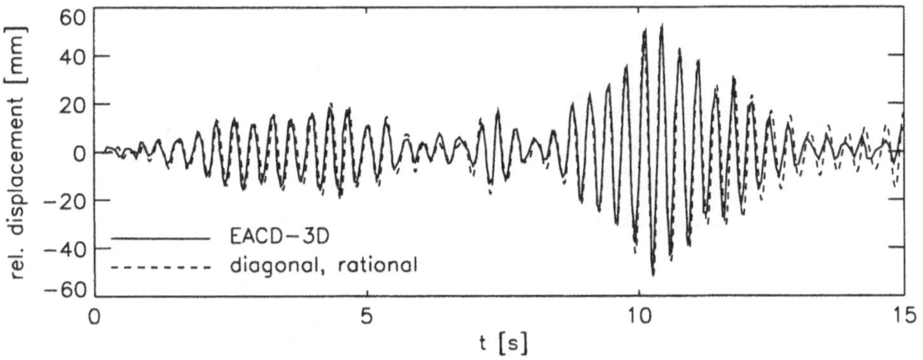

Figure 11.23: Morrow Point dam: $\alpha = 1.0$, vertical excitation

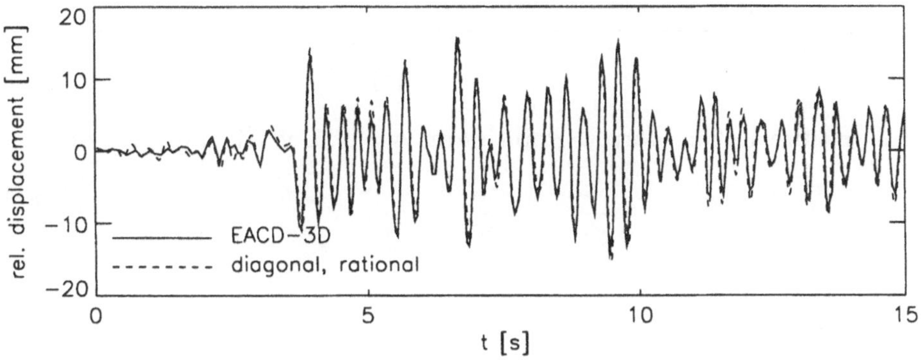

Figure 11.24: Morrow Point dam: $\alpha = 0.5$, upstream excitation

approximation. Time histories for upstream and vertical input are shown in Figures 11.24 and 11.25, respectively. Again, they compare fairly well with the results from the literature, keeping in mind that different damping models have been used.

This full-size example also shows the efficiency of the proposed method. In the example with bottom absorption the interaction of the nearfield with the farfield is captured by 24 (diagonal approximation) or 26 (full approximation) internal degrees of freedom, which are connected to the nearfield. Clearly, this is only a small number compared to the 468 degrees of freedom for the nearfield. The cross-section system, which determines the velocity input from the farfield into the nearfield, has 52 degrees of freedom. This system, however, can be solved with little effort because it is independent of the main system. The time step used is 0.005 seconds, leading to 3000 time steps for the total analysis of 15 seconds. Of course, there is some overhead to find the rational approximation. But this is really negligible compared to the time integration, as the time log shows. On a VAX 9000 using a vector processor, the frequency-domain analysis of the farfield took 1.2 seconds, the system approximation 0.3 seconds for the diagonal approximation and 15 seconds for the full approximation. This is almost negligible compared to a total computing time of about a quarter of an hour. The time-stepping analysis for the cross-section system took about 20 seconds.

Comparison of compressibilty. To complete the picture we compare the results of the incompressible and the compressible reservoir. Surprisingly, at first sight, is the fact that the difference is not too large for the displacement time histories as shown in Figure 11.26. Although the period is slightly different, the maximum values are quite similar. An explanation of this fact is found considering Figure 10.19 which compares

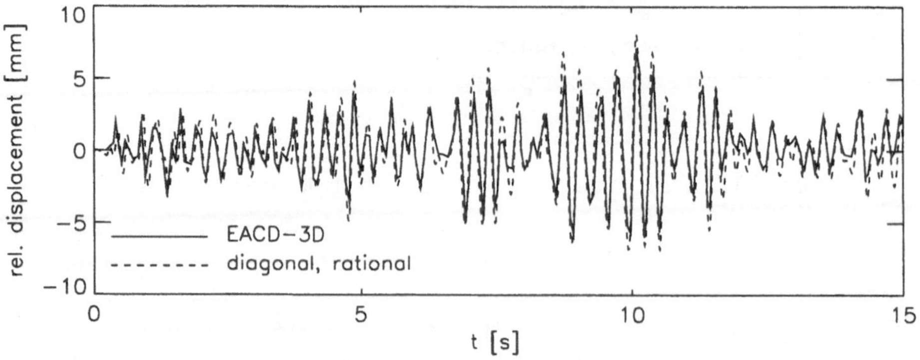

Figure 11.25: Morrow Point dam: $\alpha = 0.5$, vertical excitation

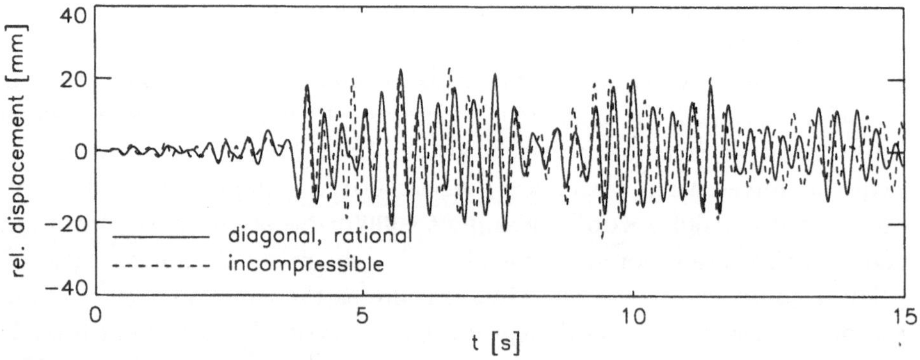

Figure 11.26: Morrow Point dam: compressible versus incompressible model, displacement due to upstream excitation

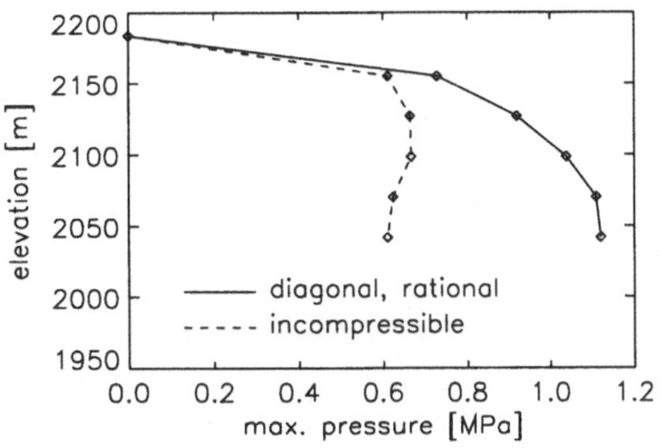

Figure 11.27: Morrow Point dam: compressible versus incompressible model, maximum dynamic pressures due to upstream excitation

the accelerations in the frequency domain. Although the maximum values of the peaks are different, their areas are quite similar, leading to similar maximum displacements in an actual earthquake analysis. The second peak of the incompressible case is of minor importance for the displacements but would, however, influence the acceleration. More pronounced is the difference between the compressible and the incompressible cases for the dynamic pressure. The maximum values of the pressure along the center-line of the upstream face of the dam are plotted in Figure 11.27. This example indicates once more the importance of compressibility effects, but shows at the same time that the results also depend strongly on the earthquake input and the kind of parameter observed.

Chapter 12

Closure

12.1 Conclusions

- Various transmitting boundaries for time-domain analysis proposed in the literature are inadequate for three-dimensional dam-reservoir interaction. Local, non-consistent boundaries and infinite elements generally give incorrect results, whereas convolution and boundary elements are highly inefficient. Also, incompressible models are not generally appropriate.

- Efficient and accurate transmitting boundaries can be formulated by linear time-invariant systems. These systems have a rational transfer function and can therefore readily be implemented in the time domain. The name rational boundaries is introduced by the author for such boundaries. Discrete-time systems lead directly to recursive difference equations, whereas continuous-time systems lead to differential equations which are solved by standard time integration algorithms. Formulating rational boundaries as continuous-time systems is preferred because the steps of approximation and discretization are separated. The approximation is simpler because both the original frequency-domain solution and the approximant are non-periodic. Because the approximation is independent of the time step, a variable time step can be used in the analysis.

- Since most numerical approximation methods are simpler for discrete-time systems, it is advantageous to map the continuous-time system to a discrete-time system by the bilinear transform. After the approximation, the system can be transformed back to the continuous-time domain. Because the infinite frequency axis is mapped onto a circle, it is possible to include the whole frequency axis without an unnecessarily large number of sampling points. Capturing the high-frequency behavior is a necessary condition to obtain a causal system. Causality, on the other hand, is equivalent to stability for rational systems.

- Simple approximation methods based on least-squares lead to reasonable results but are not quite satisfactory. The degree of the approximant has to be chosen in advance and small perturbations may lead to unstable poles. More advanced methods are based on the singular value decomposition of the Hankel matrix. The Carathéodory-Fejér (CF) method is very accurate but not easily extensible to multivariable systems. The balanced realization method is slightly less accurate but numerically preferable and equally applicable to scalar and to multivariable systems.

- For the coupling with the finite element matrices, it is convenient to transform the first-order to a symmetric second-order system. To reduce the numerical effort, a

Rayleigh-Ritz formulation can be employed. For small damping, the effort can further be reduced by diagonalizing the impedance matrix.

- Highly accurate and efficient results are obtained when compared to analytical examples and analyses from the literature. The approximation error is generally less than the one due to the finite element discretization. Only few internal degrees of freedom are necessary, making the implementation very efficient.

12.2 Future Developments

Several topics that are more or less related to the subject are not included in this work. They span from necessary complements to desirable extensions.

12.2.1 Extensions of the method

Although it is very efficient and reliable, the proposed method for constructing a rational transmitting boundary can still be improved in several directions.

- The algorithm depends on several parameters such as frequencies used for calculating the transfer function, number of frequencies for the FFT and size of the Hankel matrix, that have to be chosen by the user. A standard set of parameters has been established based on experience with the problems presented in this work, but they may not be appropriate for other examples. As experience grows some of these parameters could be chosen automatically by the computer program, while others will still need interaction by the user.

- For large transmitting boundaries the block Hankel matrix becomes quite large and may lead to problems regarding computer resources and numerical accuracy. The Ritz technique proposed in this work is one possibility to alleviate the problems. Other techniques might be necessary for different problems. One idea is to split the Hankel matrix and to approximate subregions of the transmitting boundary, so that one is somewhere in between the two extremes of approximating the whole boundary matrix at once and approximating each individual term. Another idea is to take advantage of the fact that the individual terms of the boundary matrix are related to each other. Due to linearity of the problem, the response in one point can either be calculated directly from the input at another point or indirectly by passing through a third point. More precisely, $H_{ij}(s) = H_{ik}(s)\, H_{kj}(s)$.

- The ultimate radiation boundary would be a consistent boundary directly formulated in the time domain without first calculating a frequency-domain solution. The idea is to approximate the local boundary operator Equation (3.2) similar to Engquist and Majda but to use a rational approximation including odd and even powers of $i\xi = ic\lambda/\omega$. The continued fraction approximation used by Engquist and Majda corresponds to a rational approximation with even powers of ξ, which is not able to capture the imaginary part of the local boundary operator. Because λ and ω correspond to partial derivates with respect to y and t, respectively, it should be possible to find a boundary condition that can be applied directly to the transmitting boundary without first solving an eigenvalue problem. However, at the time it is not clear whether this idea really works.

12.2.2 Extensions of the general computer program

The methods and the developed computer program are part of an ongoing research project on the dynamic behavior of dams. Several extensions are necessary in this respect.

- □ The fluid formulation with the velocity potential has several advantages including beeing a one-variable field and symmetric. On the other hand, there are also some drawbacks with this formulation. Firstly, to include the static case by a constant acceleration a linearly increasing velocity input must be specified which might quickly lead to undesirably large numbers. Secondly, if the fluid exhibits jumps in the density distribution, these jumps are also present in the pressure (recall that $p = -\rho\dot{\varphi}$) which is physically unrealistic.

  A formulation that does not have these drawbacks is the two-variable formulation using pressure and displacement potential [PA89]. The formulation is still symmetric but the number of degrees of freedom is doubled. A constant acceleration can be input directly. The variation of the pressure and of the displacement potential are specified independently and the connection between the two variables is enforced by additional equations derived from the Hu-Washizu principle.

- □ The full advantage of the procedure presented manifests itself only in a nonlinear analysis. One of the extensions will therefore be the integration of joint elements [Hoh92] allowing for opening and closing of the vertical contraction joints of the dam.

12.2.3 Applicability to other problems

The problem of transmitting boundaries occurs not only for dam-reservoir interaction. There are other fields where the proposed methods can be applied.

- □ One important application for transmitting boundaries are soil-structure-interaction problems. For the out-of-plane case only shear waves are involved and the displacement variable obeys the same differential equation and boundary conditions as the velocity potential of the fluid. Thus the problem of a soil layer on rock is identical to the problem of a two-dimensional model of a reservoir. One problem that could possibly lead to difficulties is that the usual way of including material damping by complex material properties leads to non-causal behavior.

  For the in-plane case, the soil problem is governed by a vector field of horizontal and vertical displacements, or equivalently by a shear-wave and a compression-wave potential. In the general three-dimensional case, the compression-wave potential is a scalar field and the shear-wave potential is a vector field. The two fields are uncoupled within the domain but a coupling arises at the boundaries, especially at the free surface. In principle, starting from a frequency-domain solution, the same procedure as for water can be used. One problem is that even the frequency-domain solutions are more difficult to obtain. Another problem is that the coupling at the boundaries also couples the eigenmodes and a Ritz approach or even a diagonalization as for the water is generally not appropriate. On the other hand, the block Hankel matrix of the whole transmitting boundary may become quite large and difficult to manage numerically.

- □ The geometry of the problem has a large influence on the frequency-domain solution and therefore also on the appropriate treatment of the transmitting boundaries. Typically, the existence of a cut-off frequency and two frequency ranges with completely different behavior is only encountered for prismatic geometries (layers and channels).

There are many applications in the literature that successfully apply viscous boundaries to infinite domains, for example a structure completely surrounded by fluid [OB85b]. A general classification of problems and solution behavior has not been found in the literature.

Appendix A

Program DANAID

The procedure proposed in this work has been implemented in the computer program DANAID. In this appendix we give some details of the implementation and list some of the program features. The intention is not to provide a manual but to show the principal ideas and possibilities of the program. The version described here was used to calculate the examples given in this report.

The program includes isoparametric two- and three dimensional solid, fluid and interface elements. The approximation of the farfield and the coupling to the nearfield are included. The program performs steady-state and linear dynamic analysis. However, it does not yet include static, modal and nonlinear analysis. The program has an advanced input processing including a free format with keywords and parameters, sets and mesh generation capabilities and an extensible output file structure to save the results. Post-processing has been performed with minimum effort on an ad-hoc basis.

The name DANAID stands for Dynamic ANAlysis of Infinite Domains. According to Greek mythology the Danaids were the fifty daughters of Danaus. They were punished in Hades by having to pour water perpetually into a jar with a hole in the bottom. Figure A.1 shows this scene on an amphora from Altamura, 330–320 B.C.

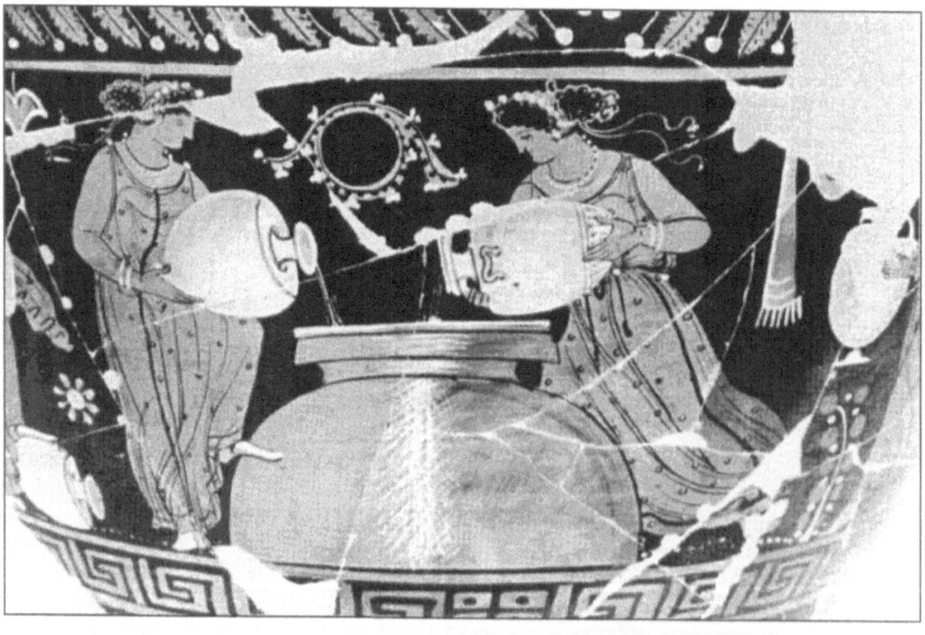

Figure A.1: Danaids on an amphora from Altamura, 330–320 B.C.

The program is written in FORTRAN and comprises about 10'000 lines of code. Choosing FORTRAN as the programming language had several reasons. One is that at the beginning (1989) the plan was to migrate the program later on to a CRAY computer. At that time supercomputing seemed to be absolutely necessary to perform the calculations within a reasonable time and FORTRAN was the only language for supercomputers. In the mean time computer power has increased faster than program development and the calculations can adequately be performed on a workstation. Generally, FORTRAN is still a good language for the numerical part, but C would perhaps be more appropriate for the input processing and the memory management. A relatively new and very promising programming language is C++. It is not widely used for finite element programs but appears to be much more suitable than either C or FORTRAN. However, important are the concepts and not the actual programming language, although some concepts might be more natural in one language than in another.

A.1 Input processing

The input processing is implemented using concepts of compiler design [ASU86, Wir86], although the procedure here is much simpler because no recursion is involved.

A.1.1 Format

For user convenience data is input format-free in a fashion similar to that in the program ABAQUS [ABA92]. The input is arranged in groups of lines, each group starting with a keyword line containing a keyword and several parameters. The keyword is preceded by an asterisk, which makes it easier to read and to interpret. For example

```
*Nodes, Nset=Top
  10  0.    0. 10.
  20  0.   20. 10.
  30  20.   0. 10.
```

Lower case characters are converted to upper case. Spaces and Tabs are considered as separators but are otherwise irrelevant. Continuation lines are given by '..' at the end of the first line. Comment is marked by '!' as shown in the following example.

```
*Elements, Type= FL3D20        ! 3-D fluid elements with 20 nodes
  1  10 20 30 40 50 60 70 80 90 100 ..     ! this line continued
       110 120 130 140 150 160 170 180 200
```

A.1.2 Scanner

In the program, the input is first processed by the scanner. The scanner reads the input stream, character by character, and groups it into symbols. Symbols are either identifiers (ID), numbers (NUM), literal strings (STR), end of line (EOL), end of file (EOF) or special characters. The symbol itself is implemented as a single-character variable SYM. The symbols ID, NUM, STR, EOL, EOF are coded as non-printable ASCII characters, special characters are directly assigned to SYM. The symbols ID, NUM, STR have an attribute, which is the associated number or string. The scanner deposits the attributes as a string in a buffer, which can be accessed by the parser by the functions GETSTR, GETINT, GETRE. These functions interpret the buffer content as string, integer or real, respectively. The interpretation as a number is done by a standard FORTRAN READ with an appropriate FORMAT specifier. The functionality of the scanner is shown schematically in Figure A.2

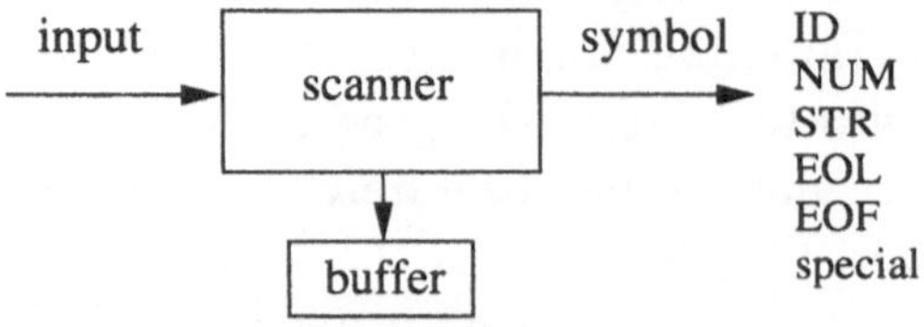

Figure A.2: Scanner

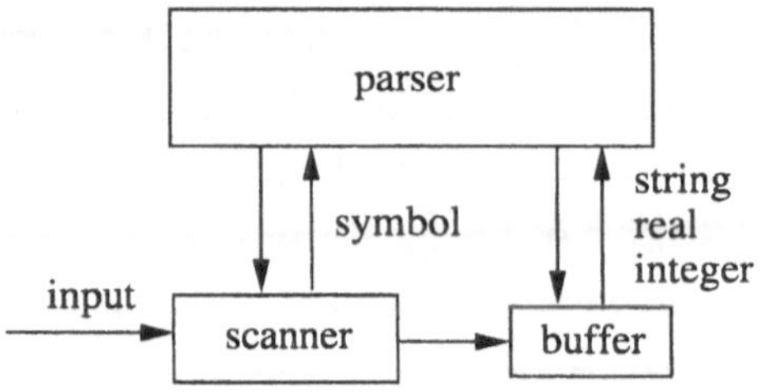

Figure A.3: Parser

A.1.3 Parser

The main program interpreting the input is the parser. The parser requests one symbol after the other from the scanner and interprets the content of the attribute buffer. Whether a number is interpreted as real or integer is determined by the parser not by the scanner. This allows to input real numbers without a decimal point. Identifiers and literal strings are interpreted as strings. This gives a certain simplification for the input because strings can be input as literal strings enclosed by apostrophes or by names of the form of an identifier without the apostrophes. The former are, for example, used for file names where a decimal point is part of the name, the latter for simple names for sets. The interaction between the parser and the scanner is depicted in Figure A.3.

A.1.4 Geometry definition

Nodes and elements are labeled by integers. They do not have to be labeled consecutively nor does their number have to be specified. The degrees of freedom of a node are only activated when elements are added. With this definition unused degrees of freedom do not have to be fixed explicitly by boundary conditions.

Nodes and elements can be grouped into sets. There are *node sets* and *element sets* allowing to use the same name for both a node set and an element set. Sets are used to address several nodes or elements with one name, for example for the definition of boundary conditions or for mesh generation. Sets are defined by setting the parameter NSET or ESET in a keyword line, either in the node or element definition in which case all nodes or elements up to the next keyword line are included in the set, or in a line containing the keyword SET followed by an explicit enumeration.

Several options are defined for the generation of nodes and elements. Only the most frequently used features are already implemented. As needs grow more features will be added. Some examples are shown in the following:

- □ For introducing nodes evenly distributed between two sets
  FILL <set1> <set2> <number of increments> <label increment>
- □ For creating new nodes by projection onto the x-plane (cross-section of channel)
  XPROJ <set> <new x-coordinate> <label increment>

- □ For creating new nodes by mirroring on the xy-plane
  XYSYM <set> <label increment>
- □ To generate new elements
  GENERATE <set> <number of new elements> <element increment> <node incr.>

Each generation line consists of a generation command and several parameters. Note, that the angular brackets are not part of the syntax. The generation lines are entered as data lines instead of direct input. The parser determines whether a line starts with a number or with an identifier. If it starts with a number it is treated as direct input, whereas if it starts with an identifier it is a generation command. Nodes or elements are generated, starting from already defined nodes or elements specified either by a number or a set. Again the parser interprets the input depending on whether the corresponding symbol is a number or an identifier. A generated set can be used to generate a new set. With this feature it easily possible to generate two- and three-dimensional meshes.

The following example shows how a row of quadrilateral elements is generated.

```
*header
  Example 2-D, 5 quads
*dimension,d=2
*nodes,nset=top
  60  0. 100.
 120 20. 100.
*nodes,nset=bot
  10 0. 0.
  70 20. 0.
*nodes
 fill bot top 5 10
*material,typ= fluid
  1  1440. 1000.
*elements, typ=fl2d4, mat=1
  1 10 70 80 20
  generate 1 4 1 10
```

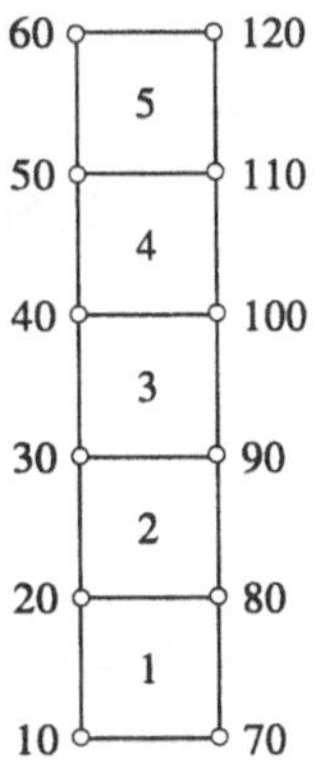

The geometry definition of a model can be stored in a separate file. This makes the input file short and clear, and allows to reuse the same geometry file for several calculations with, for example, different material properties or earthquake inputs.

A.2 Element types

The elements implemented are isoparametric solid, fluid and interface elements in two and three dimensions with linear or quadratic interpolation of the sides. The elements are drawn in Figure A.4 and described in Table A.1 The degrees of freedom are denoted by VP for the velocity potential and DX, DY, DZ for the displacement in the x, y and z-direction, respectively. Interface elements connect fluid with solid elements and have therefore the velocity potential and the displacements as variables. The global dimensionality refers to the number of x, y and z-coordinates, the local dimensionality to the number of ξ, η, ζ-coordinates in the parent element. Interface elements are surfaces curved in space or curves in a plane. There local dimensionality is one less than the global one.

Only three generic elements are programmed for solid, fluid and interface elements. All elements listed in the table are calculated by passing the appropriate parameters to these generic elements. For example, the same subroutine is used to calculate the stiffness matrix of all types of fluid elements.

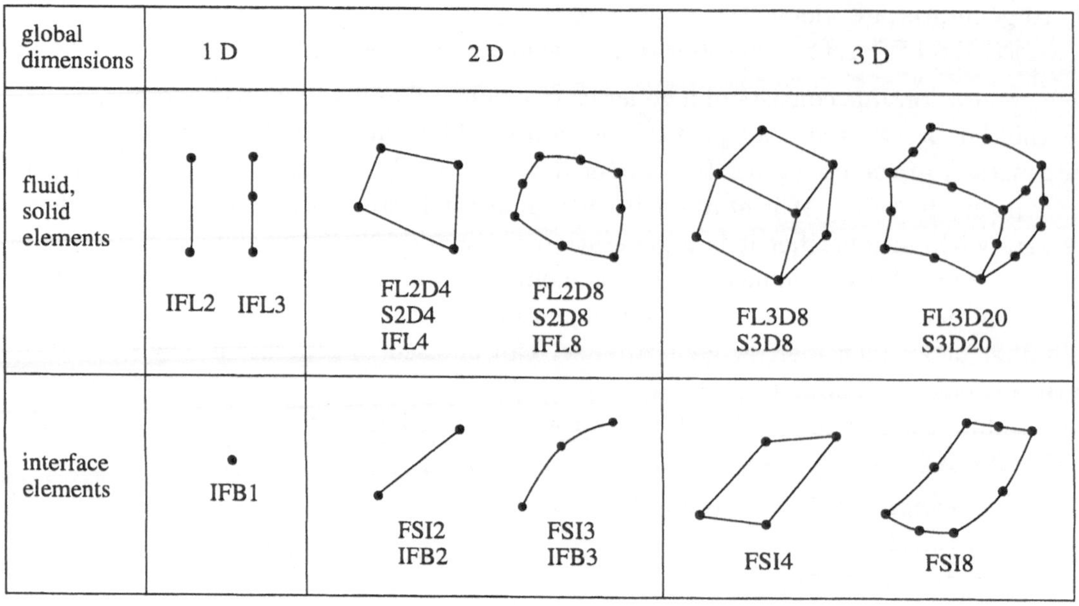

Figure A.4: Elements

Name	family	domain	global	local	nodes	dofs
S2D4	solid	nearfield	2D	2D	4	DX, DY
S2D8			2D	2D	8	DX, DY
S3D8			3D	3D	8	DX, DY, DZ
S3D20			3D	3D	20	DX, DY, DZ
FL2D4	fluid	nearfield	2D	2D	4	VP
FL2D8			2D	2D	8	VP
FL3D8			3D	3D	8	VP
FL3D20			3D	3D	20	VP
FSI2	interface	nearfield	2D	1D	2	VP, DX, DY
FSI3			2D	1D	3	VP, DX, DY
FSI4			3D	2D	4	VP, DX, DY, DZ
FSI8			3D	2D	8	VP, DX, DY, DZ
IFL2	fluid	farfield	1D	1D	2	VP
IFL3			1D	1D	3	VP
IFL4			2D	2D	4	VP
IFL8			2D	2D	8	VP
IFB1	interface	farfield	1D	0D	1	VP, DY
IFB2			2D	1D	2	VP, DY, DZ
IFB3			2D	1D	3	VP, DY, DZ

Table A.1: Elements

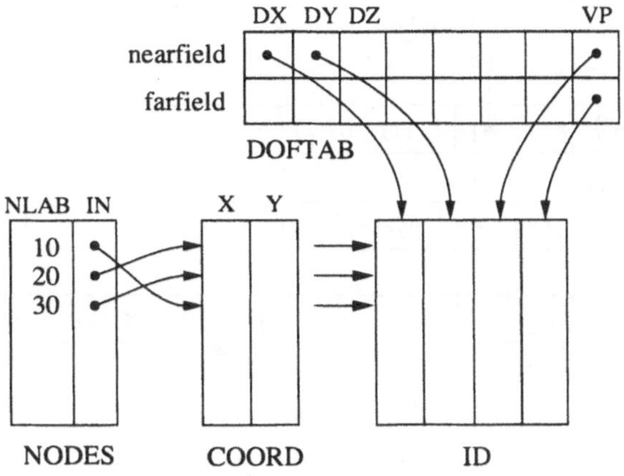

Figure A.5: Arrays for nodal data

A.3 Data structures

Data structures are implemented in standard FORTRAN. Because there are no pointers in FORTRAN, they are simulated by indirect memory addressing. The indirect addressing scheme is drawn the same way as the usual pointer addressing scheme and the word pointer is used for convenience.

A.3.1 Nodal data

Three arrays are used to store node related data, one for the node labels, one for the coordinates and one for degrees of freedom. They are depicted in Figure A.5 for the example of a two-dimensional problem. Array NODES contains the node labels NLAB and the internal node numbers IN, which are used to identify a node in the other two tables. Every node entry is inserted such that the labels are sorted. This allows to use fast binary search and to check for duplicate nodes. The array COORD contains the coordinates for each node.

The array ID contains the equation numbers pertaining to a certain node and degree of freedom. The degrees of freedom are numbered as 1, 2, 3 for the displacements in x-, y- and z-direction, and 8 for the velocity potential, respectively. Intermediate numbers are reserved for future use. Depending on the problem, there are some degrees of freedom for the nearfield and some for the farfield (cross-section). The meaning of each column in the array ID is given in the array DOFTAB. Equations are numbered separately for the nearfield and the farfield (cross-section) because these two models are treated as two separate systems of equations in the analysis. The treatment of the two models is discussed further in Section A.5.

A.3.2 Element data

Because the memory needed is different for different element types and some data may be shared among the elements, the data structure for elements is somewhat more complex. A simplified example is shown in Figure A.6. The basic entry is, as for the nodes, an array ELEMTS containing the element labels in ascending order and the corresponding element pointers. The element pointer points to an array which contains a pointer to the general element description, similar to Table A.1, a pointer to the material, and the internal node numbers pertaining to the element. In this way several elements can share

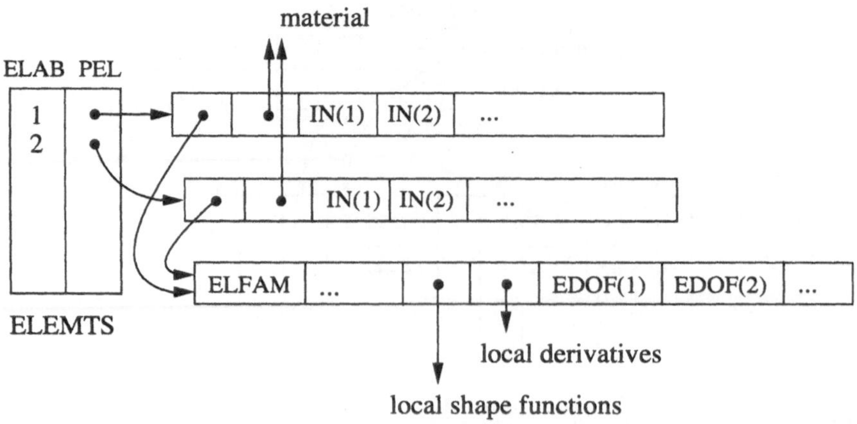

Figure A.6: Arrays for element data

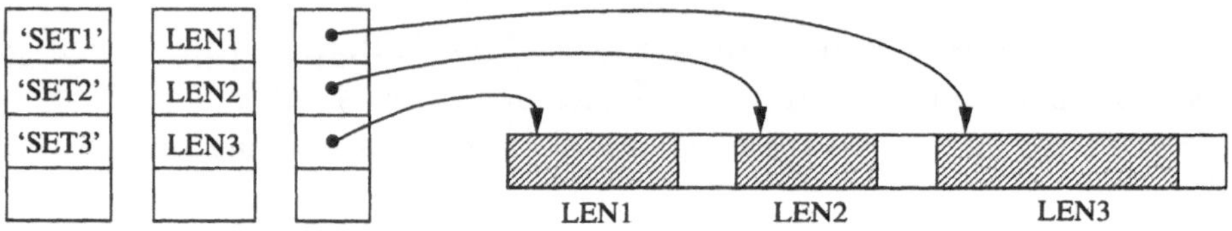

Figure A.7: Data structure for sets

the same material and the same element description. The element description is copied for each keyword line from a general table. Some of the default values may be changed by setting the corresponding parameters in the keyword line. Besides general information such as the element family (ELFAM), or the degrees of freedom involved at each node (EDOF), the element description contains pointers to the local shape function and their derivatives. These are the same for the same element type and can also be shared.

A.3.3 Sets

Sets are formulated as lists which in turn are implemented as arrays of integers. As shown in Figure A.7, each set is associated with a name ('SET1', ... in the example), the number of members in the set (LEN1, ...) and a pointer to the first member. The members of a set are stored in ascending order, which allows for fast binary search. A new member is only inserted if it is not already included in the list.

To distinguish between node sets and element sets, the letter 'N' or 'E' is prepended internally to the name. For example, 'SET1' is stored as 'NSET1' or 'ESET1' in the name table depending on whether it is a node set or an element set. This programming simplification, which avoids the need of separate tables for node sets and element sets, is not noticed by the user.

The same data structure is used for the profile optimizer (see Section A.7). (In fact, we started with the data structure for the profile optimizer when we observed that it could also be used for sets.) For each equation there is a list of equations that have a non-zero entry in the global stiffness matrix.

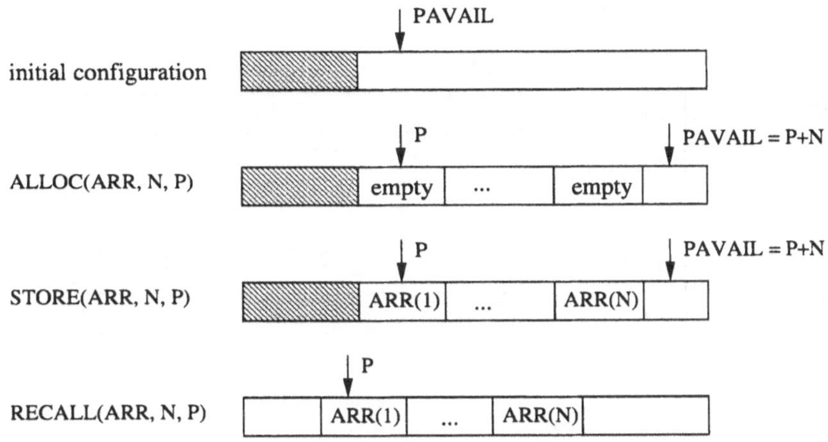

Figure A.8: Memory management routines

A.4 Memory management

Since FORTRAN has no dynamic memory allocation, the usual way of declaring a large array as a memory pool is adopted. The method is modified in the way that not one but four pools are declared, one for real and three for integer data. Having more than one memory pool makes the memory management more flexible and the separation into real and integer data has the advantage, that real data can be declared double precision without special measures.

From the three integer pools one is assigned exclusively to lists of sets. When reading nodal data, one real pool is used for the array of coordinates and one integer pool for the array of node labels. Because these two arrays belong to different pools, their size does not have to be fixed in advance. This frees the user from specifying the number of nodes in the input. The same technique is used when reading element data. Two integer pools are used for the two arrays storing the element labels and the actual element information.

To simplify memory management, several utility routines are defined for each pool. Each pool has a pointer associated with it that points to the next available address. STORE and ALLOC are used to allocate memory; STORE additionally copies values into the pool area. Data is accessed through indirect addressing or by using the routine RECALL. The operations are shown schematically in Figure A.8. The names given here for explanation are generic. In the actual program, each pool uses its specific names. The use of the memory allocation scheme is demonstrated in the following example code. The pool size is defined in the main program. Adjusting this value makes it necessary to recompile the main program.

```
INTEGER SIZE,POOL
PARAMETER (SIZE=1000000)
COMMON /CPOOL/ POOL(SIZE)
...
```

In all other routines the pool is dimensioned to 1 which is the FORTRAN artifice to declare an array of unspecified length. The address of the pointer is passed to the subroutine.

```
INTEGER POOL,N,POINTER
COMMON /CPOOL/ POOL(1)
CALL ALLOC(N,POINTER)
CALL SUB1(POOL(POINTER),N)
```

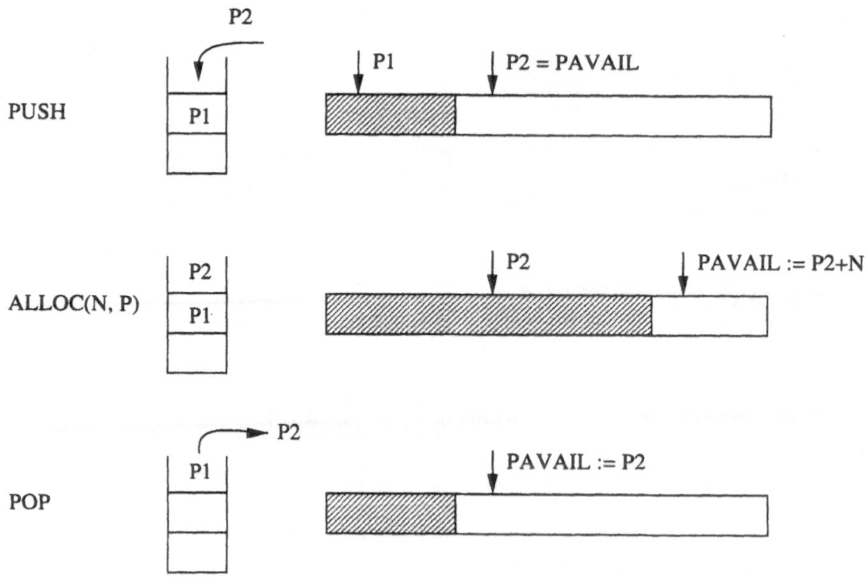

Figure A.9: Memory stack

```
...

SUBROUTINE SUB1(INTARR,N)
INTEGER INTARR(N)
...
```

To be able to reuse memory, each pool maintains a stack with pointers. A PUSH call puts the current pointer of available memory onto the stack. A POP call removes the last pushed pointer from the stack and activates it as the pointer of available memory. This effectively sets the state when PUSH was called the last time and frees the memory allocated since then. The scheme is shown in Figure A.9. The technique is simple but effective. At each level of subroutines, the calling subroutine allocates memory as workspace for the called subroutine. When the subroutine returns to the calling subroutine, the workspace can be made available for other use.

A.5 Analysis procedure

In the current version steady-state (frequency-domain) analysis and time-stepping analysis are implemented. The steady-state analysis can either be performed by using the frequency-domain solution for the transmitting boundary or by using the system matrices obtained by the balanced realization. The latter way is useful for verification of the rational approximation.

As explained in Chapter 10, there are two finite element models, the nearfield model and the cross-section (farfield) model. The cross-section model is used for two different tasks. Firstly, it is used to define the eigenvalue problem for the transmitting boundary. This part describes the interaction with the farfield and is formulated either as a transfer function matrix (only for frequency domain) or as an approximate rational system. Together with the nearfield model it makes up the total system. Secondly, the cross-section model is used to formulate the boundary-value problem of the cross-section. This part defines the loading from the farfield to the total system. The general structure of the matrices in the case of rational system approximation is shown in Figure A.10.

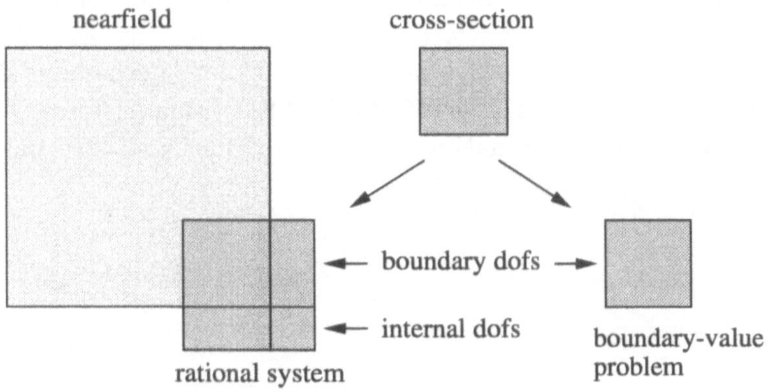

Figure A.10: Structure of matrices

As noted earlier, equations belonging to the nearfield and the cross-section are numbered separately. To make up the total system, equation numbers for the internal degrees of freedom of the rational system are appended to the the nearfield system. For the coupling of the rational system with the nearfield, equation numbers of the rational system have to mapped to the corresponding equation numbers of the nearfield. Boundary conditions on the cross-section have to be applied to both models. After the first numbering, a skyline optimization is performed for both the total model and the cross-section model and the equations are renumbered. The optimization takes place on the equations not on the nodes and includes also the internal degrees of freedom.

The main steps of the analysis procedure are the following:

1. Calculate the transfer matrix $\boldsymbol{\kappa}$ by solving the eigenvalue problem of the cross-section.

2. If desired, calculate the system matrices for the rational approximation and append them to the nearfield matrices to obtain the total system.

3. For all frequencies or time steps:

   (a) Solve the boundary-value problem for the cross-section, which is used in step 3a as the right-hand-side vector.

   (b) Calculate the response of the total (nearfield and farfield) model. The interaction with the farfield is taken into account either by the transfer matrix $\boldsymbol{\kappa}$ (step 1, only for the steady-state case) or by the system matrices of the rational system (step 2). The right-hand-side vector consists of input directly applied to the nearfield and the contribution of the farfield (step 3a).

A.6 Output file structure

Results can be saved in a file, which is then used for postprocessing. Separating calculation and postprocessing has the advantage that development and modifications of the two parts can be made independently to a certain degree. The output file serves as an interface between calculation and postprocessing. To serve that purpose, the output file structure has to fulfill certain requirements. Data should be labeled so that it can be identified by the user or by the postprocessing program. This also allows to construct a postprocessing program that lets the user select certain data by specifying node numbers, degrees of freedom, variables, etc. The other requirement is flexibility and extensibility. Extensions to the file structure should be possible without making old result files unreadable.

The file structure has so far only been used for x-y-plots, such as displacements, pressure or stresses versus time or frequency. For structural plots such as deformed shapes or stress contours a second file with a different structure should perhaps be defined.

The file is written in FORTRAN UNFORMATTED (binary) form. The general file structure is a sequence of lines alternating between key lines and data lines.

```
KEY, INDEX, N
DATA, DATA, ...
KEY, INDEX, N
DATA, DATA, ...
...
```

Key lines define the kind of information encountered in the following data line. Generally there are two classes of data lines, one being the description containing information about the type of variable (displacement, pressure, stress) and location (node number), the other containing the results at a certain time or frequency. INDEX is used to distinguish between different descriptions with the same key. Normally, N is the number of variables saved in one line. As an example, nodal displacements in time are saved as

```
201, 1, N
(NLAB(I), DOF(I), DERIV(I)) I=1,N)
221, 1, N
TIME, (VAR(I), I=1,N)
221, 1, N
TIME, (VAR(I), I=1,N)
...
```

The first line with key 201 says that the next line defines N nodal variables. The second line gives the node number (label), the degree of freedom, and the order of the derivative for each of the N variables. The number 221 in the third line says that the following line contains the time and the values of the earlier defined N variables. Definitions are numbered consecutively by an INDEX (1 in the example). This index is used to associate result lines with the corresponding definition. This is necessary because the same KEY can be used more than once, for example to define other variables at other nodes. The number N would not be necessary in lines preceding results, but all key lines have to have the same format so they can be read without knowledge of context.

Several descriptions and results can be written in interlaced sequence as long as the description precedes the results. It is also possible to have two different sets of results with different time steps. An example with two descriptions is given in the following.

```
201, 1, N1
(NLAB1(I), DOF1(I), DERIV1(I)) I=1,N1)
201, 2, N2
(NLAB2(I), DOF2(I), DERIV2(I)) I=1,N2)
221, 1, N1
TIME1, (VAR1(I), I=1,N1)
221, 2, N2
TIME2, (VAR2(I), I=1,N2)
221, 1, N1
TIME1, (VAR1(I), I=1,N1)
...
```

Other keys are defined for nodal variables versus frequency, transfer functions, stresses versus time or frequency, maximum values and general information such as date, time and title of computer run.

For the plots in this report, postprocessing has been performed by PV-WAVE programs. PV-WAVE is a high-level programming language for visualization of data, which besides the graphics routines also includes complex data structures and program flows. It is able to read many different file formats including binary FORTRAN files.

A.7 Libraries

Most of the code has been written specifically for this program. For some tasks libraries are employed. Routines to solve the eigenvalue problems and the singular value decomposition are taken from the IMSL numerical library. The skyline optimizing scheme by Gibbs, Poole and Stockmeyer [GPS76] has been obtained directly from the ACM distribution service. (ACM Transactions on Mathematical Software, Algorithm 582.)

A.8 Status of program and further developments

As pointed out at the beginning of this appendix, the program includes all features used in the examples, but it is restricted to linear problems. Within the next few years it is planed to extend the program to statics, modal analysis and nonlinear dynamics. This also includes the integration of the joint element developed at the Institute [Hoh92].

At present, the documentation is unsufficient for an external user to apply the program. Comments in the code and description on paper have been restricted to a minimum, because the code is still changing frequently and continous updating would slow down the developement too much. A more extensive documentation will not be available until the nonlinear part is also implemented. The same applies to the user manual.

For the further development of the program it would probably be worthwhile to use C++ as a programming language. One of the most difficult problems in FORTRAN, the development of data structures for all the different element types, can easily be solved in C++ by the concepts of inheritance (derived classes) and polymorphism (virtual functions). Another more general concept of C++ is encapsulation (classes). Encapsulation allows to hide implementational details while providing a public interface. This has the advantage that changes in the program can be made locally and it also provides a compact description of the program.

Bibliography

[AAK71] V. M. Adamjan, D. Z. Arov, and M. G. Krein. Analytic properties of Schmidt pairs for a Hankel operator and the generalized Schur-Takagi problem. *Math. USSR Sbornik*, 15(1):31–73, 1971.

[ABA92] *ABAQUS User's Manual*. Pawtucket, Rhode Island, 1992.

[Ach67] N. I. Achieser. *Vorlesungen über Approximationstheorie*. Akademie Velag, Berlin, 1967.

[Ana90] A. Anandarajah. Time-domain radiation boundary for analysis of plane love-wave propagation problem. *International Journal for Numerical Methods in Engineering*, 29:1049–1063, 1990.

[And93] Edoardo Anderheggen. Lineare Finite-Element-Methoden: eine Einfürung für Ingenieure. Lecture Notes, Swiss Federal Institute of Technology, Zurich, 1993.

[AO88] Mohammad T. Ahmadi and Yoshio Ozaka. A simple method for the full-scale 3-D dynamic analysis of arch dams. In *Proceedings of the Ninth World Conference on Earthquake Engineering*, volume VI, pages 373–378, Tokyo, 1988.

[ASU86] Alfred V. Aho, Ravi Sethi, and Jeffrey D. Ullman. *Compilers. Principles, Techniques, and Tools*. Addison-Wesley, Reading, Massachusetts, 1986.

[AvE87] Heinz Antes and Otto von Estorff. Analysis of absorption effects on the dynamic response of dam reservoir systems by boundary element methods. *Earthquake Engineering and Structural Dynamics*, 15:1023–1036, 1987.

[Bac93] Hugo Bachmann. Erdbebensicherung von Bauwerken. Lecture Notes, Swiss Federal Institute of Technology, Zurich, 1993.

[BB84] Peter Bettes and Jacqueline A. Bettess. Infinite elements for static problems. *Engineering Computations*, 1:4–16, 1984.

[BBE84] K. Bando, P. Bettes, and C. Emson. The effectiveness of dampers for the analysis of exterior scalar wave diffraction by cylinders and ellipsoids. *International Journal for Numerical Methods in Fluids*, 4:599–617, 1984.

[BD82] A. Bultheel and P. Dewilde. Editorial to special issue on rational approximations for systems. *Circuits, Systems, and Signal Processing*, 1(3-4):269–278, 1982.

[BP70] Charles S. Burrus and Thomas W. Parks. Time domain design of recursive digital filters. *IEEE Transactions on Audio and Electroacoustics*, AU-18:137–141, 1970.

[BT80] Alvin Bayliss and Eli Turkel. Radiation boundary conditions for wave-like equations. *Communications on Pure and Applied Mathematics*, XXXIII:707–725, 1980.

[BW76] Klaus Jürgen Bathe and Edward L. Wilson. *Numerical Methods in Finite Element Analysis.* Prentice-Hall, Englewood Cliffs, N.J., 1976.

[Cas87] Aldo Castoldi. A new criteria for the seismic monitoring of dams: A dynamic active surveillance system. In *Proceedings of the Joint China-U.S. Workshop on Earthquake Behavior of Arch Dams*, pages 258–278, Bejing, China, 1987.

[CBV76] Ruel V. Churchill, James W. Brown, and Roger F. Verhey. *Complex Variables and Applications.* McGraw-Hill, third edition, 1976.

[CC81] Anil K. Chopra and P. Chakrabarti. Earthquake analysis of concrete gravity dams including dam-water-foundation rock interaction. *Earthquake Engineering and Structural Dynamics*, 9:363–383, 1981.

[CC87] Zhang Chuhan and Zhao Chongbin. Coupling method of finite and infinite elements for strip foundation wave problems. *Earthquake Engineering and Structural Dynamics*, 15:839–851, 1987.

[CE77] Robert Clayton and Björn Engquist. Absorbing boundary conditions for acoustic and elastic wave equation. *Bulletin of the Seismological Society of America*, 67(6):1529–1540, 1977.

[CF48] George A. Campbell and Ronald M. Foster. *Fourier Integrals for Practical Applications.* Van Nostrand, Princeton, New Jersey, 1948.

[CGG+90] Bruce W. Char, Keith O. Geddes, Gaston H. Gonnet, Michael B. Monagan, and Stephen M. Watt. *Maple, First leaves. A tutorial Introduction to Maple.* Waterloo Ontario Canada, third edition, 1990.

[Che84] Chi-Tsong Chen. *Linear System Theory and Design.* Holt, Rinehart and Winston, 1984.

[Cho67] Anil K. Chopra. Hydrodynamic pressures on dams during earthquakes. *Journal of the Engineering Mechanics Division, ASCE*, 93(EM6):205–223, 1967.

[Cho68] Anil K. Chopra. Earthquake behavior of reservoir-dam systems. *Journal of the Engineering Mechanics Division, ASCE*, 94(EM6):1475–1500, 1968.

[CJ87] Martin Cohen and Paul C. Jennings. *Computational Methods for Transient Analysis*, chapter 7: Silent Boundary Methods for transient Analysis, pages 301–360. Elsevier Science Publishers, 1987.

[Dec72] Andrew G. Deczky. Synthesis of recursive digital filters using the minimum p-error criterion. *IEEE Transactions on Audio and Electroacoustics*, AU-20(4):257–263, 1972.

[Dec74] Andrew G. Deczky. Equiripple and minimax (Chebyshev) approximations for recursive digital filters. *IEEE Transactions on Acoustics, Speech, and Signal Processing*, ASSP-22(2):98–111, 1974.

[DH88] Ziyad H. Duron and John F. Hall. Experimental and finite element studies of the forced vibration response of the Morrow Point dam. *Earthquake Engineering and Structural Dynamics*, 16:1021–1039, 1988.

[DH89] Michael J. Dowling and John F. Hall. Nonlinear seismic analysis of arch dams. *Journal of Engineering Mechanics, ASCE*, 115(4):768–789, 1989.

[DM93] José Dominguez and Orlando Maseo. Earthquake analysis of arch dams. II: dam-water-foundation interaction. *Journal of Engineering Mechanics, ASCE*, 119(3):513–530, 1993.

[Dud74] Dan E. Dudgeon. Recursive filter design using differential correction. *IEEE Transactions on Acoustics, Speech, and Signal Processing*, ASSP-22(6):443–448, 1974.

[DW88] Georges R. Darbre and John P. Wolf. Criterion of stability and implementation issues of hybrid frequency-time-domain procedure for nonlinear dynamic analysis. *Earthquake Engineering and Structural Dynamics*, 16:569–581, 1988.

[EF73] Alan G. Evans and Robert Fischl. Optimal least squares time-domain synthesis of recursive digital filters. *IEEE Transactions on Audio and Electroacoustics*, AU-21(1):61–65, 1973.

[EM77] Bjorn Engquist and Andrew Majda. Absorbing boundary conditions for the numerical simulation of waves. *Mathematics of Computation*, 31(139):629–651, 1977.

[FC80] Ka-Lun Fok and Anil K. Chopra. Earthquake analysis and response of concrete arch dams. Report UCB/EERC 85/07, University of California, Berkeley, California, July 1980.

[FC83] Gregory Fenves and Anil K. Chopra. Effects of reservoir bottom absorption on earthquake response of concrete gravity dams. *Earthquake Engineering and Structural Dynamics*, 11:809–829, 1983.

[FC84] Gregory Fenves and Anil K. Chopra. Earthquake analysis and response of concrete gravity dams. Report UCB/EERC 84/10, University of California, Berkeley, California, August 1984.

[FC90] Gregory L. Fenves and Juan W. Chavez. Hybrid frequency-time domain analysis of nonlinear fluid-structure systems. In *Proceedings of Fourth U. S. National Conference on Earthquake Engineering*, volume 2, pages 97–106, Palm Springs, California, May 1990.

[Flu75] Wilhelm Fluegge. *Viscoelasticity*. Springer, 1975.

[FMR92] Gregory Fenves, S. Mojtahedi, and R. B. Reimer. Effect of contraction joints on earthquake response of an arch dam. *Journal of Structural Engineering, ASCE*, 118(4):1039–1055, 1992.

[Fun65] Y. C. Fung. *Foundations of Solid Mechanics*. Prentice-Hall, Englewood Cliffs, 1965.

[FVL88] Gregory Fenves and Luis M. Vargas-Loli. Local transmitting boundaries. *Journal of Engineering Mechanics*, 114:1011–1027, 1988.

[FWB90] Glauco Feltrin, Dieter Wepf, and Hugo Bachmann. Seismic cracking of concrete gravity dams. *Dam Engineering*, 16:279–289, 1990.

[GCP88] Keith Glover, Ruth F. Curtain, and Jonathan R. Partington. Realisation and approximation of linear infinite-dimensional systems with error bounds. *SIAM Journal of Control and Optimization*, 26(4):863–898, July 1988.

[GK81] Yves V. Genin and Sun-Yuan Kung. A two-variable approach to the model reduction problem with Hankel norm criterion. *IEEE Transactions on Circuits and Systems*, CAS-28(9):912–924, 1981.

[Glo84] Keith Glover. All optimal Hankel-norm approximations of linear multivariable systems and their L_∞-error bounds. *International Journal of Control*, 39(6):1115–1193, 1984.

[GN86] W. Gawronski and H. G. Natke. On balancing linear symmetric systems. *International Journal of Systems Science*, 17(10):1509–1519, 1986.

[GO90] Romano J. C. Gâmara and Sérgio B. M. Oliveira. Non-linear dynamic analysis of arch dams. In *Proceedings of the Ninth European Conference on Earthquake Engineering*, volume 7-B, pages 143–152, Moscow, 1990.

[GPS76] Norman E. Gibbs, William G. Poole, and Paul K. Stockmeyer. An algorithm for reducing the bandwidth and profile of a sparse matrix. *SIAM Journal of Numerical Analysis*, 13(2):236–250, 1976.

[Gra75] Karl F. Graff. *Wave motion in elastic solids.* Claredon Press, Oxford, 1975.

[GST83] Martin H. Gutknecht, Julius O. Smith, and Lloyd N. Trefethen. The Carathéodory-Féjer method for recursive digital filter design. *IEEE Transactions on Acoustics, Speech, and Signal Processing*, ASSP-31(6):1417–1426, 1983.

[GT80] Martin H. Gutknecht and Lloyd N. Trefethen. Recursive digital filter design by the Carathéodory-Féjer method. Numerical Analysis Project NA-80-01, Computer Science Department, Stanford University, Stanford, California, May 1980.

[GvL83] Gene H. Golub and Charles F. van Loan. *Matrix Computations.* The Johns Hopkins University Press, 1983.

[Hal83] H. J. Halin. The applicability of Taylor series methods in simulation. In *Proceedings of the 1983 Sommer Computer Simulation Conference*, pages 1032–1078, Vancouver, Canada, July 1983. North-Holland Publishing Company, Amsterdam.

[Hal88] John F. Hall. The dynamic and earthquake behaviour of concrete dams: Review of experimental behaviour and observational evidence. *Soil Dynamics and Earthquake Engineering*, 7(2):57–121, 1988.

[Hat65] Tadashi Hatano. An examination on the resonance of hydrodynamic pressure during earthquakes due to elasticity of water. Technical report, Technical Laboratory, Central Research Institute of Electric Power Industry, Tokyo, 1965.

[HC82] John F. Hall and Anil K. Chopra. Two-dimensional dynamic analysis of concrete gravity and embankment dams including hydrodynamic effects. *Earthquake Engineering and Structural Dynamics*, 10:305–332, 1982.

[HC83] John F. Hall and Anil K. Chopra. Dynamic analysis of arch dams including hydrodynamic effects. *Journal of Engineering Mechanics, ASCE*, 109(1):149–167, 1983.

[Hen78] Peter Henrici. Fast Fourier methods in computational complex analysis. Research Report 78-04, Seminar für Angewandte Mathematik, Swiss Federal Institute of Technology, Zurich, June 1978.

[HH92] Isaac Harari and Thomas J. R. Hughes. Analysis of continuous formulations underlaying the computation of time-harmonic acoustics in exterior domains. *Computer Methods in Applied Mechanics and Engineering*, 97:103–124, 1992.

[HHT77] Hans M. Hilber, Thomas J. R. Hughes, and Robert. L. Taylor. Improved numerical dissipation for time integration algorithms in structural dynamics. *Earthquake Engineering and Structural Dynamics*, 5:283–292, 1977.

[HJ88] J. L. Humar and A. M. Jablonski. Boundary element reservoir model for seismic analysis of gravity dams. *Earthquake Engineering and Structural Dynamics*, 16:1129–1156, 1988.

[HK66] B. L. Ho and R. E. Kalman. Effective construction of linear state-variable models from input/output functions. *Regelungstechnik*, 14(12):545–548, 1966.

[Hoh92] Jörg-Martin Hohberg. *A Joint Element for the Nonlinear Dynamic Analysis of Arch Dams.* PhD thesis, Swiss Federal Institue of Technology (ETH), Zürich, 1992. Diss. 9651.

[HT88] Laurence Halpern and Lloyd N. Trefethen. Wide-angle one-way wave equations. *Journal of the Acoustical Society of America*, 84(4):1397–1404, 1988.

[Hug87] Thomas J. R. Hughes. *The Finite Element Method.* Prentice-Hall, Englewood Cliffs, N.J., 1987.

[ISM91] ISMES. *Frist Benchmark Workshop on Numerical Analysis of Dams*, Bergamo, Italy, May 1991.

[ISM93] ISMES. *Second Benchmark Workshop on Numerical Analysis of Dams*, Bergamo, Italy, March 1993.

[JH90] A. M. Jablonski and J. L. Humar. Three-dimensional boundary element reservoir model for seismic analysis of arch and gravity dams. *Earthquake Engineering and Structural Dynamics*, 19:359–376, 1990.

[KA87] Sun-Yuan Kung and K. S. Arun. *Advances in Statistical Signal Processing*, volume 1, chapter 6: Singular-value-decomposition algorithms for linear system approximation and spectrum estimation, pages 203–250. JAI Press Inc., 1987.

[Kai80] Thomas Kailath. *Linear Systems.* Prentice-Hall, Englewood Cliffs, N.J., 1980.

[Kau88] Eduardo Kausel. Local transmitting boundaries. *Journal of Engineering Mechanics*, 114:1011–1027, 1988.

[Kau92] Eduardo Kausel. Physical interpretation and stability of paraxial boundary conditions. *Bulletin of the Seismological Society of America*, 82(2):898–913, 1992.

[KL81a] Sun-Yuan Kung and David W. Lin. Optimal Hankel-norm model reductions: Multivariable systems. *IEEE Transactions on Automatic Control*, AC-26(4):832–852, 1981.

[KL81b] Sun-Yuan Kung and David W. Lin. A state-space formulation for optimal Hankel-norm approximations. *IEEE Transactions on Automatic Control*, AC-26(4):942–946, 1981.

[KM81] R. R. Kunar and J. Marti. A non-reflecting boundary for explicit calculations. In *Computational methods for infinite domain media-structure interaction*, volume AMD-64, pages 183–204. Applied Mechanics Division, ASME, November 1981.

[Kni93] K. V. Kniffka. Investigation of the Mauvoisin concrete arch dam subjected to maximum credible earthquake. *Computers & Structures*, 47(4/5):787–800, 1993. And additional oral communication.

[Kot59] Seima Kotsubo. Dynamic water pressure on dams due to irregular earthquakes. *Memoirs of the Faculty of Enineering, Kyushu University*, 18(4):119–129, 1959.

[KT81] Eduardo Kausel and John L. Tassoulas. Transmitting boundaries: A closed-form comparison. *Bulletin of the Seismological Society of America*, 71(1):143–159, 1981.

[Kun78] Sun-Yuan Kung. A new identification and model reduction algorithm via singular value decomposition. In *Proceedings of the 12th Asilomar Conference on Circuits, Systems and Computers*, pages 705–714, Pacific Grove, CA, 1978.

[Kun80] Sun-Yuan Kung. Optimal Hankel-norm model reductions — Scalar systems. In *Proceedings of the 1980 Joint Automatic Control Conference*, San Francisco, CA, 1980.

[Lau80] Alan J. Laub. Computation for "balancing" transformations. In *Proceedings of the 1980 Joint Automatic Control Conference*, San Francisco, CA, 1980.

[Lee84] Vincent W. Lee. A new fast algorithm for the calculation of response of a single-degree-of-freedom system to arbitray load in time. *Soil Dynamics and Earthquake Engineering*, 3(4):191–199, 1984.

[Lin75] E. L. Lindman. "Free-space" boundary conditions for the time dependent wave equation. *Journal of Computational Physics*, 18:66–78, 1975.

[LK69] John Lysmer and Roger L. Kuhlemeyer. Finite dynamic model for infinite media. *Journal of the Engineering Mechanics Division, ASCE*, 95(EM4):859–877, 1969.

[LK82] David W. Lin and Sun-Yuan Kung. Optimal Hankel-norm approximation of continuous-time linear systems. *Circuits, Systems, and Signal Processing*, 1(3-4):407–431, 1982.

[LRT87] Vahid Lotfi, Jose M. Roesset, and John L. Tassoulas. A technique for the analysis of the response of dams to earthquakes. *Earthquake Engineering and Structural Dynamics*, 115:463–490, 1987.

[LW72] John Lysmer and Günter Waas. Shear waves in plane infinite structures. *Journal of the Engineering Mechanics Division, ASCE*, 98(EM1):85–105, 1972.

[LW84] Z. P. Liao and H. L. Wong. A transmitting boundary for the numerical simulation of elastic wave propagation. *Soil Dynamics and Earthquake Engineering*, 3(4):174–183, 1984.

[MB82] W. J. Mansur and C. A. Brebbia. Formulation of the boundary element method for transient problems governed by the scalar wave equation. *Applied Mathematical Modelling*, 6:307–311, 1982.

[MD93] Orlando Maseo and José Dominguez. Earthquake analysis of arch dams. I: dam-foundation interaction. *Journal of Engineering Mechanics, ASCE*, 119(3):496–512, 1993.

[Mee87] Jethro W. Meek. A recursive method of calculation for dynamics and statics: The analogy of the "simple-minded boxer". *Bautechnik*, 64:202–205, 1987. In German.

[Mei83a] Jean Meinguet. On the glover concretization of the Adamjan-Arov-Krein approximation theory. In *Modellig, Identification and Robust Control*, pages 325–334. D. Reidel Publishing Company, 1983.

[Mei83b] Jean Meinguet. A simplified presentation of the Adamjan-Arov-Krein approximation theory. In *Computational Aspects of Complex Analysis*, pages 217–248. D. Reidel Publishing Company, 1983.

[Mei88] Jean Meinguet. Once again: The Adamjan-Arov-Krein approximation theory. In *Nonlinear Numerical Methods and Rational Approximation*, pages 77–91. D. Reidel Publishing Company, 1988.

[MFH89] Isidoro Miranda, Robert M. Ferecz, and Thomas J. R. Hughes. An improved implicit-explicit time integration method for structural dynamics. *Earthquake Engineering and Structural Dynamics*, 18:643–653, 1989.

[Moo78] Bruce C. Moore. Singular value analysis of linear systems. In *Proceedings of the 1978 IEEE Conference on Decision & Control*, pages 66–73, San Diego, California, 1978.

[Moo81] Bruce C. Moore. Principal component analysis in linear systems: Controllability, observability, and model reduction. *IEEE Transactions on Automatic Control*, AC-26(1):17–32, 1981.

[MR76] Clifford T. Mullis and Richard A. Roberts. The use of second-order information in the approximation of discrete-time linear systems. *IEEE Transactions on Acoustics, Speech, and Signal Processing*, ASSP-24(3):226–238, 1976.

[MR93] J. R. Mays and L. H. Roehm. Effect of vertical contraction joints in concrete arch dams. *Computers & Structures*, 47(4/5):615–627, 1993.

[MW89] Sassan K. Mohasseb and John P. Wolf. Recursive evaluation of interaction forces of unbounded soil in frequency domain. *Soil Dynamics and Earthquake Engineering*, 8:176–188, 1989.

[Nat81] B. Nath. A novel spherical polar finite element for the solution of the steady-state scalar wave equation in three-dimensions. *Earthquake Engineering and Structural Dynamics*, 9:33–51, 1981.

[NU88] Arnold F. Nikiforov and Vasilii B. Uvarov. *Special Funtions of Mathematical Physics.* Birkhhäuser, Basel, 1988.

[OB85a] Lorraine G. Olson and Klaus Jürgen Bathe. Analysis of fluid-structure interactions. A direct symmetric coupled formulation based on the fluid velocity potential. *Computers & Structures*, 21(1/2):21–32, 1985.

[OB85b] Lorraine G. Olson and Klaus Jürgen Bathe. An infinite element for analysis of transient fluid-structure interactions. *Engineering Computations*, 2:319–329, 1985.

[OB88] J. P. F. O'Connor and J. C. Boot. A solution procedure for the earthquake analysis of arch dam-reservoir systems with compressible water. *Earthquake Engineering and Structural Dynamics*, 16:757–773, 1988.

[PA89] Peter M. Pinsky and Majib N. Abboud. Two mixed variational principles for exterior fluid-structure interaction problems. *Computers & Structures*, 33(3):621–635, 1989.

[Pap62] Athansios Papoulis. *The Fourier Integral and its Applications.* McGraw-Hill, 1962.

[Par88] Jonathan R. Partington. *An Introduction to Hankel Operators.* Cambridge University Press, Cambridge, 1988.

[PC81] Craig S. Porter and Anil K. Chopra. Dynamic analysis of simple arch dams including hydrodynamic interaction. *Earthquake Engineering and Structural Dynamics*, 9:573–597, 1981.

[Per73] Parambakatoor R. Permumalswami. Earthquake hydrodynamic forces on arch dams. *Journal of the Engineering Mechanics Division, ASCE*, 99(EM5):965–977, 1973.

[Pow81] M. J. D. Powell. *Approximation theory and methods.* Cambridge University Press, Cambridge, 1981.

[PS82] Lars Pernebo and Leonard M. Silverman. Model reduction via balanced state space representations. *IEEE Transactions on Automatic Control*, AC-27(2):382–387, 1982.

[RCH+72] Lawrence R. Rabiner, James W. Cooley, Howard D. Helms, Leland B. Jackson, James F. Kaiser, Charles M. Rader, Ronald W. Schafer, Kenneth Steiglitz, and Clifford J. Weinstein. Terminology in digital signal processing. *IEEE Transactions on Audio and Electroacoustics*, AU-20(5):322–336, 1972.

[RG75] Lawrence R. Rabiner and Bernard Gold. *Theory and Application of Digital Signal Processing.* Prentice-Hall, Englewood Cliffs, N.J., 1975.

[RHW70] F. E. Richard, J. R. Hall, and R. D. Woods. *Vibrations of Soils and Foundations.* Prentice-Hall, Englewood Cliffs, N.J., 1970.

[SB80a] Leonard M. Silverman and Maamar Bettayeb. Optimal approximation of linear systems. In *Proceedings of the 1980 Joint Automatic Control Conference*, San Francisco, CA, 1980.

[SB80b] J. Stoer and R. Bulirsch. *Introduction to Numerical Analysis.* Springer-Verlag, New York, 1980.

[SB86] Petter E. Skrikerud and Hugo Bachmann. Discrete crack modelling for dynamically loaded, unreinforced concrete structures. *Earthquake Engineering and Structural Dynamics*, 14:297–315, 1986.

[Sha76] Y. Shamash. Continued fraction methods for reduction of constant-linear multivariable systems. *International Journal of Systems Science*, 7(7):743–758, 1976.

[Sha87] S. K. Sharan. Time-domain analysis of infinite fluid vibration. *International Journal for Numerical Methods in Engineering*, 24:945–958, 1987.

[Shu87] S. G. Shul'man. *Seismic Pressure of Water on Hydrolic Structures.* A.A. Balkema, Rotterdam, 1987. Translation, originally published 1970.

[Sie86] William McC. Siebert. *Circuits, Signals, and Systems.* MIT Press, Cambridge, Massachusetts, 1986.

[Smi74] Warwick D. Smith. A nonreflecting plane boundary for wave propagation problems. *Journal of Computational Physics*, 15:492–503, 1974.

[Som65] Arnold Sommerfeld. *Vorlesungen über Theoretische Physik*, volume 4: Partielle Differentialgleichungen. Geest & Portig, Leipzig, 1965.

[ST74] C. K. Sanathanan and Hideo Tsukui. Synthesis of tranfer function from frequency response data. *International Journal of Systems Science*, 5(4):41–54, 1974.

[Sta88] Richard Stacey. Improved transparent boundary formulations for the elastic-wave equation. *Bulletin of the Seismological Society of America*, 78(6):2089–2097, 1988.

[Ste70] Kenneth Steiglitz. Computer-aided design of recursive digital filters. *IEEE Transactions on Audio and Electroacoustics*, AU-18(2):123–129, 1970.

[SW92] Tadeusz Szczesiak and Benedikt Weber. Hydrodynamic effects in a reservoir with semi-circular cross-section and absorptive bottom. *Soil Dynamics and Earthquake Engineering*, 11:203–212, 1992.

[Tak24] Teiji Takagi. On an algebraic problem related to an analytic theorem of Carathéodory and Fejér and on an allied theorem of Landau. *Japanese Journal of Mathematics*, 1:83–93, 1924.

[Tak25] Teiji Takagi. On an algebraic problem related to an analytic theorem of Carathéodory and Fejér and on an allied theorem of Landau. *Japanese Journal of Mathematics*, 2:13–17, 1925. (Continuation).

[TL87] Chong-Shien Tsai and George C. Lee. Arch dam-fluid interactions: By FEM-BEM and substructure concept. *International Journal for Numerical Methods in Engineering*, 24:2367–2388, 1987.

[TL90] Chong-Shien Tsai and George C. Lee. Method for transient analysis of three-dimensional dam-reservoir interactions. *Journal of Engineering Mechanics*, 116(10):2151–2172, 1990.

[Web90] Benedikt Weber. Fluid-structure interaction for arch dams. In *Proceedings of the European Conference on Structural Dynamics, EURODYN '90*, pages 851–858, Bochum, June 1990. A.A. Balkema, Rotterdam.

[Web92] Benedikt Weber. New method for time-domain analysis of dam-reservoir interaction. In *Proceedings of the Thenth World Conference on Earthquake Engineering*, pages 4689–4694, Madrid, Spain, July 1992. A.A. Balkema, Rotterdam.

[Wer82] Helmut Werner. A remark on the nunerics of rational approximation and the rate of convergence of equally spaced interpolation of $|x|$. *Circuits, Systems, and Signal Processing*, 1(3-4):367–377, 1982.

[Wes33] H. M. Westergaard. Water pressures on dams during earthquakes. *Transactions, ASCE*, 98:418–433, 1933.

[WHB89] Benedikt Weber, Jörg-Martin Hohberg, and Hugo Bachmann. Earthquake analysis of arch dams including joint nonlinearity and fluid-structure interaction. In *Proceedings of the International Conference on Earthquake Resistant Construction and Design*, pages 349–358, Berlin, June 1989. A.A. Balkema, Rotterdam.

[Wil83] Edward L. Wilson. Finite elements for the dynamic analysis of fluid-solid systems. *International Journal for Numerical Methods in Engineering*, 19:1657–1668, 1983.

[Wir86] Niklaus Wirth. *Compilerbau: eine Einführung.* Teubner, Stuttgart, 1986.

[WO85] John P. Wolf and Pius Obernhuber. Non-linear soil-structure interaction analysis using dynamic stiffness or flexibility of soil in the time domain. *Earthquake Engineering and Structural Dynamics*, 13:195–212, 1985.

[Wol86] John P. Wolf. A comparison of time-domain transmitting boundaries. *Earthquake Engineering and Structural Dynamics*, 14:655–673, 1986.

[Wol88] John P. Wolf. *Soil-Structure-Interaction Analysis in Time-Domain.* Prentice-Hall, Englewood Cliffs, N.J., 1988.

[Wol91] John P. Wolf. Consistent lumped-parameter models for unbounded soil: Physical representation. *Earthquake Engineering and Structural Dynamics*, 20:11–32, 1991.

[WP92] John P. Wolf and Antonio Paronesso. Lumped-parameter model and recursive evaluation of interaction forces of semi-infinite uniform fluid channel for time-domain dam-reservoir analysis. *Earthquake Engineering and Structural Dynamics*, 21:811–831, 1992.

[WS49] P. Willhelm Werner and K. J. Sundquist. On hydrodynamic earthquake effects. *Transactions of the American Geophysical Union*, 30(5):636–657, 1949.

[WS86] John P. Wolf and Dario R. Somaini. Approximate dynamic model of embeded foundation in time domain. *Earthquake Engineering and Structural Dynamics*, 14:683–703, 1986.

[WWB88] Dieter H. Wepf, John P. Wolf, and H. Bachmann. Hydrodynamic-stiffness matrix based on boundary elements for time-domain dam-reservoir-soil analysis. *Earthquake Engineering and Structural Dynamics*, 16:417–432, 1988.

[YTL90] Rihui Yang, C. S. Tsai, and G. C. Lee. Far-field modeling in 3D dam-reservoir interaction analysis. *Journal of Engineering Mechanics*, 116(10):2151–2172, 1990.

[ZBCE81] O. C. Zienkiewicz, P. Bettes, T. C. Chiam, and C. Emson. Numerical methods for unbounded field problems and a new infinite element formultation. In *Computational methods for infinite domain media-structure interaction*, volume AMD-64, pages 115–148. Applied Mechanics Division, ASME, November 1981.

[Zie77] O. C. Zienkiewicz. *The Finite Element Method.* McGraw-Hill, London, 1977.

[ZM74] H. Paul Zeiger and A. Julia McEwen. Approximate linear realization of given dimension via Ho's algorithm. *IEEE Transactions on Automatic Control*, AC-19(2):153, 1974.

Abstract

Dynamic analysis of dam-reservoir interaction is conveniently performed using finite elements. An intrinsic problem is that only the dam and the nearfield of the reservoir can be modeled directly. The farfield has to be idealized as a semi-infinite channel which is represented by a special boundary. This boundary is called *transmitting boundary* because waves are transmitted through it from the nearfield to the farfield. The rigorous frequency-domain analysis leads to two qualitatively different types of solution. Below the cut-off frequency, which corresponds to the fundamental frequency of the channel cross-section, the pressure decays exponentially with distance and no energy is transmitted. Above the cut-off frequency, pressure waves propagate to infinity taking energy out of the nearfield which results in radiation damping.

For a time-domain analysis, as required for nonlinear problems, a transmitting boundary has to be formulated in the time domain. The main difficulty is to find an algorithm that takes care of both the decaying and the wave propagation parts of the solution. Various methods proposed in the literature are shown to be inappropriate for problems involving a cut-off frequency because they either do not capture the different types of solution, as viscous boundaries, or because they are computationally inefficient as convolution or boundary elements.

The route taken here is an approach based on rational approximation and therefore is given the name *rational boundaries.* The frequency-domain solution is approximated by a linear time-invariant system expressed as a set of linear differential equations with constant coefficients. These differential equations can be solved efficiently using the same time integration algorithm as for the finite element part of the model.

Rational approximation for systems is difficult because of two main problems. Firstly, the determination of the coefficients is a nonlinear problem. Secondly, there is the additional constraint that the resulting system be stable. In a first step, simple methods approximating either the transfer function or the *Markov parameters* are explained. The approximation of the Markov parameters leads in a natural way to the *Hankel matrix* which plays the key role in more advanced methods.

An important point is the duality between discrete-time and continuous-time systems. The link between the two domains is the *bilinear transform.* The bilinear transform maps a continuous-time system to a discrete-time system and vice versa. The approximation for a continuous-time system can be done in the mapped discrete-time domain. The approximation is then transformed back to the original continuous-time domain. Because the stability regions are mapped onto each other, the bilinear transform preserves stability. This procedure has two major advantages. Firstly, the Hankel matrix can easily be calculated because the system is discrete-time and secondly, the whole frequency axis up to infinity can be included because it is mapped to the (finite) unit circle. Taking into account the high frequencies is important to preserve *causality* which in turn is necessary for *stability.*

Two methods based on the the *singular value decomposition* of the Hankel matrix are explained, the Carathéodory-Fejér (CF) method and the *balanced realization.* The CF method is the most accurate of all methods investigated and leads to an almost circular error curve in the complex plane. Unfortunately, it cannot easily be extended to the *multivariable* case. The favorite method is the balanced realization method because it is numerically robust and equally well applicable for scalar as for multivariable systems.

A special feature of the proposed algorithm is the transformation of the approximating system to a *symmetric second-order* form which is the form of the finite element matrices. The farfield can therefore be implemented by appending the second-order matrices of the farfield to the mass, damping and stiffness matrices of the nearfield.

Various examples including analyses of full-size three-dimensional models confirm the accuracy and efficiency of the method.

Keywords: dam-reservoir interaction, earthquake analysis, finite elements, transmitting boundaries, transmitting boundaries in time-domain, radiation condition, rational approximation, linear system approximation, Hankel matrix, singular value decomposition, balanced realization.

Zusammenfassung

Die dynamische Berechnung der Staumauer-Stausee-Wechselwirkung wird geeigneterweise mit Finiten Elementen durchgeführt. Ein wesentliches Problem dabei ist, dass nur die Mauer und das Nahgebiet des Stausees direkt modelliert werden können. Das Ferngebiet muss als halbunendlicher Kanal idealisiert werden und wird durch einen speziellen Rand representiert. Dieser Rand heisst *durchlässiger Rand*, weil Wellen vom Nahgebiet ins Ferngebiet durchgelassen werden. Die genaue Analyse im Frequenzbereich zeigt zwei qualitativ verschiedene Lösungen. Unterhalb der Grenzfrequenz, die der Grundfrequenz des Kanalquerschnittes entspricht, fällt der Druck exponentiell mit der Distanz ab, und es wird keine Energie abgestrahlt. Oberhalb der Grenzfrequenz wandern die Wellen ins Unendliche und transportieren Energie weg aus dem Nahbereich, was sich als Abstrahlungsdämpfung äussert.

Für eine Berechnung im Zeitbereich, wie sie für nichtlineare Probleme nötig ist, muss ein durchlässiger Rand im Zeitbereich formuliert werden. Die Hauptschwierigkeit ist dabei einen Algoritmus zu finden, der sowohl den abfallenden Teil der Lösung wie auch die Wellenausbreitung berücksichtigt. Es wird gezeigt, dass verschiedene in der Literatur vorgeschlagene Lösungen für Probleme mit einer Grenzfrequenz unzulänglich sind, da sie rechnerisch nicht effizient sind, wie zum Beispiel die Faltung oder die Randelemente.

Der hier eingeschlagene Weg ist ein Verfahren basierend auf rationaler Approximation und wird daher als *"rationale Ränder"* bezeichnet. Die Lösung im Zeitbereich wird durch ein lineares, zeitinvariantes System approximiert, das als System von linearen Differentialgleichungen mit konstanten Koeffizienten ausgedrückt wird. Diese Differentialgleichungen können mit dem gleichen Algoritmus, der für den Finite-Elemente-Teil des Modells benützt wird, effizient gelöst werden.

Die rationale Approximation von Systemen ist aus zwei Gründen schwierig. Erstens ist die Bestimmung der Koeffizienten ein nichtlineares Problem, und zweitens besteht die zusätzliche Bedingung, dass das gefundene System stabil sein muss. In einem ersten Schritt werden einfache Methoden erklärt, die entweder direkt die Uebertragungsfunktion oder dann die *Markov-Parameter* approximieren. Die Approximation der Markov-Parameter führt in natürlicher Weise zur *Hankel-Matrix*, die eine Schlüsselrolle für kompliziertere Methoden spielt.

Ein wichtiger Punkt ist die Dualität zwischen zeitlich diskreten und kontinuierlichen Systemen. Die Verbindung beider Bereiche ist durch die *bilineare Transformation* gegeben. Die bilineare Transformation ist eine Abbildung zwischen zeitlich diskreten und kontinuierlichen Systemen. Die Approximation eines zeitlich kontinuierlichen Systems kann im zeitlich diskreten Bereich ausgeführt werden. Die Approximation wird dann in den ursprünglichen, zeitlich kontinuierlichen Bereich zurücktransformiert. Da die Stabilitätsgebiete ineinander abgebildet werden, bleibt die Stabilität bei der bilinearen Transformation erhalten. Dieses Vorgehen hat zwei bedeutende Vorteile. Erstens kann die Hankel-Matrix einfach berechnet werden, da das System zeitlich diskret ist und zweitens kann die ganze Frequenzachse bis ins Unendliche erfasst werden, da sie in den (endlichen) Einheitskreis transformiert wird. Die Erfassung hoher Frequenzen ist wichtig für die Erhaltung der *Kausalität*, die ihrerseits eine Voraussetzung für die *Stabilität* ist.

Zwei Methoden werden erklärt, die auf der *Singulärwertzerlegung* der Hankel-Matrix basieren, die Carathéodory-Fejér-Methode (CF-Methode) und die "Balanced Realization". Die CF-Methode ist die genaueste von allen untersuchten Methoden und führt zu einer fast kreisförmigen Fehlerkurve in der komplexen Ebene. Leider kann diese Methode nicht einfach auf den Fall von Mehrfreiheitsgradsystemen übertragen werden. Die bevorzugte Methode ist die "Balanced Realization", da sie numerisch robust ist und sowohl auf Einfreiheitsgrad- wie auch auf Mehrfreiheitsgradsysteme angewendet werden kann.

Eine Spezialität des vorgeschlagenen Algoritmus' ist die Transformation des approximierten Systems auf ein symmetrisches System zweiter Ordnung von der gleichen Form wie die Finite-Element-Matrizen. Das Ferngebiet kann daher berücksichtigt werden, indem die Matrizen des Ferngebietes an die Massen-, Dämpfungs- und Steifigkeitsmatrizen des Nahbereichs angekoppelt werden.

Mehrere Beispiele, einschliesslich Berechnungen von grossen dreidimensionalen Modellen, bestätigen die Genauigkeit und die Effizienz der Methode.

Schlüsselwörter: Staumauer-Stausee-Wechselwirkung, Erdbebenberechnung, Finite Elemente, Durchlässige Ränder, Rationale Approximation, Approximation linearer Systeme, Hankel Matrix, Singulärwertzerlegung, Balanced Realization.

Glossary

Absorptive foundation Simplified model for reservoir-foundation interaction based on one-dimensional wave theory.

Artificial boundary Boundary between *nearfield* and *farfield.* A large domain can only be modeled to a certain extent by finite elements, introducing an artificial boundary, which is not present in the original domain. With simple boundary conditions, the artificial boundary introduces undesired reflections of waves.

Balanced realization *Realization* for which the *controllabiliy and observability Gramians* are equal and diagonal. In the present context, the term *balanced realization* also denotes a realization obtained by a specific algorithm based on the *singular value decomposition* of the *Hankel matrix.*

Bilinear transform Mapping between *continuous-time* domain and *discrete-time* domain. The left half-plane of the continuous-time domain maps onto the unit circle of the discrete-time domain. Therefore stability is preserved by this mapping. The bilinear transform allows to switch between the two domains thereby simplifying many theoretical and numerical problems.

Causal system A *system* whose *impulse response* is zero for negative times. A system is *strictly causal* if the impulse response is zero also at time zero. For a causal system the present is only affected by the past but not by the future. This is the behavior expected for physical systems. *Rational systems* that are causal are also stable.

Continuous-time system A *system* described by differential equations in time. Its counterpart is the *discrete-time system.*

Controllable system A *system* for that all states are effected by the input. The *controllability matrix* has full rank. The *controllability Gramian* is non-singular.

Cut-off frequency Frequency that separates two regimes of the wave equation. Below the cut-off frequency, the solution is decaying (*evanescent waves*) and no radiation takes place. Above the cut-off frequency, the solution consists of *traveling waves* and radiation of energy takes place. For prismatic domains, the

	cut-off frequency corresponds to the fundamental frequency of the cross-section of the domain.
Discrete-time system	A *system* described by difference equations in time. Its counterpart is the *continuous-time system.*
Evanescent waves	Exponentially decaying behavior of wave equation below the *cut-off frequency.*
Farfield	Domain beyond the finite element model, idealized as semi-infinite domain. Large domains far from the object of interest are usually not included in a finite element model. They are idealized as semi-infinite domains and solved by semi-analytical methods.
Gramian matrix	Matrix constructed by taking scalar products of all combinations of vectors from a given set. If the vectors are linearly independent, the Gramian matrix is non-singular. In system theory the *controllability* and the *observability* Gramians are used to determine controllability and observability of a system.
Hankel matrix	Matrix with constant elements along anti-diagonals. In system theory the entries of the Hankel matrix are the *Markov parameters.* The Hankel matrix can be used via the *singular value decomposition* to determine the *order* of a system and to construct the *balanced realization.*
Impulse response	Response of a system due to impulse loading. Fourier transform of *transfer function.*
Impedance	*Transfer function* expressing the local derivatives of the field variables in terms of the field variables. Dynamic stiffness for solid problems.
Lyapunov equations	Matrix equations to determine the *controllability* and *observability Gramians* from the system matrices.
Markov parameters	Coefficients of series expansion of *transfer function.* For *discrete-time systems* the Markov parameters equal the *impulse response.*
Minimal realization	*Realization* of a system with minimal *order*, that is, with the smallest possible number of *state variables.* A minimal realization is *observable* and *controllable.*
Multvariable system	*System* with several input and severval output variables.
Nearfield	Domain that is modeled by finite elements. For dam-reservoir-interaction this is usually the dam and an irregular part of the reservoir and the foundation in the vicinity of the dam.
Observable system	A *system* for which all states have an effect on the output. The *observability matrix* has full rank. The *observability Gramian* is non-singular.

Order of system Number of *state variables* in a realization of a system (= size of matrix **A**).

Radiation condition Boundary condition for unbounded domains, excluding waves originating at infinity.

Rational boundary *Transmitting boundary* obtained by approximating the original transfer function by a rational transfer function. The rational transfer function is usually formulated using the state-variable description with constant system matrices. The system can readily be applied in the time domain.

Realization Specific form of a system. The same system can have many different realizations.

Scalar system System with one input and one output variable.

Singular values Characteristic values of a matrix found by the singular value decomposition. The singular values are non-negative. The number of non-zero values equals the rank of the matrix. Applied to the *Hankel matrix*, the number of non-zero singular values equals the minimal *order* of a realization.

State variables Internal variables used in the state-variable description of systems. The state variables fully describe the condition of the system at any time. They contain all necessary information from the past to proceed to the future. They need not describe any physical quantity. The number of state variables is called the *order* of a system.

System Idealized mathematical description of a physical phenomenon by an input-output relationship. Frequently used descriptions are *transfer function*, *impulse response* and *state-variable* description.

Transfer function Response of a system due to steady-state input. Fourier transform of *impulse response*.

Transmitting boundary *Artificial boundary* with special boundary condition avoiding reflection of waves.

Traveling waves Sinusoidal behavior of wave equation above the *cut-off frequency*.

Velocity potential Field variable for describing irrotational fluids. Velocity and dynamic pressure can be derived from the velocity potential. For coupled fluid-solid systems, the formulation with the velocity potential leads to symmetric equations.

Viscous boundary Simplified *transmitting boundary* consisting of dashpots. Also known as the Lysmer boundary.

Brief List of Notations

Wave problems, Chapters 2, 3, 9

c	Wave velocity of rod, 16 Wave velocity of fluid, 26
c_P	Compression wave velocity, 18
c_S	Shear wave velocity, 18
λ	Eigenvalue of channel cross-section, 149 Material parameter for rod on elastic foundation, 21
q	Parameter for bottom reflection, 149
α	Bottom reflection coefficient, 149
k	Wave number, 21, 25, 27
κ	Normalized impedance, 25
u	Displacement, 16
ψ	Displacement potential, 17
φ	Velocity potential, 26, 148
p	Dynamic pressure, 26, 148
v	Fluid velocity, 26, 148
Φ	Velocity potential over channel cross-section, 149
P	Continued fraction approximation, 30
R	Reflection coefficient, 32

Signal processing, Chapters 4, 5

z	z-tansform variable, 54
s	Laplcae transform variable, 62
ω	Fourier transform variable, circular frequency, 19
α	Coefficient for bilinear transform, 72
a_k	Denominator coefficients of rational function, 54, 67
b_k	Numerator coefficients of rational function, 54, 67
$H(z)$	Discrete-time transfer function, 54
$H(s)$	Continuous-time transfer function, 64, 67
h_k	Discrete-time impulse response (Markov parameters), 54 Continuous-time Markov parameters, 69
$h(t)$	Continuous-time impulse response, 63, 69

$\mathbf{\Gamma}$	Hankel matrix, 58
Γ	Hankel integral operator, 70
$\mathbf{U}, \mathbf{\Sigma}, \mathbf{V}$	Singular value decomposition matrices, 90

Linear Systems, Chapters 6, 7, 8

$\mathbf{A}, \mathbf{B}, \mathbf{C}, \mathbf{D}$	System matrices, discrete-time or continuous-time, 102, 114
$\hat{\mathbf{A}}, \hat{\mathbf{B}}, \hat{\mathbf{C}}$	System matrices obtained by similarity transform, 104
$\tilde{\mathbf{A}}, \tilde{\mathbf{B}}, \tilde{\mathbf{C}}, \tilde{\mathbf{D}}$	Discrete-time system matrices obtained by bilinear transform, 118
$\bar{\mathbf{A}}, \bar{\mathbf{B}}, \bar{\mathbf{C}}, \bar{\mathbf{D}}$	System matrices obtained balanced realization, 122
$\mathbf{A}_0, \mathbf{A}_1, \mathbf{A}_2, \mathbf{B}_1, \mathbf{B}_2, \mathbf{D}_1, \mathbf{D}_2$	Second-order system matrices, 146
$\mathbf{H}(z)$	Discrete-time transfer matrix, 103
$\mathbf{H}(s)$	Continuous-time transfer matrix, 114
$\mathbf{h}_k$	Discrete-time impulse response matrices, 103 Continuous-time Markov paramters, 114
$\mathbf{h}(t)$	Continuous-time impulse response matrix, 114
$\mathbf{T}$	Similarity transformation matrix, 104
$\mathbf{\Gamma}$	Block Hankel matrix, 110
$\check{\mathbf{\Gamma}}$	Shifted block Hankel matrix, 110
$\mathbf{U}, \mathbf{\Sigma}, \mathbf{V}$	Singular value decomposition matrices, 122
$\mathbf{W}_o$	Observability matrix, 106
$\mathbf{W}_c$	Controllability matrix, 107
$\mathbf{P}$	Controllability Gramian, 107
$\mathbf{Q}$	Observability Gramian, 106

Finite Elements, Chapters 10, 11

$\mathbf{M}_{ff}, \mathbf{C}_{ff}, \mathbf{K}_{ff}$	Fluid mass, damping and stiffness matrices, 174
$\mathbf{M}_{ss}, \mathbf{C}_{ss}, \mathbf{K}_{ss}$	Solid mass, damping and stiffness matrices, 175
$\mathbf{M}_I, \mathbf{C}_I, \mathbf{K}_I, \mathbf{A}_I$	Mass, damping, stiffness and weighting matrices of channel cross-section, 178
$\mathbf{M}_I^*, \mathbf{C}_I^*, \mathbf{K}_I^*$	Reduced mass, damping, stiffness matrices of channel cross-section, 184
$\mathbf{C}_{fs}$	Fluid-solid interface matrix, 176
$\mathbf{V}_f$	Fluid velocity input vector, 174
$\mathbf{V}_I$	Velocity input vector of channel cross-section,
$\mathbf{F}_s$	Force vector of solid domain, 175
$\mathbf{S}_{ff}$	Dynamic stiffness for fluid, 182
$\mathbf{S}_{ss}$	Dynamic stiffness for solid, 182
$\mathbf{S}_{sf}$	Dynamic stiffness for fluid-solid coupling, 182
$\mathbf{N}$	Fluid shape functions, 173
$\mathbf{N}_s$	Solid shape functions, 175 178
$\mathbf{u}$	Vector of nodal displacements, 175

$\boldsymbol{\varphi}$	Vector of nodal velocity potentials, 174
$\boldsymbol{\zeta}$	Vector of internal variables, 201
$\boldsymbol{\Psi}$	Eigenvectors of channel cross-section, 179
$\boldsymbol{\Psi}_0$	Zero-frequency eigenvectors of channel cross-section, used as Ritz vectors, 183
$\boldsymbol{\Phi}^h$	Nodal velocity potential due to excitation normal to channel cross-section, 178
$\boldsymbol{\Phi}^p$	Nodal velocity potential due to excitation in the plane of channel cross-section, 178
$\mathbf{k}$	Transfer matrix for transmitting boundary, 179
$\boldsymbol{\kappa}$	Normalized transfer matrix for transmitting boundary, 200

Index

α-method, 204
Absorptive foundation, 149, 171
Added mass, 8, 185
Analytic functions, 66
Angle of incidence, 27
Approximation
  of mapped discrete-time system, 85
  of Markov parameters, 82
  of transfer function, 77
Artificial boundary, 1

Balanced
  realization, 121
    algorithm, 121
    examples, 128, 130, 131
  system, 111
Bilinear transform, 72
  of system matrices, 119
Block
  column, 122
  Hankel matrix, 110
  row, 122
Boundary elements, 38
  in frequency domain, 40
  in time domain, 41

Causality, 24
  continuous-time system, 69
  discrete-time system, 56
Consistent, non-local boundaries, 3
Controllability
  continuous-time systems, 114
  discrete-time systems, 107
  Gramian
    continuous-time systems, 115
    discrete-time systems, 107
Convolution integral, 20
Coupled system
  approximated farfield, 200
  frequency domain, 180
  incompressible fluid, 185, 203
  viscous boundary, 186, 203
Criterion for numerical rank, 127

Damping matrix for fluid, 174
DANAID, 228
Data structures, 233
Diagonal approximation, 203, 210
Digital filter, recursive, 11, 52, 81, 92, 99
Dispersion
  due to discretization, 46
  of waves, 22
Doubly asymptotic approximation, 33

Engquist-Majda boundaries, 30
Error curve, 86, 87
Evanescent waves, 22
Extrapolation algorithm, 35

Farfield, 1
Fast Fourier Transform, 60
FFT, 60
Finite elements, 43, 170
  fluid, 173
  infinite fluid domain, 177
  interface, 175
  isoparametric, 172
  solid, 175
  types, 231
Fluid model, 148
Fourier transform, 19
  discrete, 60, 71
  Fast, 60
  numerical, 71
Frequency sampling, 77
Fundamental solution, 40

Generation commands, 230

Hankel
  integral operator, 70, 116
  matrix, 58, 110
    block, 110
    finite dimensional, 83
High-frequency behavior, 200
Hilbert transform, 66
Hybrid time-frequency method, 50

Identification, 12
Impedance, 17
Impulse response
  approximation, 83, 84
  causal, 56
  continuous-time, 114
  discrete-time, 54, 101, 103
  multivariable, 103, 114
Incompressible fluid, 185, 203
Infinite
  elements, 47
  mapping, 48
  shape function, 47
Input processing, 229
Isoparametric elements, 45

Laplace transform, 62
Least-squares Padé approximation, 84
Lindman boundary, 34
Local, non-consistent boundaries, 3
Lyapunov equation
  continuous-time, 115
  discrete-time, 109

Markov parameters, 69
  approximation, 82, 85
  continuous-time, 69, 114
Mass matrix
  for fluid, 174
  for rod, 46
Memory management, 235
Minimal realization, 108
Model reduction, 11
Morrow Point dam, 195, 218
Multivariable systems, 100
  continuous-time, 114
  discrete-time, 102

Nearfield, 1
Non-recursive system, 77
Null space, 92
Numerical rank, 90, 127
Nyquist frequency, 61

Observability
  continuous-time systems, 114
  discrete-time systems, 106
  Gramian
    continuous-time systems, 115
    discrete-time systems, 106
Output file, 237

Padé approximation, 83
Paraxial boundaries, 34
Parser, 230
Partial realization, 83
Periodic transfer function, 79
Pine Flat, 217
Plane waves, 27
Propagating waves, 21
Proper transfer function, 54

Radiation
  condition, 2, 16
  damping, 1
Range of a matrix, 91
Rational
  boundaries, 3
  systems, 51
Rectangular channel, 162
  cross-stream excitation, 163
  time domain, 212
  upstream excitation, 162
  vertical excitation, 163
Ritz vectors, 183

Sampling
  in time, 60
  points, 77
Scalar systems, 100
Scaling, 199
Scanner, 229
Semi-circular channel, 164
  cross-stream excitation, 168
  time domain, 214
  upstream excitation, 165
  vertical excitation, 167
Semi-infinite rod, 15
  on elastic foundation, 21
  on visco-elastic foundation, 25, 130
Sets, 230, 234
Shape function, 44, 172
Similarity transform, 104
Simplified models
  analytical, 156
  finite elements, 187
Singular value decomposition, 59, 90
  of Hankel integral operator, 71, 116
Smith boundaries, 37
Sommerfeld, 2
Spherical cavity, 17
  by balanced realization, 128

Stability
continuous-time system, 67
multivariable, 105
of discrete-time system, 56
State-Variable Description, 100
Stiffness matrix
for fluid, 174
for rod, 46
for solid, 175
Strictly
causal impulse response, 57
proper transfer function, 54
Superposition boundaries, 37
System, 4, 51
linear, time-invariant, 4
matrices
multivariable, 104
scalar, 101
rational, 4
theory, 4

Taft earthquake, 216
Talvacchia dam, 190
Time
integration, 204
stepping, 204
Time-domain implementation, 199
Transfer function
approximation, 77
continuous-time, 67, 114
discrete-time, 52, 101
in complex plane, 95
multivariable, 103, 114
rational, 54, 67
Transmitting boundary, 1, 16
Two-degree-of-freedom system, 131
Two-dimensional fluid problem, 149
fixed boundary, 157, 159
incompressible fluid, 158, 160
upstream excitation, 150
vertical excitation, 155
viscous boundary, 158, 160

Viscous boundary, 186, 203

Wave equation, 148
Weak form for fluid, 170

z-transform, 54